Biological Invasions in Changing Ecosystems: Volume 3

Biological Invasions in Changing Ecosystems: Volume 3

Aubree Anderson

Callisto Reference
www.callistoreference.com

Callisto Reference,
118-35 Queens Blvd., Suite 400,
Forest Hills, NY 11375, USA

Visit us on the World Wide Web at:
www.callistoreference.com

ISBN: 978-1-64116-776-5 (Hardback)

Cataloging-in-Publication Data

Biological invasions in changing ecosystems: Volume 3 / Aubree Anderson.
p. cm.
Includes bibliographical references and index.
ISBN 978-1-64116-776-5
1. Biological invasions--Environmental aspects. 2. Ecological disturbances.
3. Introduced organisms--Environmental aspects. I. Anderson, Aubree.
QH353 .B56 2023
577.18--dc23

Table of Contents

	Preface	**VII**
Chapter 16	**Invasive Alien Plants and its Effect on South African Rangelands** Timothy G. O'Connor and Brian W. van Wilgen	**1**
Chapter 17	**Assessing the Influence of Alien Species on Biodiversity** Tsungai A. Zengeya, Sabrina Kumschick, Olaf L. F. Weyl and Brian W. van Wilgen	**29**
Part V	**Handling the Invasions**	**53**
Chapter 18	**Developing and Enforcing Policies and Legislation** Peter Lukey and Jenny Hall	**55**
Chapter 19	**Invasive Alien Plants and their Biological Control over the Years** Martin P. Hill, Vincent C. Moran, John H. Hoffmann, Stefan Neser, Helmuth G. Zimmermann, David O. Simelane, Hildegard Klein, Costas Zachariades, Alan R. Wood, Marcus J. Byrne, Iain D. Paterson, Grant D. Martin and Julie A. Coetzee	**91**
Chapter 20	**Biological Invasions and their Risk Analysis** Sabrina Kumschick, Llewellyn C. Foxcroft and John R. Wilson	**111**
Chapter 21	**Alien Plant Control Projects and their Assessment** Brian W. van Wilgen, John R. Wilson, Andrew Wannenburgh and Llewellyn C. Foxcroft	**134**
Chapter 22	**Major Takeaways from Alien and Invasive Animal Control Projects** Sarah J. Davies, Martine S. Jordaan, Minette Karsten, John S. Terblanche, Andrew A. Turner, Nicola J. van Wilgen, Ruan Veldtman, Tsungai A. Zengeya and John Measey	**166**

Chapter 23 **Biological Invasion and Restoration Interventions** **201**
Patricia M. Holmes, Karen J. Esler, Mirijam Gaertner, Sjirk Geerts, Stuart A. Hall, Mlungele M. Nsikani, David M. Richardson and Sheunesu Ruwanza

Chapter 24 **Biological Invasions: Social and Human Dimensions** **237**
Ross T. Shackleton, Ana Novoa, Charlie M. Shackleton and Christian A. Kull

Permissions

Index

Preface

The purpose of the book is to provide a glimpse into the dynamics and to present opinions and studies of some of the scientists engaged in the development of new ideas in the field from very different standpoints. This book will prove useful to students and researchers owing to its high content quality.

Biological invasion is a process wherein certain organisms are introduced into a new ecosystem, where they did not exist previously. It may be a natural process or may be done deliberately. These organisms are called invasive species. They may have an adverse or positive impact on the native inhabitants and natural resources of the ecosystem they invade. Some common characteristic features of invasive species include rapid reproduction and growth, phenotype plasticity, high dispersal ability, ability to tolerate a broad spectrum of environmental conditions, and prior successful invasions. They also have an ability to survive by consuming various types of foods. Some invasive species have favorable effects, such as, providing a suitable habitat or food for the native species. Invasive species can play a crucial role in restoring the native place, where the indigenous species have reached a point of extinction. The topics included in this book on biological invasions in changing ecosystems are of utmost significance and bound to provide incredible insights to readers. It is appropriate for students seeking detailed information in this area of study as well as for experts.

At the end, I would like to appreciate all the efforts made by the authors in completing their chapters professionally. I express my deepest gratitude to all of them for contributing to this book by sharing their valuable works. A special thanks to my family and friends for their constant support in this journey.

Aubree Anderson

16

Invasive Alien Plants and its Effect on South African Rangelands

Timothy G. O'Connor and Brian W. van Wilgen

Abstract Rangeland covers >70% of the land surface of South Africa, and includes grassland, savanna, thicket, and karroid shrubland vegetation. These rangelands support domestic livestock and wildlife whose economic value is around ZAR 30 billion annually. They are invaded by hundreds of alien plant species, of which 71 have been identified as being of special importance in South African rangelands. These species are able to proliferate in response to disturbances, of which grazing and fire are the two most important for South African rangelands; changes to fire and grazing regimes can therefore promote invasion, especially by alien trees. These trees replace palatable grasses and are generally unpalatable themselves. At a national scale, invasive alien plants are estimated to reduce the value of livestock production by ZAR 340 million annually, but this is expected to increase dramatically as plant invasions spread, and as additional alien species become invasive. Invasive species that have increased their range dramatically by up to 671% between 2006 and 2016 include *Campuloclinium macrocephalum* (Pompom Weed), *Opuntia engelmannii* (Small Round-leaved Prickly Pear), *Opuntia humifusa* (Large-Flowered Prickly Pear), *Parthenium hysterophorus* (Parthenium Weed), *Trichocereus spachianus* (Torch Cactus), and *Verbena bonariensis* (Wild Verbena). Studies that document the impacts of individual species on rangeland composition and structure cover only a few species, including *Prosopis* species (Mesquite), *Acacia mearnsii* (Black Wattle) and *Parthenium hysterophorus*, all of which can dramatically reduce grass cover and the capacity to support livestock, especially at high densities. Invasive plants in rangelands can also be beneficial as sources of firewood, fodder, shade and medicinal products. Some species may offer considerable value when at low abundance but

T. G. O'Connor (✉)
South African Environmental Observation Network, Pietermaritzburg, South Africa
e-mail: tim@saeon.ac.za

B. W. van Wilgen
Centre for Invasion Biology, Department of Botany and Zoology, Stellenbosch University, Stellenbosch, South Africa

their detrimental impact far outweighs advantages as their abundance increases, resulting in a net negative value. An escalating threat of alien plants to rangelands demands innovative responses in addition to biological control and clearing programmes.

16.1 Introduction

Rangelands are areas of natural or semi-natural vegetation that support large mammalian herbivores. They cover >70% of the land surface of South Africa, and include grassland, savanna, thicket, and karroid shrubland vegetation. This diversity of vegetation supports a richness of domestic and native large mammalian herbivores that consume graminoid plants, herbaceous dicotyledons, and woody plants in different proportions depending on whether they are grazers, browsers, or mixed feeders. Ecosystems dominated by woody plants therefore support more browsers, while grasslands support only grazers. Production or subsistence rangelands are used for the rearing of 40.7 million head of livestock (cattle, 12.8 million; sheep, 22.5 million; goats, 5.4 million; Department of Agriculture, Forestry and Fisheries 2018). In addition, over 60,000 km^2 is set aside either as protected areas (Department of Environmental Affairs 2018a), or for farming game animals (~3 million head, Meissner et al. 2013).

Alien plants can provide novel foodstuffs for herbivores, but they also alter vegetation structure and composition in ways that can negatively affect herbivores. Despite their prominence, there have been few studies of the impact of alien plants on South African rangelands (Richardson and van Wilgen 2004). Impacts of invasive alien species on the ecosystem services provided by rangelands (livestock production and conservation), catchments (water supply), and biodiversity (harvested products and non-use activities) are of particular concern (van Wilgen et al. 2008).

In this chapter we assess the impact of invasive alien plants on South African rangelands. We review the types, extent, economic importance, and key ecological processes relevant to invasions of South African rangelands, and identify the status of important alien species in each biome. We address the question of whether rangeland management has influenced the success of alien plants, and consider their ecological and economic impact, illustrated by case studies. We conclude with a distillation of key features concerning the ingress of alien plants into rangelands and how effective rangeland management could be used for their control.

16.2 Rangelands in South Africa

16.2.1 Types of Rangelands in South Africa

South African biomes serving as rangeland include parts of the Grassland, Nama-Karoo, Succulent Karoo, Savanna, and Albany Thicket Biomes, and the Indian Ocean Coastal Belt, but not the Fynbos Biome (although lowland fynbos was once rangeland it has largely been transformed), Desert or Forest Biomes (Table 16.1). Dominant plant growth forms characterising each biome reflect climatic differences between the biomes (Mucina and Rutherford 2006). Grassland is dominated by perennial C_4 grasses at low altitude and C_3 grasses above 2700 m above sea level, a mixture of woody plants and perennial C_4 grasses characterise the Savanna Biome, the Karoo Biomes are dominated by dwarf shrubs, especially by succulent dwarf shrubs in the Succulent Karoo Biome, with greater abundance of grasses in the eastern Nama-Karoo Biome, a mix of succulent and woody trees dominate the Albany Thicket Biome, whilst the Indian Ocean Coastal Belt Biome is a mosaic of grassland, wetland, forest and savanna.

Climate, vegetation, and historical human occupation have influenced the manner in which biomes have been used as rangeland. The San peoples who originally occupied parts of southern Africa were hunter-gatherers who did not keep domestic livestock. The first livestock in South African biomes, principally the Nama-Karoo and Succulent Karoo, arrived when the Khoekhoen immigrated into South Africa together with their herds of fat-tailed sheep (*Ovis aries*) about 4000 years ago (Elphick 1985). Next, Bantu peoples with their herds of cattle (*Bos indicus*) commenced colonising South Africa from about 1700 years ago (Huffman 2007), and Khoekhoen had also acquired cattle by 500 AD (Orton et al. 2013). The Bantu peoples occupied well-watered and wooded areas along the eastern and northern regions of the country at the time European settlement commenced during the mid-seventeenth century. Commencing in the south-west (Karoo) region, European colonists decimated wildlife by hunting and by the rapid introduction of barbed-wire fencing in the late 1800s that led to population collapses of migratory animals such as springbuck (*Antidorcas marsupialis*; Dean et al. 2018). Following European colonisation, a spatial pattern of land access based on race was initiated that developed into the template for land tenure and use which remains evident today. Most land was under private tenure for commercial agricultural use by white people; other ethnic groups were consolidated into blocks of land of which non-arable portions were used as communal rangeland for livestock. Currently, commercial producers use grazing systems with fenced paddocks that constrain movement of livestock (O'Reagain and Turner 1992).

Biomes differ in the type of livestock carried and the importance of wildlife. Small stock (sheep and goats) have been farmed in watered regions of the Succulent and Nama-Karoo Biomes for 4000 years, and intensely across this region over the past two centuries, ostrich farming was developed in the Little Karoo during the late nineteenth century, and springbuck farming developed from the 1960s onward (Beinart 2003). Communal land accounts for 3% of the Succulent and Nama

Table 16.1 Biomes serving as rangelands in South Africa

Biome	Total area (km^2)	% of South Africa	Area transformed (km^2)	Remaining natural area (km^2)	Area under conservation (km^2)	Range in mean annual precipitation (mm)	Mean grazing capacity (ha/LSU)
Indian Ocean Coastal Belt	11,529	0.9	7381	4148	825.88	577–1467	4
Grassland	330,860	27.1	132,803	198,057	14,844.78	197–1925	6
Savanna	394,158	32.3	75,065	319,093	52,863.43	120–1807	12
Nama-Karoo	249,353	20.4	4827	244,526	3901.66	54–638	25
Succulent Karoo	78,203	6.4	3595	74,607	6077.25	41–676	65
Albany Thicket	35,250	2.9	3124	32,125	4212.38	86–1179	14

Biomes are described in detail in Mucina and Rutherford (2006). Data on areas and rainfall are from Skowno et al. (2019); grazing capacity is from Department of Agriculture, Forests and Fisheries (2018a); *LSU* large stock unit

Karoo Biomes (Walker et al. 2018), on which continuous grazing by small-stock is practised (Todd and Hoffman 2009). More than 30% of the Grassland Biome has been transformed, <20% is under communal tenure, and the remainder is used mainly for commercial production of sheep and cattle (Mucina and Rutherford 2006). The Savanna Biome contains approximately 100,000 km^2 of communal area, the backbone of state wildlife conservation areas including the Kruger National (19,000 km^2) and Kalahari Gemsbok National (9590 km^2) Parks, commercial ranching based mainly on cattle but including sheep and goats in semi-arid regions, and private wildlife properties that have emerged as an increasingly important land use since the 1960s (Carruthers 2008). Albany Thicket was used mainly for goat farming following European settlement but commercial wildlife enterprises are now well established. The overall value of rangeland in South Africa is reflected by the annual value of commercial livestock production estimated to be ZAR 18 billion (Meissner et al. 2013), that of the commercial wildlife industry ZAR 10 billion (Department of Environmental Affairs 2018b), and that 19 million people reside in rural areas.

Here we discuss four components of change within the rangelands described above—shifts in the composition of the herbivores, increases in the intensity of grazing, changes to fire regimes, and fencing—and their consequences for invasions (see also Box 16.1 for the evidence of the impact of grazing and fire).

Box 16.1 Experimental Evidence for the Role of Fire and Grazing in Promoting or Preventing Invasion

South Africa has a rich legacy of long-term experiments in which regimes of fire and grazing management have been manipulated in order to improve our understanding of their ecological effects. In addition, the ecological outcomes of different long-standing management regimes can commonly be compared across the fence line separating two properties. While the effects of these regimes on invasion by alien plants were not included in the formal experimental designs, the experiments or fence-line comparisons themselves offer an opportunity to examine outcomes in this regard, and to look for generalities. These experiments and contrasts have been collated in Supplementary Appendix 16.1 to provide a novel data base about the influence of management on the success of alien plants.

Exclusion of fire and grazing was investigated at eight locations ranging from semi-arid savanna through to moist grassland; investigation of fire return-interval was similarly represented along a climate gradient. Moist and mesic areas grow sufficient fuel that can support frequent fires, semi-arid systems do not. The effects of complete fire and grazing exclusion in semi-arid systems may therefore be attributed mainly to grazing exclusion; those in mesic or moist systems are usually attributed mainly to exclusion of fire. The effects of fire have not been experimentally investigated in the semi-arid Albany

(continued)

Box 16.1 (continued)

Thicket, Nama-Karoo or Succulent Karoo Biomes. Long-term exclusion of fire has promoted the success of woody alien species throughout the Grassland Biome including the Drakensberg, Athole in Mpumulanga, nThabamhlope and Ukulinga in KwaZulu-Natal, and Döhne in the Eastern Cape. All these grasslands are high rainfall or at least mesic in character. By contrast for the Savanna Biome, woody alien species were successful only in the mesic savannas of the Kruger National Park and Towoomba; woody alien species did not establish in semi-arid savannas of Kruger National Park or of Mopani, Limpopo Province. A long fire-return interval (relative to the management norm of the system) also allowed woody alien species to increase in high rainfall Drakensberg grassland and in mesic savanna in Kruger National Park. Woody alien species which gained greatest prominence when fire was excluded were *Acacia mearnsii* (Black Wattle) and *Rubus cuneifolius* (American Bramble) across the Grassland Biome, and *Lantana camara* (Lantana) in the Savanna Biome. Exclusion of fire also appeared to promote herbaceous alien species but this was not consistently evident.

Differences in grazing management had far less of an effect on the success of alien plant species than differences in fire management. Neither woody nor herbaceous aliens showed a response to grazing differences in mesic grassland at Athole, Kokstad, or Ermelo, in semi-arid savanna of the southern Kalahari, at the Sundays River Valley in the Albany Thicket Biome, in the Nama-Karoo Biome at Grootfontein, Carnarvon or Sundays River Valley, or in the Succulent Karoo Biome at Tierberg or Hoekdoorn. An increase in alien plants in response to heavy grazing pressure was recorded only for the Ikwezi district in the Albany Thicket Biome. The success of the shrub *Atriplex lindleyi* (Australian Saltbush) observed at Tierberg in the Succulent Karoo in response to grazing treatment was the result of invasion onto nutrient-enriched mounds ("heuweltjies", see Moore and Picker 1991) (Milton and Dean 2010), a response also observed following nitrogen addition. Similarly, alien herbaceous plant species are also evident in the grazing experiments in the Nama-Karoo Biome at Carnarvon and Grootfontein on accumulated dung piles at livestock resting spots (<1% of paddock area) that were absent from the rangeland matrix.

Rangelands that are managed by prescribed grazing and fire within acceptable limits therefore appear to be largely resistant to invasive alien plants.

16.2.2 Shifts in the Composition of Herbivores

Over large parts of South Africa, free-ranging herds of grazing and browsing wild herbivores have been replaced by domestic livestock. Domestic animals do not have the same effect as native herbivores on vegetation because there are far fewer livestock species than wildlife species and, consequently, less of a range in body size and feeding ecology. However, the foraging ecology of individual livestock species is similar to that of individual wildlife species. In the Karoo biomes, the feeding ecology and impact of caprines is comparable with that of the springbuck they have replaced. Within the Savanna Biome, cattle are comparable with buffalo (*Syncerus caffer*) in terms of body size and social aggregation, but there is no natural counterpart for cattle in the Grassland Biome. Moreover, mesic montane grasslands are stocked at 35-fold higher biomass by commercial producers than that occurring within natural wildlife ecosystems (Rowe-Rowe and Scotcher 1986).

16.2.3 Increases in the Intensity of Grazing

Grazing systems involving paddocks are the norm for commercial livestock production (O'Reagain and Turner 1992), with rotation among paddocks meaning that a paddock is subjected to heavy grazing then rested. Communal areas are more commonly grazed continuously but at up to three times higher stocking rates than corresponding commercial ranches (Tapson 1993; Rutherford et al. 2012b). O'Reagain and Turner (1992) concluded that stocking rate rather than grazing system was the most important management variable affecting rangeland vegetation; chronic (consistent and heavy) grazing over time has been responsible for depressed herbaceous biomass that may offer establishment opportunities for alien plants. Livestock may also create piospheres, a gradient of grazing impact diminishing with distance from water points (Thrash 1998), and nutrient-enriched patches where they repeatedly rest.

16.2.4 The Role of Fire

Fire is integral to the management of grassland and savanna, and is employed in grassland mainly for removing moribund material and in savanna for controlling bush encroachment (Tainton 1999). Fire-return interval across savanna and grassland is inversely related to mean annual precipitation (Archibald et al. 2009), which varies from 200 mm to 1200 mm; fire-return interval correspondingly varies from >10 years to 1 year (van Wilgen and Scholes 1997). Woody plants establish in moist grasslands within 6 years if fire is excluded (Titshall et al. 2000). Accordingly, a biennial fire regime is commonly imposed for their control. Savanna trends toward dry thicket if fire is excluded (O'Connor et al. 2014). In semi-arid savanna or grassland, fire may occur as infrequently as once a decade, or be completely excluded if herbaceous fuel production is lessened by chronic grazing (Archibald et al. 2009). Alien plants invading grassland or savanna would therefore be expected to be fire-adapted, or to exploit fire refugia, localised areas of low fire intensity, or areas in which fire has been excluded for long enough to allow for the development of a stand self-protected against fire. The Karoo Biomes and Albany Thicket do not normally experience fire although a few fires have recently occurred within the transition between the Nama-Karoo and Grassland Biomes following an increase in grass biomass resulting from years of above-average rainfall (du Toit et al. 2015). Fire also interacts strongly with grazing pressure to influence vegetation responses (Box 16.1).

16.2.5 The Influence of Fencing

Fencing facilitates invasion of alien species through offering a perch site for birds to disperse the seeds of alien plants, as crows appear to do for *Opuntia* species (Dean and Milton 2000), and because fire breaks along fences are commonly prepared by breaking soil which contributes to establishment success of ruderal including alien plants (Roux 1969).

16.3 Invasive Plants in Rangelands

South African rangelands harbour hundreds of alien plant species, ranging from an estimated 129 species in the Nama Karoo Biome to 404 species in the Savanna Biome (van Wilgen and Wilson 2018), but $<10\%$ of these have become seriously invasive to date (Fig. 16.1). Many alien species remain restricted to land transformed by crop agriculture, timber plantations, and urban settlements. The wetter biomes (Indian Ocean Coastal Belt, Savanna, Grassland) and riparian rather than dryland habitats within each biome have been impacted most by invasive alien plants. Well-managed dryland rangeland has proven to be largely resistant to invasion although a few alien species in riparian habitats have also invaded adjacent dryland; a set of species have invaded only dryland rangeland. Several invasive alien species have transformed vegetation structure owing to a larger growth form than those of native plants, especially evident for invasion by woody plants into grassland or karroid shrubland. Patches of soil disturbance or nutrient enrichment, often created by livestock, offer additional pathways of ingress into dryland rangelands. Episodic proliferation of annual grasses that may fuel fires would threaten transformation of affected areas of the Succulent Karoo Biome, whose vegetation is not adapted to fire (Rahlao et al. 2009). Many alien species establish and proliferate in response to disturbances (Hobbs and Huenneke 1992), of which grazing and fire are the two most important types in South African rangelands.

To assess biome-level patterns, we focused on species of concern identified by van Wilgen et al. (2008). We defined these as species which occupy $>10\%$ of quarter degree grid cells of at least one biome (QDGC; 15′ latitude by 15′ longitude), species with limited distributions but severe local impacts (Rouget et al. 2004), and species which are rapidly becoming major threats to rangelands based on the extent

Fig. 16.1 Examples of alien plants that invade rangelands in South Africa. (**a**) *Chromolaena odorata* (Triffid Weed) in the Indian Ocean Coastal Belt in KwaZulu-Natal; (**b**) *Cylindropuntia fulgida* var. *mamillata* (Boxing Glove Cactus) in the Northern Cape; (**c**) *Campuloclinium macrocephalum* (Pompom Weed) in grasslands, Gauteng; and (**d**) *Acacia mearnsii* (Black Wattle) trees invading grasslands, KwaZulu-Natal. Photographs courtesy of: (**a**) Plant Protection Research Institute; (**b**) Travor Xivuri; (**c**) Lesley Henderson; (**d**) John Hoffmann

of their increase between 2000 and 2016 (Henderson and Wilson 2017). Affected biomes and habitats (riparian or dryland) were determined from the literature (e.g. Henderson 2001). We separated riparian from other habitats, as riparian areas are usually critical for large mammalian herbivores because they provide key forage resources during seasonal bottlenecks.

Of the 773 species listed in the Southern African Plant Invader Atlas data base (Henderson and Wilson 2017), 71 species were identified as being of especial importance for rangelands (i.e., are widespread, or have severe local impacts, or are rapidly increasing in extent). These included 11 tall trees, 15 medium-height trees, 1 short tree, 13 shrubs, 2 low shrubs, 5 woody climbers, 10 succulent shrubs, 1 succulent tree, 1 herbaceous climber, 3 perennial and 5 annual herbaceous species, 1 annual herbaceous shrub, and 3 perennial grasses (Table 16.2). The greatest number of important species were found within the biomes receiving higher mean annual rainfall, specifically the Grassland and Savanna Biomes and the Indian Ocean Coastal Belt, whereas the three more arid biomes (Albany Thicket, Nama-Karoo, Succulent Karoo) each contained about a half to a third of the average number of species recorded for a moist biome (Fig. 16.2). Based on a climate-envelope modelling exercise of the potential of 71 invasive species to invade further, Rouget et al. (2004) determined that the rank order (in terms of susceptibility to invasion)

Table 16.2 Widespread, locally severe, or rapidly spreading invasive alien plant species in South African rangelands[a]

Species	Growth form	QDGCs occupied (%) in 2000	QDGCs occupied (%) by 2016	Increase (%) in QDGCs occupied 2000–2016	Affected biomes[b]	Affected habitat (riparian, dryland)	Biological control[c]
Acacia baileyana	Medium tree	4.4	5.2	16.5	G	Both	Negligible
Acacia cyclops	Medium tree	8.4	8.9	5.4	AT, SK	Both	Substantial
Acacia dealbata	Medium tree	13.0	15.4	18.0	G	Both	Negligible
Acacia decurrens	Medium tree	5.1	6.4	24.8	G, IOCB	Both	Negligible
Acacia longifolia	Medium tree	4.8	4.9	2.1	AT, IOCB, NK, Sv	Both	Substantial
Acacia mearnsii	Medium tree	21.8	23.6	8.2	AT, G, IOCB, Sv	Both	Substantial
Acacia melanoxylon	Tall tree	6.8	8.7	27.6	G	Both	Substantial
Acacia saligna	Medium tree	8.0	8.5	6.3	AT, SK	Both	Substantial
Achyranthes aspera	Perennial herb	3.9	5.0	22.2	G, IOCB, Sv	Dryland	None
Agave americana	Succulent shrub	21.9	26.9	22.7	AT, G, NK, SK, Sv	Dryland	None
Ageratum conyzoides	Annual herb	1.9	2.2	18.9	G, IOCB, Sv	Riparian	None
Araujia sericifera	Woody climber	1.8	2.8	52.8	G, Sv	Riparian	None
Arundo donax	Tall grass	18.9	23.1	22.4	G, IOCB, NK, SK, Sv	Riparian	Under investigation
Atriplex inflata	Low shrub	8.3	9.9	18.9	NK, SK	Dryland	None
Atriplex nummularia	Low shrub	8.7	12.1	37.8	NK, SK	Both	None
Caesalpinia decapetala	Shrub	6.5	7.4	15.0	IOCB, Sv	Both	Negligible
Campuloclinium macrocephalum	Perennial herb	0.7	5.5	671.4	G, IOCB	Dryland	Not determined
Cardiospermum grandiflorum	Woody climber	2.2	3.1	41.9	G, IOCB, Sv	Riparian	Not determined
Cereus jamacaru	Succulent tree	6.3	10.1	60.5	AT, IOCB, Sv	Dryland	Complete
Cestrum laevigatum	Shrub	3.6	3.5	0.0	AT, IOCB, Sv	Both	Under investigation
Chromolaena odorata	Shrub	4.7	6.1	28.0	G, IOCB, Sv	Both	Not determined
Cosmos bipinnatus	Annual herb	2.4	7.0	185.4	G	Riparian	None
		0.05	0.9	1700.0	Sv	Riparian	None

Cryptostegia grandiflora	Woody climber						
Cuscuta campestris	Parasitic herb	4.2	4.7	11.8	G, IOCB	Both	None
Cylindropuntia imbricata	Succulent shrub	6.7	10.6	58.8	G, NK, Sv	Dryland	Substantial
Dolichandra unguis-cati	Woody climber	1.1	2.5	122.8	Sv	Both	Negligible
Eucalyptus camaldulensis	Tall tree	6.2	9.9	61.2	AT, G, SK	Riparian	None
Eucalyptus grandis	Tall tree	5.1	5.9	16.0	G, IOCB	Riparian	None
Gleditsea triancanthos	Medium tree	5.6	11.0	94.6	G, Sv	Both	Not determined
Ipomoea indica	Herb climber	1.2	1.7	47.8	AT, G, IOCB, Sv	Both	None
Jacaranda mimosifolia	Tall tree	9.9	11.9	19.5	G, IOCB, Sv	Both	None
Lantana camara	Shrub	15.9	15.9	26.3	G, IOCB, Sv	Both	Variable
Melia azedarach	Tall tree	28.0	34.3	22.5	AT, G, IOCB, Sv	Both	Research discontinued
Morus alba	Medium tree	6.7	9.5	45.0	G, IOCB, Sv	Riparian	None
Nasella trichotoma	Perennial grass	0.6	0.6	0.0	G	Dryland	None
Nicotiana glauca	Shrub	19.5	22.2	13.8	AT, NK, SK, Sv	Both	None
Opuntia aurantiaca	Succulent shrub	3.1	3.3	4.9	AT, G, NK, Sv	Dryland	Substantial
Opuntia engelmannii	Succulent shrub	0.5	3.3	550.0	G, NK, Sv	Dryland	Negligible
Opuntia ficus-indica	Succulent shrub	43.8	51.0	16.4	AT, G, IOCB, NK, SK, Sv	Dryland	Substantial
Opuntia humifusa	Succulent shrub	1.3	5.0	296.0	G, Sv	Dryland	Complete

(continued)

Table 16.2 (continued)

Species	Growth form	QDGCs occupied (%) in 2000	QDGCs occupied (%) by 2016	Increase (%) in QDGCs occupied 2000–2016	Affected biomes[b]	Affected habitat (riparian, dryland)	Biological control[c]
Opuntia monocantha	Succulent shrub	2.4	3.9	58.3	AT, IOCB, Sv	Both	Complete
Opuntia robusta	Succulent shrub	11.4	18.0	57.3	G, NK, Sv	Dryland	None
Opuntia stricta	Succulent shrub	5.4	8.4	55.7	AT, G, Sv	Dryland	Substantial
Parthenium hysterophorus	Annual herb	0.8	4.5	493.3	IOCB, Sv	Dryland	Not determined
Pennisetum setaceum	Perennial grass	3.4	8.9	163.6	G, NK, SK, Sv	Both	None
Pinus halepensis	Tall tree	4.3	6.1	40.0	G, NK	Dryland	None
Pinus patula	Tall tree	4.3	4.9	14.1	G	Dryland	None
Populus alba	Tall tree	0.8	1.2	60.0	G	Riparian	None
Populus canescens	Tall tree	18.9	24.6	6.5	AT, G, SK	Riparian	None
Populus nigra	Tall tree	4.6	5.8	26.7	G	Riparian	None
Prosopis glandulosa / velutina	Medium tree	19.8	24.5	23.3	NK, SK	Both	Negligible
Prunus persica	Medium tree	16.2	19.8	22.3	Sv	Dryland	None
Psidium guajava	Medium tree	8.1	9.7	19.4	IOCB, Sv	Both	None
Pyracantha angustifolia	Shrub	7.2	9.9	37.3	G	Dryland	None
Ricinus communis	Annual shrub	23.2	26.0	12.1	AT, G, IOCB, Sv, NK	Riparian	None
Robinia pseudoacacia	Tall tree	5.6	8.5	37.3	G	Both	None
Rubus cuneifolius	Shrub	3.8	5.5	44.0	G, IOCB	Both	None
Rubus fruticosus	Shrub	4.5	4.7	3.4	G	Both	None
Salix babylonica	Medium tree	24.2	27.1	12.0	G	Riparian	None
Schinus molle	Medium tree	11.7	14.1	20.3	G, NK, SK	Both	None
Schinus terebinthifolius	Small tree	1.5	2.0	33.3	IOCB	Both	None
Senna didimobotrya	Shrub	7.1	8.6	21.6	G, IOCB, Sv	Both	None
Senna occidentalis	Shrub	2.8	4.4	53.6	G, IOCB, Sv	Both	None

Sesbania punicea	Shrub	16.4	19.2	16.7	AT, G, IOCB, Sv	Riparian	Complete
Solanum mauritianum	Medium tree	13.5	16.8	24.5	G, IOCB, Sv	Both	Negligible
Solanum seaforthianum	Woody climber	1.5	2.6	70.0	IOCB, Sv	Both	None
Solanum sisymbriifolium	Shrub	2.0	4.5	135.0	G, IOCB	Both	Substantial
Tecoma stans	Shrub	2.9	7.1	143.9	IOCB, Sv	Both	Not determined
Trichocereus spachianus	Succulent tree	2.9	6.3	115.8	G, NK, Sv	Dryland	None
Verbena bonariensis	Annual herb	3.0	13.6	360.3	G, IOCB	Riparian	None
Xanthium strumarium	Annual herb	7.6	11.9	57.0	G,, IOCB, Sv	Both	None

[a]Adapted from van Wilgen et al. (2008), Rouget et al. (2004) and Henderson and Wilson (2017)
[b]Biomes: *AT* Albany Thicket, *G* Grassland, *IOCB* Indian Ocean Coastal Belt, *NK* Nama-Karoo, *SK* Succulent Karoo, *Sv* Savanna. Data on occupancy of QDGCs (quarter degree grid cells) are from Henderson and Wilson (2017);
[c]Estimates of the effectiveness of biological control are from Klein (2011) and Zachariades et al. (2017)

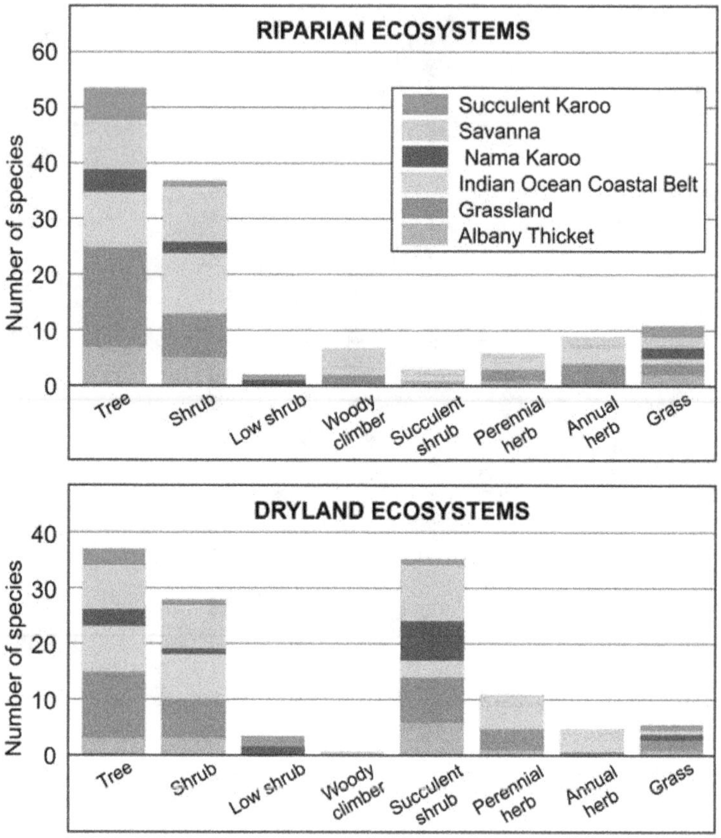

Fig. 16.2 The number of prominent alien plant species in different growth forms that impact on rangelands in six biomes in South Africa. The upper panel shows species present in riparian ecosystems, and the lower panel shows those present in dryland areas (away from riparian ecosystems). The figure is based on the 71 species listed in Table 16.2

was the Grassland Biome, Albany Thicket, Savanna Biome (including the Indian Ocean Coastal Belt), Nama-Karoo and Succulent Karoo Biomes, which described a fourfold difference in the proportion of 71 species to which a biome was vulnerable. Invasion potential of moist grassland and moist savanna vegetation types was particularly high (>25 of the 71 species).

Trees and shrubs contribute the most invasive alien species in all biomes except the Nama-Karoo, in which succulent shrubs contribute the most (Fig. 16.2). Succulent shrubs were, however, also well-represented in the Grassland and Savanna Biomes. These large growth forms may therefore easily transform the host vegetation that is naturally dominated by growth forms of smaller stature, such as grasses or karroid shrubs. The tall reed *Arundo donax* (Giant Reed) also transforms riparian vegetation owing to its greater stature than any native riparian vegetation except trees. The extent or severity of transformation of native vegetation has not been well described, with a few exceptions (e.g. *Prosopis* spp.; van den Berg 2010; Shackleton et al. 2015), but impacts are widely evident, especially in riparian environments. Growth forms other than trees, shrubs, and succulent shrubs are not usually important invasive alien species (Table 16.2). Despite an updated national list of 256 alien grass species (Visser et al. 2017), only three grass species were prominent. However, a number of species which have more than doubled their distribution between 2000 and 2016 were herbaceous; these include *Campuloclinium macrocephalum* and *Parthenium hysterophorus* (Table 16.2; Fig. 16.1). Although many species occurred in more than one biome, a biome supports a distinctive community of alien plants, which suggests that a biome-specific approach to the control of alien plants would be appropriate (Rouget et al. 2015; see also Richardson et al. 2020, Chap. 3). Half of the important species impact both riparian and dryland habitat, whilst 19 and 16 species tend to be restricted to riparian or dryland habitat, respectively (Table 16.2). In the dry Karoo biomes, however, 65% of the 101 alien species are found in riparian habitat (Milton and Dean 2010).

16.4 Benefits of Alien Plants to Rangelands

Many of the alien species that have become invasive were introduced for a specific purpose (Richardson et al. 2020, Chap. 3). For example, Milton and Dean (2010) determined that 54% of the 101 alien species recorded in arid parts of South Africa (<250 mm mean annual precipitation, MAP) are cultivated for utility products or services, 34% as ornamentals, and only 12% were accidental introductions. Valuable services provided by alien plants include forage and shelter for livestock, soil protection, fuel and building materials, barrier plants, medicines, and pollination. However, their relative value depends on what is displaced. A species may provide multiple services, for example *Acacia dealbata* (Silver Wattle) in the Eastern Cape Province is used by 100% of households for firewood, 73% for fencing, 18% as a medicine, and 79% for livestock fodder (Agripa 2016).

Alien plants may contribute to soil stability in the face of chronic grazing if they can withstand trampling and grazing, as observed for the small, mat-growing alien herb *Richardia brasiliensis* (Tropical Richardia), a sub-dominant plant (13% cover) on communal grassland in the moist southern Drakensberg (O'Connor 2005). Its mat-like growth serves to intercept rainfall energy, promote infiltration, reduce lateral flow, and contain soil wash. This widespread species only becomes conspicuous in degraded swards but can be displaced by improved grazing management. By contrast, the widespread invasive alien perennial grass *Pennisetum clandestinum* (Kikuyu Grass) fulfils the same function but is not easily displaced once established.

Fuelwood is the primary source of energy in rural areas under communal tenure; a rural household consumes on average 5.3 tonnes of firewood annually (Shackleton and Shackleton 2004). Australian *Acacia* species are the primary source of fuel wood for many rural communities located in the Grassland Biome (Agripa 2016), in the absence of which native forest may be harvested for fuel wood (Geldenhuys and Cawe 2011). By contrast, *Prosopis* species are sparingly used as fuel wood, owing to most infestations being distant from settlements (van den Berg 2010), and also due to the inferior quality of mesquite wood compared with that of native species (Shackleton et al. 2015).

The medicinal plant trade in South Africa is a commercial enterprise valued at ZAR 2.9 billion annually and it involves 27 million consumers who consume 20,000 tonnes of plant material annually comprising 771 native species of which most are derived from rangeland (von Ahlefeldt et al. 2003). By contrast, only a few alien species are used for medicinal purposes. Examples include common use of *Acacia dealbata* in the Eastern Cape (Agripa 2016), *Senna occidentalis* (Stinking Weed) (Henderson 2001), and the naturalised perennial wetland herb *Acorus calamus* (Sweet Flag) in the Indian Ocean Coastal Belt (von Ahlefeldt et al. 2003). In most cases, rangeland lost to invasive aliens is expected to result in a loss of medicinal plants. Alien species introduced for providing edible fruits for humans include *Opuntia* species, *Rubus cuneifolius* (American Bramble), and *Psidium guajava* (Guava), all now very successful invaders, but all these are used incidentally as a food source. *Eucalyptus* species are important for supporting native bee species which pollinate most of the insect-pollinated crops in South Africa that are worth an estimated ZAR 10.3 billion annually (SANBI 2018). In extensively transformed landscapes, *Eucalyptus* plantations and stands serve as their primary source of food.

16.5 Negative Impacts of Invasions on Rangelands

16.5.1 Physical and Economic Impacts

This section assesses the influence of invasive alien plants on animal production through affecting the amount and quality of forage, producing harmful chemicals that may be ingested, reducing accessibility to forage, and inflicting physical harm to animals. Their effect on forage abundance depends on whether alien plants are more, equally, or less desirable for animals than the native vegetation they supplant, with the severity of this effect depending on the total area invaded. Van Wilgen et al. (2008) estimated that (at the time of their study) alien plants had caused a 1% decrease in livestock production, but predicted that this may increase to as much as 71% in the future. The forage value of a grass sward invaded by alien grasses is expected to be maintained in most instances because the majority of the 256 alien grass species in the country, of which 122 species have become naturalised and 41 species have become invasive or have had an economic impact, were introduced at pasture research stations for screening of their potential mainly as fodder plants

(Visser et al. 2017). Examples include invasive *Paspalum* and *Bromus* species, *Pennisetum clandestinum* (Kikuyu Grass) and *Lolium multiflorum* (Ryegrass), all palatable species. By contrast, grazing capacity has been reduced by the unpalatable perennials *Nasella trichotoma* (Nasella Tussock) and *Nasella tenuissima* (White Tussock) in mountain grassland in the Eastern Cape Province, *Pennisetum setaceum* (Fountain Grass) in grassland and the Nama-Karoo, and by the annual *Stipa capensis* (Mediterranean Needle Grass), which damages wool, in the Succulent Karoo. However, even the normally unpalatable *P. setaceum* can serve as fodder during drought. We found no studies concerning the impact of alien species on wildlife systems, or which had quantified the impact of alien grasses on livestock production or on wildlife in southern Africa.

Invasion of grassland or savanna by alien trees and shrubs will markedly depress grass production as occurs with encroachment by native woody species (O'Connor and Stevens 2017). This negative impact is offset to some extent when the woody species offer palatable forage for browsers or mixed feeders. Most of the woody invasive species (legumes in the genera *Acacia* and *Prosopis*, and *Chromolaena odorata*, Triffid Weed) (see below) do not offer palatable foliage, although some of the legume species were introduced for their palatable pods. Cattle and browsers consume and effectively disperse the pods of, for example, *Gleditsea triancanthos* (Honey Locust), *Leucaena leucocephala* (Leucaena) and *Prosopis* species, and this has contributed to their recent increase in distribution (Table 16.2; 94.4% increase for *L. leucocephala*) although these foodstuffs can be harmful when large quantities are consumed (Binggeli 2001). The foliage of all woody species listed in Table 16.2 is not considered as quality forage, although a few of these species [e.g., *Acacia mearnsii*, *Morus alba* (White Mulberry), *Psidium guajava* (Guava), *Robinia pseudoacacia* (Black Locust), *Salix babylonica* (Weeping Willow)] may be browsed incidentally. *Schinus molle* (Pepper Tree) seedlings are palatable and consequently may experience potentially high levels of mortality when exposed to sheep grazing in semi-arid savanna (Iponga et al. 2009), and this offers an approach for controlling this species which threatens to expand considerably its distribution (Rouget et al. 2004). However, even species providing poor-quality forage may be used by livestock, especially goats, under intensive stocking; examples include *Acacia dealbata* (Silver Wattle), *Populus X canescens* (Grey Poplar) (Du Toit 2016), and *Lantana camara* (Lantana) (T. O'Connor, pers. obs.). *Atriplex* species and the succulent species *Agave americana* (American Agave), *Opuntia ficus-indica* (Mission Prickly Pear) and *Opuntia monocantha* (Cochineal Prickly Pear) are commonly used as forage (Henderson 2001), and *Melia azedarach* (Syringa) has been made into silage.

Woody alien species that possess deterrents such as thorns may form impenetrable thickets that prevent access to forage, as is evident for drainage lines in the Succulent Karoo Biome, which harbours key forage resources that have been invaded by *Prosopis* spp. (van den Berg 2010). Dense infestations of *Opuntia ficus-indica*, a succulent shrub protected by thorny cladodes, had occurred in the

Eastern Cape by the early 1900s that prevented livestock access to forage and ultimately seriously perturbed the pastoral economy of the region (van Sittert 2002). Physical damage to livestock by well-armed alien plants has been recorded in East Africa for *Opuntia stricta* (Australian Pest Pear) (Shackleton et al. 2017) and *Prosopis* spp. (Bekele et al. 2018), most commonly by thorns damaging hooves.

Animal products are the most important ecosystem service provided by rangelands. De Lange and van Wilgen (2010) estimated the economic impact of alien plant invasions on this service, using van Wilgen et al.'s (2008) estimates of livestock reduction due to invasive alien plants, and assuming a value of ZAR 2500 per large livestock unit. The estimated annual value of livestock production from rangelands in South Africa would have been ZAR 33 billion (2010 values) had they not been invaded. Current levels of invasion have reduced this by an estimated ZAR 340 million annually. These authors also attempted to estimate what the reduction would have been in the absence of historical control efforts, and concluded that current losses could have amounted to ZAR 12.7 billion had there not been historical control. However, because this estimate was based on an approach that required large assumptions, there is a low level of confidence in estimates of avoided cost (van Wilgen and Wilson 2018).

Rangelands are also critical for the provision of water services, and they provide native medicines, fuel wood, wildlife-based tourism, pollination services, and maintain the biodiversity which underpins these. Water resources provided by the higher-rainfall Grassland Biome, for example, were estimated to be worth ZAR 50 billion per year under a scenario of no invasion, and this had been reduced to ZAR 48.9 billion per year due to invasions in 2010 (De Lange and van Wilgen 2010). Biodiversity can provide economic value from harvested products and non-use sources (such as tourism). The annual value of biodiversity-based services from all terrestrial biomes was estimated at ZAR 22.1 billion per year (with no invasions) and ZAR 21.7 billion per year at current levels of invasion (De Lange and van Wilgen 2010). Again, these reductions would almost certainly have been far more severe had there been no attempt at control in the past, but the magnitude of the economic value of historical control cannot be estimated with high confidence due to a lack of data.

A cost-benefit threshold is expected to apply in that alien plants may offer considerable value when at low abundance but their detrimental impact far outweighs advantages as their abundance increases (Shackleton et al. 2007; van Wilgen and Richardson 2014). For example, *Prosopis* trees delivered positive benefits in the form of edible pods when first introduced to South Africa, but as they spread the negative impacts grew, eventually exceeding the benefits and resulting in net negative impacts (Richardson et al. 2000; Wise et al. 2012; Fig. 16.3).

In the following paragraphs we illustrate some consequences of invasion by individual alien species on rangeland quality and condition.

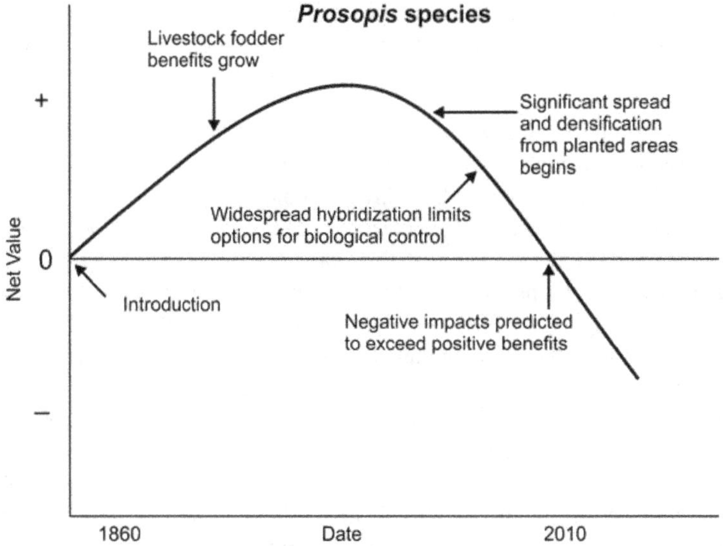

Fig. 16.3 Conceptual illustration showing changing net values (sum of benefits minus sum of impacts) over time associated with *Prosopis* species (economic information from Wise et al. 2012; Figure redrawn from van Wilgen and Richardson 2014)

16.5.2 Prosopis *Species (Mesquite)*

Prosopis is one of the few invasive alien taxa whose ecological and economic impacts have been reasonably well studied in South Africa (van den Berg 2010; Milton and Dean 2010; Ndhlovu et al. 2011, 2016; Shackleton et al. 2015; Zimmermann and Pasiecznik 2005; Dzikiti et al. 2018; see also Bekele et al. (2018) for information from Ethiopia). The genus *Prosopis* includes several tree species and their hybrids that have successfully invaded the Nama-Karoo, Succulent Karoo, and semi-arid Savanna Biomes. In the Northern Cape Province, the cover of *Prosopis* spp. has increased from 0.35% (127,821 ha) in 1974 to 4.06% (1.4 million ha) by 2007 (van den Berg 2010); between 2000 and 2016 the occupation country-wide of quarter degree grid cells by *Prosopis glandulosa/velutina* and hybrids increased by 23% (Henderson and Wilson 2017). Mesquite trees were introduced to South Africa from the Americas to provide fodder and shade for livestock. Their pods provide fodder; foliage is not utilised by livestock.

The success of *Prosopis* species as invaders in South African karroid shrubland is paralleled by the encroachment of *P. glandulosa* into savanna grassland in its home range. This has been attributed to chronic grazing, resulting in low fuel loads and a consequent reduction in fire, as well as dispersal of seeds by cattle and creation of suitable micro-sites for establishment (Archer et al. 1988; Brown and Archer 1989). Although the climate and vegetation of North American C_4 savanna grassland and South African karroid shrubland differ, they display a similar environmental history of chronic grazing and an absence of fire. In the Karoo biomes, livestock (mainly sheep) have similarly facilitated invasion of dryland through dispersal of seed from riparian areas.

In terms of impacts, *Prosopis* species have had a disproportionate impact on riparian areas, which serve as key resource areas for livestock during the dry season. About 30% of riparian areas have been invaded compared with 3.3% of dryland area (van den Berg 2010), reducing their grazing capacity (Milton and Dean 2010). For dryland rangeland, moderate levels of *Prosopis* invasion on a heavily grazed Nama Karoo rangeland site (approximately 15% canopy cover) reduced grazing capacity by 34% from 3.87 to 2.56 LSU per 100 ha (Ndhlovu et al. 2011). The negative impacts of *Prosopis* invasion on native vegetation (especially grasses) become apparent above a threshold of around 30% *Prosopis* cover (Ndhlovu et al. 2011, 2016). Shackleton et al. (2015) found that invasion by *Prosopis* reduced perennial grass cover from above 15% where the basal area of *Prosopis* was below 2 m^2/ha, to 0 where the *Prosopis* basal area was above 4.5 m^2/ha. Similarly, the cover of native perennial herbaceous plants decreased from above 20% to zero with similar increases in *Prosopis* cover. In contrast, the cover of annual grasses or annual herbaceous plants persisted at quite high levels of invasion.

16.5.3 Acacia mearnsii *(Black Wattle)*

Acacia mearnsii was introduced from Australia to provide tannins from bark and for wood products. This tree species has extensively invaded the relatively humid parts of South Africa, especially riparian areas, and it has occupied most of its suitable climatic envelope (Rouget et al. 2004). It showed only an 8% increase in distribution range between 2000 and 2016 (Henderson and Wilson 2017), but further increases in density are possible. Grass cover decreases in a negative exponential manner with an increase in wattle cover (Gwate et al. 2016). Yapi et al. (2018) investigated the impacts of *A. mearnsii* invasion on rangeland condition, and rangeland recovery after clearing. They found a 72% reduction in grazing capacity in densely invaded sites, and this improved by 66% 5 years after clearing. In densely invaded sites, the basal cover of grasses was reduced by up to 42%, resulting in a reduction in grazing capacity of 75%, from 50 to 12.5 large stock units per 100 ha in uninvaded and densely invaded sites, respectively. These results suggest that, if left uncontrolled, wattle species can substantially reduce livestock carrying capacity within montane grasslands in South Africa. Exclusion of fire, by intention or as a consequence of chronic grazing that removes the grass fuels necessary for fires, promotes an increase of wattle, whereas in grasslands, this species can be controlled using repeated burning because it does not resprout. Fire-return intervals under repeated burning in mesic grasslands are usually around 2 years. Although burning stimulates germination, seedlings would not develop into a reproductive plant because the fire-return interval is too short. *Acacia mearnsii* possesses many traits typically associated with an aggressive invader species including prolific seed set, seeds capable of remaining dormant for up to 50 years, and a rapid sapling growth rate that promotes overtopping of native vegetation (Richardson and Kluge 2008).

16.5.4 **Parthenium hysterophorus** *(Parthenium)*

Parthenium hysterophorus is an annual herb native to tropical and subtropical America that has increased alarmingly between 2000 and 2016 (Table 16.2). It has negative impacts on livestock production, and is also a serious threat to environmental and human health because of its ability to produce chemicals that cause severe dermatitis, allergy and toxicity in humans and animals (Terblanche et al. 2016). In a study of impacts on subsistence and commercial farmers, Wise et al. (2007) found that unchecked spread reduced the economic returns to small-scale farmers by between 26 and 41%, but that these losses could be offset if the levels of invasion were reduced by at least 50%. However, although 80% mortality can be achieved using chemical or manual control, density of *P. hysterophorus* will return to pre-control levels within a year owing to the size of the seed bank, but density will decline if follow-up clearing is maintained (Goodall et al. 2010).

16.5.5 **Chromolaena odorata** *(Triffid Weed)*

Chromolaena odorata is a tall, bushy, scrambling shrub that was introduced to Africa from the Caribbean and has spread rapidly to occupy most of the climatically suitable habitats for the species in the eastern part of South Africa (Rouget et al. 2004). It has negative impacts on agricultural practices and on biodiversity. Studies of the magnitude of impacts are rare, but Wise et al. (2007) found that the introduction of a mechanical control programme saw annual returns from cattle sales increase by between 7 and 34%, depending on the area of the initial invasion, in the Ntambanana district in KwaZulu-Natal, South Africa.

16.5.6 **Opuntia ficus-indica** *(Mission Prickly Pear)*

The invasive cactus *Opuntia ficus-indica* provides an example of the devastating impact that invasive cacti can have on arid rangelands. The species was responsible for severe degradation in parts of the Karoo, but has now been dramatically reduced in density over most of its distribution by effective biological control (Hill et al. 2020, Chap. 19). At the height of its dominance at the start of the twentieth century, *O. ficus-indica covered* almost 1 million ha of semi-arid rangelands. There were no formal ecological studies that assessed its impact at the time, but several historical overviews have recorded that these impacts were dramatic. Annecke and Moran

(1978) noted that "*Prickly pears ... are problem plants from two points of view: firstly, they form impenetrable thickets over wide areas, overwhelming the native vegetation and completely preventing normal agricultural activities: and secondly, livestock eat the fruit and cladodes, the spines and glochids of which may lead to severe or fatal gastro-intestinal disorders*". Van Sittert (2002) notes that in the 1890s, farmers "*urged the colonial state take immediate action against opuntia to save the pastoral economy from the threatened ruination of its stock holdings, labour and pastures*", and that "*opuntia cut deep into the material base of the settler economy and rendering it unfit for pastoral farming*." Now that the species is under effective biological control, these impacts have largely been forgotten (Hill et al. 2020, Chap. 19).

16.6 Management of Rangelands

Rangelands are managed using manipulations of herbivory and fire that may influence the success of alien plants. Sources of information for considering these factors were studies which have contrasted the influence on botanical composition of commercial ranching, communal tenure, or conservation estate, and the results of field trials or fence-line contrasts investigating fire and grazing management.

Communal rangeland tends to support a greater diversity and abundance of alien plants than commercial rangeland in moist but not in semi-arid or arid environments. Across KwaZulu-Natal [670–1120 mm mean annual precipitation (MAP)], three alien *Paspalum* grass species dominated wetter communal sites but not wetter commercial or semi-arid communal sites (O'Connor et al. 2003). In the southern Drakensberg region of KwaZulu-Natal, the alien dicotyledon *Richardia brasiliensis* (Tropical Richardia) was a sub-dominant species on dryland communal rangeland (13% aerial cover) but this and other alien species were almost absent from commercial or conservation grassland (0.03% cover of *R. brasiliensis*) despite the proximity of transformed land (cultivated grass pastures, maize fields, plantations) that were dominated by a richness (50) of alien herbaceous species (O'Connor 2005). Scott-Shaw and Morris (2015) describe grazing gradients in the KwaZulu-Natal midlands along which alien forb species replace native forb species lost to chronic grazing. This evidence suggests that well-managed rangeland is resistant to invasion. Wetlands were more severely invaded than drylands in this region; their total of 47 alien species included 6 of the 30 most abundant species, with wetlands in communal areas more densely invaded than those in commercial or protected areas (Walters et al. 2006). By contrast, alien species were absent or very scarce on both sides of the fence in semi-arid areas. Such fence-line contrasts included chronically-grazed communal rangeland versus moderately-stocked commercial rangeland or nature reserves in the Succulent Karoo (Todd and Hoffman 2009), the transition from the Nama-Karoo to the Grassland Biome (Rutherford et al. 2012a), Thornveld Savanna (Rutherford and Powrie 2011), and Mopane Savanna (Rutherford et al. 2012b). Within Albany Thicket, chronic continuous grazing by domestic livestock, when compared to a protected area, resulted in the dominant native shrubs *Euclea undulata* (Common Guarri) and *Portulacaria afra* (Spekboom) being replaced by the native *Putterlickia pyracantha* (False Spike-thorn) and the invasive alien *Opuntia ficus-indica* (Mission Prickly Pear) (Hoffman and Cowling 1990).

Graziers commonly implement a long-term burning regime (where applicable), and a system of grazing management involving various combinations of stocking rate, animal type, and grazing intensity. The effect of these on the abundance of alien plants was assessed by compiling a novel data set of the results of long-term management experiments and long-standing fence-line contrasts of management systems (Supplementary Appendix 16.1). Box 16.1 describes the value of this database in illustrating some key management principles. First, exclusion or very infrequent fire promotes the establishment and often dominance of alien woody species in moist grassland or mesic savanna, an effect which is not apparent in regularly-burnt rangeland. By contrast, alien species are not conspicuous in semi-arid savanna despite the fact that they seldom burn. Secondly, commercial dryland rangeland under prescribed grazing and fire management was largely resistant to invasive alien plants with only small amounts of a few usually herbaceous species being recorded. Alien plants increased in response to heavy grazing pressure.

Available evidence indicates that effective use of fire in particular can be used for controlling invasive plants. However, riparian habitats are vulnerable in part because they are not usually subjected to planned fires.

Biological control has been implemented for a small proportion of the 71 prominent species; specifically, 3 are considered to be completely controlled, 16 are under substantial control, 16 are under negligible control, 4 are under investigation, 5 have not been determined, and the remainder are subject to no biological control (Table 16.2) (Hill et al. 2020, Chap. 19).

16.7 Prognosis

The threat posed by invasive alien plants to rangelands is expected to increase with time because many species have not yet spread to occupy their potential climatic range, whereas others that have, such as Australian *Acacia* species, can be expected to increase in density (Rouget et al. 2004). Many currently quiescent species are expected to become invasive once an exponential phase of population growth commences (Hui and Richardson 2017). An example of this is the recent expansion of *Campuloclinium macrocephalum* (Pompom Weed) following a long period where it was only localised (Wilson et al. 2013). Despite increased effort to control the introduction of alien plant species into the country, the pool of alien species is growing (van Wilgen and Wilson 2018) and the irruption of new invasive species is almost inevitable.

Direct actions such as clearing and biological control will continue to provide a foundation for containing invasive alien plants, but management of rangelands and the animals they support have been complicit in the success of many alien species. Refinement of rangeland management in order to suppress or contain invasive alien plants deserves closer attention. Withdrawal of fire, or an increase in fire-return interval, appears to have been an important management action promoting woody alien species in the fire-dependent Grassland and Savanna Biomes. The effect of chronic grazing on the availability of fuel, especially in communal areas, has contributed to increased fire-return intervals. Appropriate resting and burning programmes should be adopted, which would further serve to suppress native bush (O'Connor et al. 2014). Fire exclusion, not uncommon with commercial producers, will lead to invasion by alien woody species into savanna and grassland. Management options in the Karoo biomes, Albany Thicket, or riparian habitats, which are not prone to fire, are limited because grazing management has less of an effect on the success of alien species, although critical study awaits. Heavy grazing intensity apparently promotes invasive alien plants. Effective seed dispersal of certain woody species by livestock can be reduced by appropriate management of their movement; palatable alien species may be contained by stocking with additional herbivore species. Rangelands, on account of covering >70% of South Africa and influencing the delivery of the bulk of the country's ecosystem services, warrant closer investigation of the manner in which management can be refined in order to contain invasive alien plants.

The large archive of phytosociological data in South Africa awaits analysis for defining the effect of land tenure and environment on the richness and abundance of alien plants.

No panacea for preventing further increases in invasive alien plants has been unearthed. Rather, a diversity of alien species displaying a wide range in their invasion dynamics demands further innovative response to identify economically sustainable means of preventing or limiting their increase. Biocontrol has been successful in some cases, but the sheer number of invasive alien species militates against this avenue providing overall success. Achievements realised through government-sponsored clearing have been noteworthy, but this approach will require additional funding in order to maintain the current level of threat of treated species.

Acknowledgements BvW acknowledges support from the DSI-NRF Centre for Invasion Biology and the South African National Research Foundation (Grant 109467).

References

Agripa N (2016) Perceptions and livelihood uses of an invasive alien tree (*Acacia dealbata*) by rural communities in the Eastern Cape. MSc thesis, Rhodes University, Grahamstown

Annecke DP, Moran VC (1978) Critical reviews of biological pest control in South Africa. 2. The prickly pear, *Opuntia ficus-indica* (L.) Miller. J Entomol Soc S Afr 41:161–188

Archer S, Scifres C, Bassham CR et al (1988) Autogenic succession in a subtropical savanna: conversion of grassland to thorn woodland. Ecol Monogr 58:111–127. https://doi.org/10.2307/1942463

Archibald SA, Roy DP, van Wilgen BW et al (2009) What limits fire? An examination of drivers of burnt area in sub-equatorial Africa. Glob Chang Biol 15:613–630. https://doi.org/10.1111/j.1365-2486.2008.01754.x

Beinart W (2003) The rise of conservation in South Africa: settlers, livestock, and the environment 1770–1950. Oxford University Press, Oxford

Bekele K, Haji J, Legesse B et al (2018) Impacts of woody invasive alien plant species on rural livelihood: generalised propensity score evidence from *Prosopis* spp. invasion in Afar region in Ethiopia. Pastor Res Policy Pract 8:28. https://doi.org/10.1186/s13570-018-0124-6

Binggeli P (2001) The human dimensions of invasive woody plants. In: McNeely JA (ed) The great reshuffling – human dimensions of invasive alien species. IUCN, Gland, pp 145–159

Brown JR, Archer S (1989) Woody plant invasion of grasslands: establishment of honey mesquite (*Prosopis glandulosa* var. *glandulosa*) on sites differing in herbaceous biomass and grazing history. Oecologia 80:19–26. https://doi.org/10.1007/BF00789926

Carruthers J (2008) "Wilding the farm or farming the wild?" The evolution of scientific game ranching in South Africa from the 1960s to the present. Trans R Soc S Afr 63:160–181. https://doi.org/10.1080/00359190809519220

De Lange WJ, van Wilgen BW (2010) An economic assessment of the contribution of biological control to the management of invasive alien plants and to the protection of ecosystem services in South Africa. Biol Invasions 12:4113–4124. https://doi.org/10.1007/s10530-010-9811-y

Dean WRJ, Milton SJ (2000) Directed dispersal of *Opuntia* species in the Karoo, South Africa: are crows the responsible agents? J Arid Environ 45:305–314. https://doi.org/10.1006/jare.2000.0652

Dean WRJ, Seymour CL, Joseph GS (2018) Linear structures in the Karoo, South Africa, and their impacts on biota. Afr J Range For Sci 35:223–232. https://doi.org/10.2989/10220119.2018.1514530

Department of Agriculture, Forestry and Fisheries (2018) Conservation of Agricultural Resources Act, Act 43 of 1983. New long-term grazing capacity map for South Africa 2017. Government Gazette No. 41870, 31 August 2018, Pretoria

Department of Environmental Affairs (2018a) EGIS protected areas database. https://egis.environment.gov.za/protected_areas_database

Department of Environmental Affairs (2018b) Unlocking the socio-economic potential of South Africa's biodiversity assets through sustainable use of wildlife resources. Available at: https://pmg.org.za/files/180227DEA.pdf

Du Toit JCO (2016) Control of the invasive grey poplar (*Populus* x *canescens*) using mechanical removal and grazing by livestock. Research report of the Grootfontein Agricultural Development Institute, Middelburg, pp 74–75

Du Toit JCO, van den Berg L, O'Connor TG (2015) Photographic evidence of fire-induced shifts from dwarf-shrub- to grass-dominated vegetation in Nama-Karoo. S Afr J Bot 101:148–152. https://doi.org/10.1016/j.sajb.2015.06.002

Dzikiti S, Bugan R, Le Maitre D et al (2018) Comparison of water use by *Prosopis* spp and the co-occurring *Vachellia karroo* trees before and after clearing the invasions: implications on ground water. Water Research Commission Report No 2256/1/18, Pretoria

Elphick R (1985) Khoikhoi and the founding of white South Africa. Ravan Press, Johannesburg

Geldenhuys CJ, Cawe SG (2011) Matching resource use needs with resource status and population dynamics of target species in Transkei coastal forests to sustain resource use, Port St Johns Forest Estate, South Africa. In: Geldenhuys CJ, Ham C, Ham H (eds) Sustainable forest management in Africa: some solutions to natural forest management in Africa. Proceedings of the Sustainable Forest Management in Africa Symposium, Stellenbosch, 3–7 November 2008, Department of Forest and Wood Science, Stellenbosch University, pp 264–278

Goodall J, Braack M, de Klerk J et al (2010) Study on the early effects of several weed-control methods on *Parthenium hysterophorus* L. Afr J Range For Sci 27:95–99

Gwate O, Mantel SK, Finca A et al (2016) Exploring the invasion of rangelands by *Acacia mearnsii* (black wattle): biophysical characteristics and management implications. Afr J Range For Sci 33:265 273

Henderson L (2001) Alien invasive weeds and invasive plants. Plant Protection Research Institute

Handbook No. 12, Agricultural Research Council, Department of Agriculture, Pretoria
Henderson L, Wilson JRU (2017) Changes in the composition and distribution of alien plants in South Africa: an update from the Southern African Plant Invaders Atlas. Bothalia 47(2):a2172
Hill MP, Moran VC, Hoffmann JH et al (2020) More than a century of biological control against invasive alien plants in South Africa: a synoptic view of what has been accomplished. In: van Wilgen BW, Measey J, Richardson DM et al (eds) Biological invasions in South Africa. Springer, Berlin, pp 549–568. https://doi.org/10.1007/978-3-030-32394-3_19
Hobbs RJ, Huenneke LF (1992) Disturbance, diversity, and invasion: implications for conservation. Conserv Biol 6:324–337. https://doi.org/10.1046/j.1523-1739.1992.06030324.x
Hoffman MT, Cowling RM (1990) Desertification in the Lower Sundays River Valley, South Africa. J Arid Environ 19:105–117. https://doi.org/10.1016/S0140-1963(18)30834-6
Huffman TN (2007) Handbook to the Iron Age. University of KwaZulu-Natal Press, Pietermaritzburg
Hui C, Richardson DM (2017) Invasion dynamics. Oxford University Press, Oxford. https://doi.org/10.1093/acprof:oso/9780198745334.001.0001
Iponga DM, Milton SJ, Richardson DM (2009) Soil type, microsite, and herbivory influence growth and survival of *Schinus molle* (Peruvian pepper tree) invading semi-arid African savanna. Biol Invasions 11:159–169. https://doi.org/10.1007/s10530-008-9221-6
Klein H (2011) A catalogue of the insects, mites and pathogens that have been used or rejected, or are under consideration, for the biological control of invasive alien plants in South Africa. Afr Entomol 19:515–549. https://doi.org/10.4001/003.019.0214
Meissner HH, Scholtz MM, Palmer AR (2013) Sustainability of the South African livestock sector towards 2050 Part I: worth and impact of the sector. S Afr J Anim Sci 43:282–297. https://doi.org/10.4314/sajas.v43i3.5
Milton SJ, Dean WRJ (2010) Plant invasions in arid areas: special problems and solutions: a South African perspective. Biol Invasions 12:3935–3948. https://doi.org/10.1007/s10530-010-9820-x
Moore JM, Picker MD (1991) Heuweltjies (earth mounds) in the Clanwilliam district, Cape Province, South Africa: 4000-year-old termite nests. Oecologia 86:424–432. https://doi.org/10.1007/BF00317612
Mucina L, Rutherford MC (eds) (2006) The vegetation of South Africa, Lesotho and Swaziland. Strelitzia 19. South African National Biodiversity Institute, Pretoria
Ndhlovu T, Milton-Dean SJ, Esler KJ (2011) Impact of *Prosopis* (mesquite) invasion and clearing on the grazing capacity of semiarid Nama Karoo rangeland, South Africa. Afr J Range For Sci 28:129–137. https://doi.org/10.2989/10220119.2011.642095
Ndhlovu T, Milton SJ, Esler KJ (2016) Effect of *Prosopis* (mesquite) invasion and clearing on vegetation cover in semi-arid Nama Karoo rangeland, South Africa. Afr J Range For Sci 33:11–19. https://doi.org/10.2989/10220119.2015.1036460
O'Connor TG (2005) Influence of land use on plant community composition and diversity in Highland Sourveld grassland in the southern Drakensberg, South Africa. J Appl Ecol 42:975–988. https://doi.org/10.1111/j.1365-2664.2005.01065.x
O'Connor TG, Stevens N (2017) Bush encroachment. Oxford online library. https://doi.org/10.1093/OBO/9780199363445-0069
O'Connor TG, Morris CD, Marriott DJ (2003) Change in land use and botanical composition of KwaZulu-Natal's grasslands over the past fifty years: Acocks' sites revisited. S Afr J Bot 69:105–115. https://doi.org/10.1016/S0254-6299(15)30365-3
O'Connor TG, Puttick JR, Hoffman MT (2014) Bush encroachment in southern Africa: changes and causes. Afr J Range For Sci 31:67–88. https://doi.org/10.2989/10220119.2014.939996
O'Reagain PJ, Turner JR (1992) An evaluation of the empirical basis for grazing management recommendations for rangeland in southern Africa. J Grassl Soc South Afr 9:38–49. https://doi.org/10.1080/02566702.1992.9648297
Orton J, Mitchell P, Klein R et al (2013) An early date for cattle from Namaqualand, South Africa: implications for the origin of herding in southern Africa. Antiquity 87:108–120. https://doi.org/10.1017/S0003598X00048651
Rahlao SJ, Milton SJ, Esler KJ et al (2009) Effects of invasion of fire-free arid shrublands by a

fire-promoting invasive alien grass (*Pennisetum setaceum*) in South Africa. Austral Ecol 34:920–928. https://doi.org/10.1111/j.1442-9993.2009.02000.x

Richardson DM, Kluge RL (2008) Seed banks of invasive Australian Acacia species in South Africa: role in invasiveness and options for management. Perspect Plant Ecol 10:161–177. https://doi.org/10.1016/j.ppees.2008.03.001

Richardson DM, van Wilgen BW (2004) Invasive alien plants in South Africa: how well do we understand the ecological impacts? S Afr J Sci 100:45–52

Richardson DM, Bond WJ, Dean WRJ et al (2000) Invasive alien organisms and global change: a South African perspective. In: Mooney HA, Hobbs RJ (eds) Invasive species in a changing world. Island Press, Washington, DC, pp 303–349

Richardson DM, Foxcroft LC, Latombe G et al (2020) The biogeography of South African terrestrial plant invasions. In: van Wilgen BW, Measey J, Richardson DM, Wilson JR, Zengeya TA (eds) Biological invasions in South Africa. Springer, Berlin, pp 65–94. https://doi.org/10.1007/978-3-030-32394-3_3

Rouget M, Richardson DM, Nel JL et al (2004) Mapping the potential ranges of major plant invaders of South Africa, Lesotho and Swaziland using climatic suitability. Divers Distrib 10:475–484

Rouget M, Hui C, Renteria J et al (2015) Plant invasions as a biogeographical assay: vegetation biomes constrain the distribution of invasive alien species assemblages. S Afr J Bot 101:24–31. https://doi.org/10.1016/j.sajb.2015.04.009

Roux E (1969) Grass: a story of Frankenwald. Oxford University Press, Cape Town

Rowe-Rowe DT, Scotcher JSB (1986) Ecological carrying capacity of the Natal Drakensberg for wild ungulates. S Afr J Wildl Res 16:12–16

Rutherford MC, Powrie LW (2011) Can heavy grazing on communal land elevate plant species richness levels in the Grassland Biome of South Africa? Plant Ecol 212:1407–1418. https://doi.org/10.1007/s11258-011-9916-0

Rutherford MC, Powrie LW, Husted LB (2012a) Plant diversity consequences of a herbivore-driven biome switch from grassland to Nama-Karoo shrub steppe in South Africa. Appl Veg Sci 15:14–25

Rutherford MC, Powrie LW, Thompson DI (2012b) Impacts of high utilisation pressure on biodiversity components in *Colophospermum mopane* savanna. Afr J Range For Sci 29:1–11

SANBI (2018) Gums and bees. A roadmap for landowners in South Africa. South African National Biodiversity Institute, Pretoria

Scott-Shaw R, Morris CD (2015) Grazing depletes forb species diversity in the mesic grasslands of KwaZulu-Natal, South Africa. Afr J Range For Sci 32:21–31. https://doi.org/10.2989/10220119.2014.901418

Shackleton CM, Shackleton SE (2004) The importance of non-timber forest products in rural livelihood security and as safety nets: evidence from South Africa. S Afr J Sci 100:658–664

Shackleton CM, McGarry D, Fourie S et al (2007) Assessing the effects of invasive alien species on rural livelihoods: case examples and a framework from South Africa. Hum Ecol 35:113–127. https://doi.org/10.1007/s10745-006-9095-0

Shackleton RT, Le Maitre DC, Richardson DM et al (2015) The impact of invasive alien *Prosopis* species (mesquite) on native plants in different environments in South Africa. S Afr J Bot 97:25–31

Shackleton RT, Witt A, Merinyi F, van Wilgen BW (2017) Distribution and socio-ecological impacts of the invasive alien cactus *Opuntia stricta* in eastern Africa. Biol Invasions 19:2427–2441. https://doi.org/10.1007/s10530-017-1453-x

Skowno AL, Poole CJ, Raimondo DC et al (2019) National biodiversity assessment 2018: the status of South Africa's ecosystems and biodiversity. Synthesis report. South African National Biodiversity Institute, Pretoria

Tainton NM (1999) Veld management in South Africa. University of Natal Press, Pietermaritzburg

Tapson D (1993) Biological sustainability in pastoral systems: the KwaZulu case. In: Behnke RH, Scoones I, Kerven C (eds) Range ecology at disequilibrium. Overseas Development Institute, London, pp 118–135

Terblanche C, Nänni I, Kaplan H et al (2016) An approach to the development of a national strategy for controlling invasive alien plant species: the case of *Parthenium hysterophorus* in South Africa. Bothalia 46(1):a2053. https://doi.org/10.4102/abc.v46i1.2053

Thrash I (1998) Impact of water provision on herbaceous vegetation in Kruger National Park, South Africa. J Arid Environ 38:315–324. https://doi.org/10.1006/jare.1997.0331

Titshall LW, O'Connor TG, Morris CD (2000) Effect of long-term exclusion of fire and herbivory on the soils and vegetation of sour grassland. Afr J Range For Sci 17:70–80. https://doi.org/10.2989/10220110009485742

Todd SW, Hoffman MT (2009) A fence line in time demonstrates grazing-induced vegetation shifts and dynamics in the semiarid Succulent Karoo. Ecol Appl 19:1897–1908. https://doi.org/10.1890/08-0602.1

van den Berg E (2010) Detection, quantification and monitoring *Prosopis* spp. in the Northern Cape Province of South Africa using remote sensing and GIS. MSc thesis, North-West University, Potchefstroom

van Sittert L (2002) 'Our irrepressible fellow-colonist': the biological invasion of prickly pear (*Opuntia ficus-indica*) in the Eastern Cape c.1890-c.1910. J Hist Geog 28:397–419. https://doi.org/10.1006/jhge.2001.0457

van Wilgen BW, Richardson DM (2014) Challenges and trade-offs in the management of invasive alien trees. Biol Invasions 16:721–734. https://doi.org/10.1007/s10530-013-0615-8

van Wilgen BW, Scholes RJ (1997) The vegetation and fire regimes of southern hemisphere Africa. In: van Wilgen BW, Andreae MO, Goldammer GJ et al (eds) Fire in southern African savannas: ecological and atmospheric perspectives. Wits University Press, Johannesburg, pp 27–46

van Wilgen BW, Wilson JR (2018) The status of biological invasions and their management in South Africa 2017. South African National Biodiversity Institute, Kirstenbosch and DST-NRF Centre of Excellence for Invasion Biology, Stellenbosch

van Wilgen BW, Reyers B, Le Maitre DC et al (2008) A biome-scale assessment of the impact of invasive alien plants on ecosystem services in South Africa. J Environ Manag 89:336–349. https://doi.org/10.1016/j.jenvman.2007.06.015

Visser V, Wilson JRU, Canavan K et al (2017) Grasses as invasive plants in South Africa revisited: patterns, pathways and management. Bothalia 27(2):a2169. https://doi.org/10.4102/abc.v47i2.2169

von Ahlefeldt D, Crouch NR, Nichols G et al (2003) Medicinal plants traded on South Africa's eastern seaboard. Ethekwini Parks Department and University of KwaZulu-Natal, Porcupine Press

Walker C, Milton SJ, O'Connor T et al (2018) Drivers and trajectories of social and ecological change in the Karoo region of South Africa. Afr J Range For Sci 35:157–177. https://doi.org/10.2989/10220119.2018.1518263

Walters DJ, Kotze DC, O'Connor TG (2006) Impact of land use on vegetation composition, diversity, and selected soil properties of wetlands in the southern Drakensberg mountains, South Africa. Wetl Ecol Manag 14:329–348. https://doi.org/10.1007/s11273-005-4990-5

Wilson JRU, Ivey P, Manyama P et al (2013) A new national unit for invasive species detection, assessment and eradication planning. S Afr J Sci 109(5/6):#0111, 13 p. https://doi.org/10.1590/sajs.2013/20120111

Wise RM, van Wilgen BW, Hill MP et al (2007) The economic impact and appropriate management of selected invasive alien species on the African continent. CSIR report number: CSIR/NRE/RBSD/ER/2007/0044/C, Council for Scientific and Industrial Research, Pretoria, 64 p

Wise RM, van Wilgen BW, Le Maitre DC (2012) Costs, benefits and management options for an invasive alien tree species: the case of mesquite in the Northern Cape. J Arid Environ 84:80–90

Yapi TS, O'Farrell PJ, Dziba LE et al (2018) Alien tree invasion into a South African montane grassland ecosystem: impact of Acacia species on rangeland condition and livestock carrying capacity. Int J Biodivers Sci 14:105–116

Zachariades C, Paterson ID, Strathie LW et al (2017) Assessing the status of biological control as a management tool for suppression of invasive alien plants in South Africa. Bothalia 47(2):a2142. https://doi.org/10.4102/abc.v47i2.2142

17

Assessing the Influence of Alien Species on Biodiversity

Tsungai A. Zengeya, Sabrina Kumschick, Olaf L. F. Weyl and Brian W. van Wilgen

Abstract Studies of the impact of alien species on the environment are increasingly being carried out, and there has been ongoing debate about how to standardise the description of these impacts. This chapter evaluates the state of knowledge on the impacts of alien species on biodiversity in South Africa based on different assessment methods. Despite South Africa being one of the most biologically diverse countries in the world, there have been very few studies that formally document the impacts of alien species on biodiversity. Most of what is known is based on expert opinion, and consequently the level of confidence in the estimates of the magnitude of these impacts is low. However, it is clear that a significant number of alien species cause major negative impacts, and that there is cause for serious concern. There is a

T. A. Zengeya (✉)
South African National Biodiversity Institute, Kirstenbosch Research Centre, Cape Town, South Africa

Centre for Invasion Biology, Department of Zoology and Entomology, University of Pretoria, Pretoria, South Africa
e-mail: t.zengeya@sanbi.org.za

S. Kumschick
South African National Biodiversity Institute, Kirstenbosch Research Centre, Cape Town, South Africa

Centre for Invasion Biology, Department of Botany and Zoology, Stellenbosch University, Stellenbosch, South Africa

O. L. F. Weyl
DSI-NRF South African Research Chair in Inland Fisheries and Freshwater Ecology, South African Institute for Aquatic Biodiversity, Makhanda, South Africa

Centre for Invasion Biology, South African Institute for Aquatic Biodiversity, Makhanda, South Africa

B. W. van Wilgen
Centre for Invasion Biology, Department of Botany and Zoology, Stellenbosch University, Stellenbosch, South Africa

growing global effort to assess all alien species with standardised protocols to alleviate the problem of comparing impacts measured using different approaches. Formal assessments have been done for a few alien species in South Africa, but most naturalised and invasive species have not been evaluated, and, we suspect, for most alien species there has been no attempt, as yet, to document their impacts. However, red-listing processes found that alien species were frequently included as a significant extinction risk for several native species of fish, amphibians, and plants. There are very few studies that cover the combined impacts of co-occurring alien species in particular areas, and these studies could provide the rationale for regulation and management, which is often absent. While reductions due to alien species in the value of ecosystem services, the productivity of rangelands, and biodiversity intactness are relatively low at present these impacts are expected to grow rapidly as more invasive species enter a stage of exponential growth.

17.1 Introduction

South Africa occupies only 2% of the world's land area but it is one of the most biologically diverse countries globally (Mittermeier et al. 1997). For example, it is estimated that 7% of the world's vascular plants, 5% of mammals and 7% of birds are found in South Africa (Le Roux 2002). In addition, more than half of the plants, butterflies, amphibians and reptiles native to South Africa are endemic (Colville et al. 2014). This is partly because of the country's diverse environmental conditons that have resulted not only in high species diversity and endemism, but also in a diversity of terrestrial, freshwater and marine ecosystems (van Wilgen et al. 2020a, Chap. 1; Wilson et al. 2020, Chap. 13). The country's terrestrial vegetation types can be broadly grouped into nine biomes that range from deserts and Mediterranean-climate shrublands (fynbos) in the west to grasslands and savanna in the eastern interior, and evergreen forests in high rainfall areas along the east coast (Mucina and Rutherford 2006). Three areas within these biomes—the Cape Floristic Region, the Succulent Karoo, and the Maputaland-Pondoland-Albany region—have been designated global biodiversity hotspots (Myers et al. 2000; Mittermeier et al. 2004). The marine realm includes several ecosystems in the surrounding Indian and Atlantic oceans and offshore islands (van Wilgen et al. 2020a, Chap. 1). The marine ecosystems are also diverse with over 12,000 species that represent 15% of all known coastal marine species worldwide (Le Roux 2002; Griffiths et al. 2010).

Alien species can change the composition, structure and functioning of biodiversity in many ways. Examples include hybridisation with native species (e.g. Tracey et al. 2008), extirpation of native species through predation (e.g. Rodda and Fritts 1992) or disease transmission (Hulme 2014), alterations to the structure of vegetation by introducing novel growth forms (van Wilgen and Richardson 1985), and changes to ecosystem processes like fire (Brooks et al. 2004) and hydrology (Le Maitre et al. 2015). A recent study concluded, based on data from the IUCN Red List, that alien species were the most common threat associated with extinctions of mammals, amphibians and reptiles, and for vertebrate extinctions overall (Bellard et al. 2016a).

In South Africa, a total of 107 alien species are suspected to have major negative impacts on biodiversity, and most (75%) of these are plants (van Wilgen and Wilson 2018). Despite these concerns, there have been relatively few studies that have formally quantified the impacts of alien species on different facets of biodiversity in South Africa. In a review focussing on alien plants, Richardson and van Wilgen (2004) concluded that information on the ecological impacts of alien plants was generally poor, and that the consequences of invasions for the delivery of ecosystem goods and services were, with the notable exception of their influence on water resources, inadequately studied. Since then, the situation has improved to some degree.

This chapter evaluates the state of knowledge on the impacts of alien species on biodiversity in South Africa based on different assessment methods. It focuses mainly on explicit knowledge of impacts of alien species on compositional and structural aspects of biodiversity at a species scale, and less at genetic, community, and ecosystem scales (cf. Noss 1990). Some references are presented that discuss the cumulative impact of all invasions on biodiversity, but how these impacts interact with other global change drivers is not covered.

17.2 Assessing the Impact of Alien Species on Biodiversity

Different approaches have been used to quantify, assess and summarise the impacts caused by alien species on native biodiversity. Here, we focus on three main methods: impact-scoring schemes, expert opinion assessments, and impacts identified during red-listing processes. Impact scoring schemes such as the Environmental Impact Classification for Alien Taxa (EICAT; Blackburn et al. 2014; Hawkins et al. 2015; see also Box 17.1) and the Socio-Economic Impact Classification of Alien Taxa (SEICAT; Bacher et al. 2018), amongst others, can be used to formally assess and quantify impacts of individual alien species. Impact-scoring schemes are based on published evidence, and aim to make impacts comparable between taxa and regions by assigning them to semi-quantitative classes which are clearly defined to minimise assessor bias. We focus here on the EICAT scheme as it is most relevant for impacts of a particular alien species on native species. Expert opinion studies (e.g. Zengeya et al. 2017) also assess impacts of specific alien species, but use experts rather than published impact evidence to gauge impacts. Expert opinion studies can be done at a species level (impact of a particular alien species on native species) and or on the entirety of invasions on an ecosystem (e.g. ecosystem services; van Wilgen et al. 2008). Lastly, the International Union for the Conservation of Nature (IUCN) Red List of Threatened Species reports threats from a combination of invasions on a particular native species (Mace et al. 2008).

Box 17.1 The Environmental Impact Classification for Alien Taxa (EICAT)

EICAT is a simple, objective and transparent method to classify alien taxa according to the magnitude of environmental impacts caused in their introduced ranges (Blackburn et al. 2014; Hawkins et al. 2015). It assesses impacts caused by a specific alien taxon (mostly a species) on native species in the recipient area and can help identify taxa with different levels of impacts as well as facilitate comparisons of impacts between taxa and regions. Based on published records of impacts, EICAT classifies detrimental impacts based on 12 mechanisms, namely:

- Competition
- Predation
- Hybridisation
- Transmission of diseases
- Parasitism
- Poisoning/toxicity
- Bio-fouling
- Grazing/herbivory/browsing
- Chemical, physical or structural impact on ecosystem
- Interaction with other taxa

Furthermore, if records of impact are available, EICAT distinguishes between five impact levels based on the organisational level of the species affected, as follows:

Minimal Concern (MC)—the alien taxon does not affect the performance of any native species it interacts with through any of the above mentioned mechanisms

Minor (MN)—the alien taxon affects the performance of at least one native species through any of the above mentioned mechanisms, but does not affect population size

Moderate (MO)—the alien taxon reduces the population size of at least one native species through any of the above mentioned mechanisms, but the native species is still present in the community

Major (MR)—the alien taxon leads to the local or metapopulation extinction of at least one native species changing the native community, with the impacts being reversible when the alien taxon is no longer present

Massive (MV)—irreversible community changes caused by the alien taxon through the local, sub-population or global extinction of at least one native species

If no data on impacts of the taxon in the alien range is available, it is classified as Data Deficient (DD).

(continued)

Box 17.1 (continued)
EICAT is in the process of being adopted by the IUCN as a standard system for classifying invasive alien species based on the nature and magnitude of their impacts (see https://www.iucn.org/theme/species/our-work/invasive-species/eicat for more details).

17.2.1 Impact-Scoring Schemes

The Environmental Impact Classification for Alien Taxa has been applied to various taxa that are known to occur as alien species in South Africa including grasses (Visser et al. 2017; Nkuna et al. 2018; Canavan et al. 2019), amphibians (Kumschick et al. 2017), birds (Evans et al. 2016), mammals (Hagen and Kumschick 2018), fish (Marr et al. 2017), gastropods (Kesner and Kumschick 2018), and some other invertebrates (Nelufule 2018). Most EICAT assessments performed to date have been done at a global scale, i.e. including all records of impact of a given species in its global alien range. It is important to note that although global assessments are comprehensive, there will need to be regularly updated as further assessments are done at national scales. In South Africa, national-level EICAT assessments have been done for some alien grasses (Visser et al. 2017) and fishes (Marr et al. 2017) (Table 17.1).

A global effort is in progress to use EICAT to assess all species in their alien ranges, and researchers from South Africa have contributed many assessments to date. Evidence-based assessments are also needed in South Africa for regular reporting on the status of biological invasions (van Wilgen and Wilson 2018; Wilson et al. 2018; van Wilgen et al. 2020a, Chap. 1). Moreover, the risk analysis framework currently used as evidence to support listing of alien species in South Africa (Kumschick et al. 2018) requires impact assessments analogous to EICAT (Kumschick et al. 2020, Chap. 20). However, while some progress has been made, there is still a very large number of species that have to be formally assessed in South Africa.

17.2.1.1 Grasses

There are six grass species that have had EICAT assessments done at a national scale for South Africa, with *Arundo donax* (Giant Reed) and *Glyceria maxima* (Reed Meadow Grass) evaluated as having major impacts (Visser et al. 2017; Nkuna et al. 2018; VK Nkuna, unpublished data). These two species have been implicated in competitively displacing native species; *A. donax*, for example, dominates riparian areas and can locally exclude native plants (Holmes et al. 2005; Guthrie 2007), while *G. maxima* has locally displaced some native wetland species (Mugwedi 2012).

Table 17.1 Alien species that have been assessed by the Environmental Impact Classification for Alien Taxa (EICAT) scheme to have major (MR) and massive (MV) impacts on biodiversity in South Africa[a]

Taxon	Species	Assessment category	Mechanisms	Impacts	Source
Fish	*Oreochromis niloticus* (Nile Tilapia)	MV	Hybridisation	Native *Oreochromis* species such as *O. mossambicuss* are under threat from hybridisation with *O. niloticus*	Marr et al. (2017)
Grass	*Arundo donax* (Giant Reed)	MR	Competition	Dominates riparian areas and can locally exclude native species	Visser et al. (2017), Nkuna et al. (2018), VK Nkuna (unpublished data)
Grass	*Glyceria maxima* (Reed Meadow Grass)	MR	Competition	Outcompetes native wetland species	Visser et al. (2017), Nkuna et al. (2018), VK Nkuna (unpublished data)
Fish	*Micropterus dolomieu* (Smallmouth Bass)	MR	Competition and predation	Causes changes in the community structure of invertebrates and local extirpations and fragmentation of native fish populations through predation	Marr et al. (2017)
Fish	*Micropterus salmoides* (Largemouth Bass)	MR	Competition and predation	Causes changes in the community structure of invertebrates and local extirpations and fragmentation of native fish populations through predation	Marr et al. (2017)
Fish	*Oncorhynchus mykiss* (Rainbow Trout)	MR	Competition and predation	Causes declines, and in some cases local extirpation, of native invertebrates, frogs and fish through predation	Marr et al. (2017)
Fish	*Salmo trutta* (Brown Trout)	MR	Competition and predation	Causes declines, and in some cases local extirpation, of native invertebrates, frogs and fish through predation	Marr et al. (2017)

[a]Major impacts lead to community changes, which are reversible; and massive impacts leads to irreversible community changes and extinctions

17.2.1.2 Gastropods

Thirty-four gastropod species alien to South Africa have formal global EICAT assessments. Four of these—*Helisoma duryi* (Seminole Rams-Horn), *Tarebia granifera* (Quilted Melania Snail), *Oxychilus draparnaudi* (Drapanaud's Glass Snail) and *Theba pisana* (White Garden Snail)—are known to change community structures in their global invaded range (Kesner and Kumschick 2018). For example, *H. duryi* outcompetes and displaces native snail species (e.g. Christie et al. 1981), *T. granifera* has been implicated in causing local extinctions of native snails in wetlands (e.g. Karatayev et al. 2009) and *O. draparnaudi* is a predator that causes the local disappearance of prey species (e.g. Frest and Sanders Rhodes 1982).

In South Africa, the only documented impacts are for *Tarebia granifera* and *Theba pisana*. The former was accidentally introduced into South Africa through the aquarium trade and has since invaded several rivers, lakes, wetlands and estuaries in the eastern and northern parts of the country (Appleton et al. 2009). It can reach high population densities, and has been implicated in displacing native snails and becoming the dominant component of invertebrate assemblages in invaded areas (Miranda and Perissinotto 2014). *Theba pisana* can reach high densities on the west and south coast of South Africa, with the potential to impact on native fauna and flora (Odendaal et al. 2008).

17.2.1.3 Fish

No fish species have been globally assessed with EICAT to date, but an assessment of alien fish species naturalised in South Africa identified five species that have negative impacts (Marr et al. 2017). *Micropterus dolomieu* (Smallmouth Bass), *M. salmoides* (Largemouth Bass), *Oncorhynchus mykiss* (Rainbow Trout), *Salmo trutta* (Brown Trout) and *Oreochromis niloticus* (Nile Tilapia) are known to have adverse impacts in South Africa that span all levels of biological organisation from gene to ecosystem (Ellender and Weyl 2014). On a genetic level, the integrity of native *Oreochromis mossambicus* (Mozambique Tilapia) is under threat from hybridisation with *O. niloticus*, a species introduced for aquaculture (Firmat et al. 2013). Evidence of species and population-level impacts are mainly linked to direct predation by *O. mykiss*, *S. trutta* and *Micropterus* species (Black Basses), which have resulted in local extirpations of native fishes and invertebrates (Ellender and Weyl 2014; van der Walt et al. 2016; Shelton et al. 2018).

Oncorhynchus mykiss and *S. trutta* are cold water species that, as a result of stocking for sport fishing, have naturalised in many cool, well-oxygenated headwater streams in the country (Ellender et al. 2017; Weyl et al. 2017a). They are generalist predators that feed primarily on aquatic invertebrates, but will also opportunistically prey on terrestrial invertebrates, fish and amphibians. Their impacts therefore span numerous trophic levels, and in South Africa include the decline, and in some cases local extirpation, of native invertebrates, frogs and fish

(Karssing et al. 2012; Rivers-Moore et al. 2013; Jackson et al. 2016). *Oncorhynchus mykiss* and *S. trutta* have also been implicated in the decline of populations of *Hadromophryne natalensis* (Natal Cascade Frog) in a streams in the uKhahlamba Drakensberg Park (Karssing et al. 2012). The two trout species have aslo been implicated in the extirpation of the endangered fish *Amatolacypris trevelyani* (Border Barb) in headwater streams of the Keiskamma River in the Eastern Cape. In the Breede River system, Shelton et al. (2015b) demonstrated that the abundance (mean density and biomass) of native fish species—*Pseudobarbus burchelli* (Breede River Redfin), *Sandelia capensis* (Cape Kurper) and *Galaxias zebratus* (Cape Galaxias), were 5–40 times lower where *O. mykiss* was present, and that invertebrate species assemblages in streams with *O. mykiss* were distinctly different from those found in streams without *O. mykiss*. In addition, Shelton et al. (2015a) observed that *O. mykiss* had a weaker predation effect on aquatic invertebrates relative to the native fishes that it had replaced. As a result, there was also evidence of cascading effects, with algal biomass being significantly lower when *O. mykiss* was present because of higher abundance of herbivorous invertebrates relative to uninvaded sites. In the Drakensberg, and the Keiskamma River system, *O. mykiss* and *S. trutta* also share emerging insects as a food resource with riparian spiders, causing a decline in spider abundance at invaded sites (Jackson et al. 2016).

Black bass is the collective name for species of the genus *Micropterus* which, in South Africa, includes *M. salmoides*, *M. dolomieu*, *M. punctulatus* (Spotted Bass) and *M. floridanus* (Florida Bass) (Weyl et al. 2017b). As warm water species, Black Bass were introduced for sport fishing in the lower reaches of rivers and impoundments (see Khosa et al. 2019). Their impacts are similar to those documented for *O. mykiss* and *S. trutta*, and include the alteration of invertebrate communities (Weyl et al. 2010) and local extirpations and fragmentation of native fish populations (e.g. Kimberg et al. 2014; van der Walt et al. 2016; Ellender et al. 2018). A quantitative assessment of aquatic invertebrates in the Wit River, Eastern Cape, found that *M. salmoides* changed relative abundance and community structure (Weyl et al. 2010). In sections with *M. salmoides*, several members of large or conspicuous taxa (Odonata, Hemiptera and Coleoptera) were significantly reduced or even absent, while members of cryptic/inconspicuous taxa (Trichoptera, Leptoceridae, Mollusca, and Physidae) were significantly more abundant. In the headwaters of the Swartkops River system in the Eastern Cape, for example, *M. salmoides* predation has fragmented populations of the now endangered *Pseudobarbus afer* (Eastern Cape Redfin) (Ellender et al. 2018). Similarly, in the Olifants River system in the Western Cape, *M. dolomieu* and *M. punctulatus* invasions have restricted native *Cedercypris calidus* (Clanwilliam Redfin) and *Pseudobarbus phlegethon* (Fiery Redfin) to headwater reaches above physical barriers to invasion (van der Walt et al. 2016) and in the Marico River in the North West, *M. salmoides* and *M. punctulatus* have depleted mainstream *Enteromius motebensis* (Marico Barb) populations (Kimberg et al. 2014).

Trout and black bass often act synergistically on native fish populations. Trout are cold water species that are established in cooler headwater reaches of rivers (Ellender et al. 2016; Shelton et al. 2018), from which they exclude native fishes through predation and competition. Downstream trout populations are limited by temperature, which also mediates their predatory impacts (Shelton et al. 2018). Black bass, on the other hand, are warm water species that invade up rivers from mainstream source populations, and their penetration of headwaters is limited only by physical barriers such as waterfalls or gradients (see van der Walt et al. 2016; Ellender et al. 2018). This has resulted in an increasing compression of native fish populations between these two invaders, a situation which has resulted in the loss of more than 80% of habitat in some catchments (van der Walt et al. 2016).

17.2.1.4 Amphibians

A total of 104 species of alien amphibians globally have been assessed with EICAT (Kumschick et al. 2017), of which 21 are alien species in South Africa (van Wilgen and Wilson 2018). *Duttaphrynus melanostictus* (Asian Common Toad) is the only one of these species with major impacts globally that also occurs as an alien in South Africa, but there is no evidence for it having any impact in the country to date, probably due to its lack of establishment within the country (Measey et al. 2017). The only cases of documented impacts in the country are restricted to native species. *Xenopus laevis* (African Clawed Toad), native to South Africa but traded intensively globally, hybridises with the endemic *X. gilli* (Cape Platanna) (Picker 1985), but importantly there is no evidence of introgression (Furman et al. 2017). Even though these two species might have likely overlapped for millennia (Schreiner et al. 2013), densities of *X. leavis* have probably been artificially increased in the last 400 years (Measey et al. 2017), leading to intense competition and predation (Vogt et al. 2017; de Villiers et al. 2016). *Sclerophrys gutturalis* (Guttural Toad), native to much of the country but introduced to a peri-urban area of Cape Town (Vimercati et al. 2017), could potentially threaten the native endangered *Sclerophrys pantherina* (Western Leopard Toad), but no evidence of the extent of this threat has been reported to date (Measey et al. 2017). *Hyperolius marmoratus* (Painted Reed Frog) is also native to some parts of South Africa but has become invasive in other areas of the country that are outside its natural range (Tolley et al. 2008; Davies et al. 2013). No studies on its impacts on biodiversity have been conducted to date, but it is suspected that it may impact the endemic *Hyperolius horstocki* (Arum Lily Frog) (Measey et al. 2017).

17.2.1.5 Birds

There are 415 alien bird species that have been assessed with EICAT at a global scale (Evans et al. 2016), of which 37 occur as aliens in South Africa (van Wilgen and Wilson 2018). Two species, *Anas platyrhynchos* (Mallard) and *Pycnonotus jocosus* (Red-Whiskered Bulbul) are known to cause major impacts in their global invasive range (Evans et al. 2016). *Anas platyrhynchos* threatens the genetic integrity of native congeners through hybridisation (e.g. Tracey et al. 2008). In southern Africa *A. platyrhynchos* hybridises with endemic species such as *Anas undulata*

(Yellow-billed Duck), but currently there are low levels of introgression of *A. platyrhynchos* genes into *A. undulata* (Stephens et al. 2020). Introgression could become more extensive in the future if *A. platyrhynchos* populations are not controlled (Stephens et al. 2020; Davies et al. 2020, Chap. 22). *Pycnonotus jocosus* have been shown to damage crops, spread weeds, prey on native species and compete with them elsewhere (Mo 2015), but their impact in South Africa is unknown. Similarly, *Psittacula krameri* (Ring-necked Parakeet) is thought to compete for breeding holes with bats and other birds in other alien ranges leading to population declines of affected species. It has a very limited but expanding distribution in South Africa, and only ancedotal observations of impacts have been reported so far (Hart and Downs 2014). The species is a conflict species because it has both societal benefits (as a pet) and negative impacts (see Zengeya et al. 2017; Shackleton et al. 2020, Chap. 25).

17.2.1.6 Mammals

There are 42 alien mammal species that have been recorded outside of captivity in South Africa (van Wilgen and Wilson 2018), eleven of which have formal global-scale EICAT assessments. All of these species cause large impacts in their invasive range globally (Hagen and Kumschick 2018). For example, several alien rodent species such as *Rattus rattus* (Black Rat), *Rattus norvegicus* (Norwegian Rat) and *Mus musculus* (House Mouse) have caused local extinctions of native species of invertebrates, birds, bats and rodents on several islands through predation, competition for food, and disease transmission (e.g. Steadman 1995; Marris 2000; Courchamp et al. 2003). The feral species *Capra hircus* (Goat), *Equus asinus* (Donkey) and *Sus scrofa* (Pig) cause massive impacts through competition with native species for food, altering the structure and composition of plant communities by grazing and rooting (e.g. Kurdila 1998; Campbell and Donlan 2005; Means and Travis 2007). This has led to habitat loss, resulting in local extinction of some native species and accelerated soil erosion. Other domestic species such as *Felis catus* (Cat) can cause major impacts through predation leading to population declines, and in some cases local extirpation, of native mammals, reptiles, and birds (Fitzgerald and Veitch 1985; Winter and Wallace 2006).

For a few species, impacts have been recorded in South Africa. For example, *M. musculus* and *F. catus* have caused major impacts on offshore islands (Berruti 1986; Greve et al. 2017, 2020, Chap. 8). *Mus musculus* was introduced on Marion Island before 1818, likely as a stowaway on ships whose crews were engaged in seal hunting (Watkins and Cooper 1986). On the island, it preys on invertebrates (Jones and Ryan 2010; Dilley et al. 2016) and this changed the population densities, reproduction strategies and growth rates of some invertebrates on the island (Treasure and Chown 2014). Similarly, declines in albatross populations have been ascribed to predation of chicks by *M. musculus* (Dilley et al. 2016). *Felis catus* was introduced onto Marion Island in 1949 to control *M. musculus* (Bester et al. 2002). Although *F. catus* fed on *M. musculus*, it also preyed on seabirds and this severely affected seabird populations, especially burrowing species such as petrels (Procellariidae), leading to decreased breeding success, population declines and the local extinction of at least one species (van Rensburg 1983; Bester et al. 2002). *Felis catus* was eventually eradicated from the island in 1991 (Bester et al. 2002; Davies et al. 2020, Chap. 22), but the population of summer-breeding burrowing petrels is still to recover (Cerfonteyn and Ryan 2016).

Rattus norvegicus, *R. rattus*, *S. scrofa* and *C. hircus* have all caused massive impacts in other alien ranges (Hagen and Kumschick 2018), but their impacts in South Africa have not been studied apart from a few unpublished reports on wild boar impacts on vegetation and soil erosion (as mentioned in Spear and Chown 2009). The two rat species were unintentionally introduced into South Africa through the shipping trade (Long 2003). These two species are amongst the most invasive *Rattus* species with a global distribution in urban areas (Aplin et al. 2011). *Rattus rattus* is believed to be widely distributed but restricted by the drier parts of South Africa, while *R. norvegicus* is assumed to be limited to coastal areas of the country (De Graaf 1981). The two rat species are widely regarded as pests; in South Africa, specifically, they damage infrastructure, contaminate foodstuffs, and act as reservoirs of zoonotic diseases (e.g. Jassat et al. 2013; Julius et al. 2018; Potgieter et al. 2020, Chap. 11). *Sus scrofa* damages some critically endangered plants in the Western Cape, affecting succession and facilitating alien plant spread (Picker and Griffiths 2011). *Capra hircus* grazing has reduced the cover and density of endemic geophytes and succulents shrubs in thicket vegetation, and conservation of this endemic-rich flora is seriously threatened (Moolman and Cowling 1994).

17.2.2 Expert Opinion Assessments

Given that few taxa have been formally evaluated for the impacts using EICAT in South Africa, a classification based on expert opinion rather than on published literature was used as an interim measure in South Africa's first national status report on biological invasions (van Wilgen and Wilson 2018). Here, experts scored species according to their perceived ecological impacts and their socio-economic impacts (separately for negative and positive impacts) (see Zengeya et al. 2017 for more details). Using this scheme, 25 species were assessed as having severe impacts, and 82 as having major impacts (Table 17.2), the majority of these being terrestrial and freshwater plants (80 species). Here, using selected examples we highlight impacts on biodiversity of some of these alien species.

17.2.2.1 Plants

There are 893 alien plants species that are known to occur outside of cultivation in South Africa (van Wilgen and Wilson 2018). For a majority (56%) of these plant species, their impact on biodiversity has not been evaluated (Table 17.2). Of the 379

Table 17.2 The number of species that have been assessed by means of expert opinion as having impacts at different levels in South Africa[a]

Taxon	Not evaluated	Data deficient	Negligible	Few	Some	Major	Severe
Amphibians	0	15	1	2	1	2	0
Birds	73	0	5	5	8	1	0
Freshwater fish	6	1	0	5	9	4	1
Freshwater invertebrates	4	0	7	9	4	1	4
Mammals	3	0	4	16	11	8	0
Marine invertebrates	4	73	2	1	4	1	0
Marine plants	0	8	0	0	0	0	0
Microbial species	103		6	0	1	0	0
Reptiles	80	18	11	11	8	0	0
Terrestrial and freshwater plants	514	2	48	116	133	63	17
Terrestrial invertebrates	460	5	94	16	20	2	3
Totals	1247	122	178	181	199	82	25

[a]Adapted from Zengeya et al. (2017) and van Wilgen and Wilson et al. (2018)
Assignment to impact levels was based on an assessment of impact on both ecological or socio-economic aspects, with the assignment to a level being determined by the highest impact for either ecological or socio-economic aspects. Species that were identified by the experts as having insufficient information to formulate an opinion on its impacts were classified as "Data deficient" and a species was classified as "Not evaluated" if it was not evaluated against the criteria

species that have been assessed by expert opinion, the majority (33%) were categorised as low-impact species (i.e. species with negligible, few or some impacts). Only 80 plants were classified as having major and severe impacts, 17 of which had severe impacts, most of them Australian wattles (genus *Acacia*) and mesquite trees (genus *Prosopis*).

Prosopis are some of the few alien plants whose impacts (ecological and economic) have been examined in some detail in South Africa. The genus consists of several species and their hybrids that are regarded among some of the world's most damaging invasive plants. They were introduced to South Africa to provide supplemaneraty feed and shade for livestock, but similar to other countries where they have been introduced in the world, they have become invasive, generating negative impacts. In South Africa they have been found to have pronounced effects on insects, birds, and plants. Steenkamp and Chown (1996) found that invasion reduced the abundance and diversity of dung beetles. Native bird communities in invaded sites were found to be less diverse; some feeding guilds such as raptors were eliminated, and there was a marked decline in the abundance of frugivores and insectivorous species (Dean et al. 2002). However, other guilds such as nectarivores and seedeaters were less affected. Invasion also reduced the abundance and diversity of native plants. For example, in some plots the number of native tree species declined by 37% ($n = 8$) when the denstity of *Prosopis* doubled, while native perennial grasses and herbaceous plants were eliminated (Shackleton et al. 2015).

Invasive *Prosopis* competes with the native *Vachellia erioloba* (Camel Thorn) for groundwater, increasing the likelihood of competive exclusion of Camel Thorn trees when water resources are limiting (Schachtschneider and February 2013).

Other than for the genus *Prosopis*, there are very few studies that have quantified the impact of alien plants on biodiversity in South Africa. Richardson et al. (1989) reviewed published and unpublished data on plant species richness in the Fynbos Biome with different levels of invasion by alien trees and shrubs in the genera *Pinus*, *Hakea* and *Acacia*. They found significant declines in native plant species richness at the scale of the sample quadrats used in their study (4–256 m^2), notably when the cover of alien plants exceeded 50%. Similarly, Yapi et al. (2018) recorded marked declines in the cover of native grass species with increases in the canopy cover of alien *Acacia mearnsii* (Black Wattle) trees in the Eastern Cape. Such declines in the abundance and richness of native plant species, and associated faunal species, are likely to take place where any alien plant species becomes dominant. Given that many alien plant species have recently entered a phase of rapid spread, it can be expected that these kinds of impacts will increase. For example, Henderson and Wilson (2017) showed that the number of quarter degree grid cells occupied by alien plants in South Africa increased by ~50% between 2000 and 2016, with at least nine species having expanded rapidly over the past decade (see also van Wilgen et al. 2020a, Chap. 21).

Besides trees, some grasses have been shown to affect native biodiversity in South Africa. In addition to the examples mentioned earlier, *Ammophila arenaria* (European Beach Grass) has changed native dune communities (Hertling and Lubke 1999). *Avena barbata* (Slender Wild Oat) has invaded some lowland fynbos and can dominate old field habitats (Heelemann et al. 2013), and a recent study on *Paspalum quadrifarium* (Tussock Paspalum) showed its ability to affect native species communities (Nkuna 2018).

17.2.2.2 Invertebrates

Many alien invertebrates are major pests of agriculture, but here we focus on taxa that have negatively impacted native biodiversity. There are 629 alien freshwater and terrestrial invertebrates that are known to occur outside of captivity in South Africa (van Wilgen and Wilson 2018). For a majority (74%) of these, their impacts on biodiversity have not been evaluated. Of the remainder, 25% were assessed using expert opinion as low-impact species, and less than 2% had major to severe impacts (Table 17.2). Five species of terrestrial invertebrates that purportedly have major to severe impacts include *Cornu aspersum* (Common Garden Snail), *Deroceras invadens* (Tramp Slug), *Linepithema humile* (Argentine Ant), *Thebia pisana* and *Trogoderma granarium* (Khapra Beetle). *Linepithema humile* has invaded fynbos, where it displaces native ants. The displacement of native ants is a direct impact on biodiversity, but the role that native ant species play in the dispersal of the seeds of fynbos plants has also been disrupted. Bond and Slingsby (1984) found that *L. humile* differed from native ants in moving seeds shorter distances, and in failing to store them in nests below the soil. The seeds were left on the soil surface, where they were eaten by vertebrate and invertebrate seed predators. Bond and Slingsby (1983) estimated that *L. humile* displacement of native ants could threaten the long-term survival of approximately 1000 fynbos plant species that were dependent on native ants for the dispersal and protection of their seeds.

Among the marine invertebrates, *Mytilus galloprovincialis* (Mediterranean Mussel) has had significant impacts in South African marine environments (Robinson et al. 2020, Chap. 9). First recorded in South Africa in the late 1970s, it presently occurs along the whole of the west coast and as far east as East London (approx. 2000 km of coastline) (Robinson et al. 2005). Within its invaded range, it has been implicated in causing impacts on native species and altering the structure of rocky shore communities. Along the west coast, *M. galloprovincialis* competively excludes native species such as mussels and limpets from prime habitas (Robinson et al. 2007), while along the south coast it co-exists with the native *Perna perna* (Brown Mussel) (Bownes and McQuaid 2006). Interestingly, *M. galloprovincialis* forms complex mussel beds that have increased habitat availability leading to an increase in the diversity and abundance of native fauna on invaded shores (Sadchatheeswaran et al. 2015).

17.2.2.3 Mammals

Of the 42 alien mammal species that have been recorded in South Africa, 8 species have been assessed using expert opinion as causing major impacts in South Africa (van Wilgen and Wilson 2018). Six of these (*R. rattus*, *R. norvegicus*, *F. catus*, *M. musculus*, *S. scrofa* and *C. hircus*) cause massive impacts in the country, in similar ways to those identifed by formal global EICAT assessments (see Sect. 17.2.1.6). For the remainder, *Hippotragus equinus koba* (Western Roan) has been implicated in hybridising with local sub-species (Mathee and Robinson 1999; Alpers et al. 2004; van Wyk et al. 2019) and the impacts of *Macaca fascicularis* (Crab-Eating Macaque) are not well documented in South Africa, so their potential impacts can only be inferred from their global invasive range.

17.2.2.4 Fishes

Five fish species were classified by experts as causing major to severe impacts in invaded catchments in South Africa (Zengeya et al. 2017). *Micropterus dolomieu*, *M. salmoides, O. mykiss* and *S. trutta* had major impacts through mainly competition and predation. *Oreochromis niloticus* was assessed as having severe impacts mainly through hybridisation with congeneric native species.

17.2.3 Impacts Identified During Red-Listing Processes

Some recent studies have used data from the IUCN Red List of Threatened Species (Mace et al. 2008) to assess the role of alien species as drivers of recent extinctions (Bellard et al. 2016a), and global patterns of threats to vertebrates posed by biological invasions (Bellard et al. 2016b). In South Africa, several taxa have undergone formal red list assessments and these include plants, dragonflies, freshwater fishes, amphibians, reptiles, birds and mammals (Child et al. 2015; Taylor et al. 2015; IUCN 2018; SANBI 2019). Following a similar approach as Bellard et al. (2016b), we assessed whether and how species listed under the IUCN Red List in South Africa were affected by alien species. We calculated the proportion of threatened native species, i.e. those in the upper tier of extinction risk (Vulnerable, Endangered, and Critically endangered), where alien species were indicated as a threat (Table 17.3).

A total of 23,609 native species were assessed, of which 17% had alien species as a major threat to their extinction risk. The proportion of threatened species that are imperilled by alien species varied across threat categories, being higher for Endagered (60%, $N = 609$ out of 1007) and Vulnerable species (48%, $N = 788$ out of 1641) and lower for Critically Endangered (40%, $N = 276$ out of 688) (Table 17.3). This trend is also reflected when comparions are made across taxa in each threat category (Table 17.3). More than half of the taxa assessed as Endangered (five out of eight), Critically Endangered (three out of five) and Vulnerable (three out of five) were threatened by alien species. Across all three threat categories, the proportion of species that are being threatened by alien species was highest for fishes, amphibians and plants. The Red List assessments identified and classified 12 major threats to the persistence of a species, and alien species were rarely considered to be the sole extinction risk for most species. For example, most of the fish and amphibian species listed as Critically Endangered have small distributional ranges, experience a continuous decline in habitat quality through several anthropogenic activities (e.g. pollution, excessive water abstraction and altered flow regimes), and are additionally threatened by invasive species through predation, competition and physical alteration of ecosystems (van Wilgen et al. 2020b, Chap. 29).

17.2.4 Impacts on Biodiversity at a Biome Scale

A limited number of studies have quantifed the impact of invasive species on ecosystem services at a landscape scale. Several studies have either focused on a particular ecosystem service (e.g. surface water supplies, Le Maitre et al. 2000), or on a single species (e.g. *A. mearnsii*, De Wit et al. 2001). This has led to problems in extraplolating the results generated by the different studies to make broad conclusions about the entirety of impacts of invasions on an ecosystem. Van Wilgen et al. (2008), however, assessed current and potential impacts of alien plants on selected ecosystem services (surface water runoff, ground water recharge, livestock production and biodiversity) in

Table 17.3 Numbers of of threatened taxa found in South Africa that have formal IUCN Red list assessments and where alien species are indicated as a threatening process[a]

Threat category	Taxa	Number of assessed species	Proportion (%) of species threatened by alien species
Critically endangered	Freshwater Fishes	7	100
	Amphibians	6	67
	Reptiles	4	50
	Plants	632	40
	Butterflies	20	40
	Mammals	5	20
	Birds	13	15
	Dragonflies	1	0
	All taxa combined	688	40
Endangered	Amphibians	9	100
	Freshwater Fishes	24	92
	Dragonflies	5	80
	Plants	874	62
	Butterflies	30	57
	Reptiles	7	29
	Mammals	20	25
	Birds	38	21
	All taxa combined	1007	60
Vulnerable	Amphibians	1	100
	Freshwater Fishes	11	73
	Plants	1516	50
	Reptiles	12	42
	Dragonflies	14	29
	Butterflies	23	26
	Mammals	31	13
	Birds	33	12
	All taxa combined	1641	48

[a]Threatened taxa are species that were assessed as experiencing the highest extinction risk (Vulnerable (VU), Endangered (EN), and Critically endangered (CR). Data sources: Child et al. (2015), Taylor et al. (2015), IUCN (2018), SANBI (2019)

five terrestrial biomes (Savanna, Fynbos, Grasslands, Succulent Karoo and Nama Karoo) in South Africa. With the exception of the fynbos, the current invasions had no measurable impact on biodiversity intactness (an estimate of impact of land-use changes on populations of various taxa such as plants, mammals, birds, reptiles and amphibians in a particular area, see Scholes and Biggs 2005) at a landscape scale. While the current impacts of alien plants were relatively low, the future impacts were predicted to be very high. In addition, while the errors in these estimates are likely to be substantial, the predicted impacts were sufficiently large to suggest that there is cause for serious concern. De Lange and van Wilgen (2010) used the above estimates of intactness to estimate the economic value of ecosystem services based on biodiversity. It was estimated that the reduction in value of selected ecosystem services due to invasive alien plants was the highest for fynbos, but still substantial for others like savanna, thicket, grassland, succulent karoo and Nama karoo. Overall, terrestrial ecosystems in South Africa would deliver biodiversity-based ecosystem services valued at ZAR 22.1 billion per annum if no invasions were present. The estimated value of the annual flow of these services was reduced by 2% (ZAR 428 million) each year due to alien plant invasions at current levels; under a scenario where alien plants are allowed to invade all available habitat, these losses were estimated to increase to more than 50% (ZAR 12.9 billion) annually. Other studies have also estimated the total value of ecosystem services in South Africa (Anderson et al. 2017; Turpie et al. 2017). Anderson et al. (2017) estimated this value to be about ZAR 8 trillion per year—this is nearly double that of South Africa's gross domestic product of ZAR 4.7 trillion for the same period. Turpie et al. (2017) estimated the value of ecosystem services provided by terrestrial, freshwater and estuarine ecosystems to be at ZAR 245 billion annually. The differences in the estimates between these assessments reflect methodological differences and the general challenges associated with attempts to monetise the values of biodiversity. Nevertheless, they highlight that ecosystem services make a substantial contribution to the economy.

17.3 Synthesis

The issue of quantifying the impacts of alien species on biodiversity remains a major challenge, both globally and in South Africa. For the majority of species found in South Africa, there are no studies documenting impacts, and there has been no formal assessment of impacts at a national scale. Evidence-based impact assessments are needed to support calls for management interventions. For example, in South Africa 556 taxa are listed as invasive, and landowners have an obligation to manage them (van Wilgen et al. 2020a, Chap. 1, Box 1.2). However, the regulations should arguably focus on priority species because not all of the listed species are necessarily harmful to the extent that would justify management especially when the current capacity to manage and to regulate them is limited (Zengeya et al. 2017). The use of expert opinion and or formal assessments such as EICAT could help to identify and priotise those species that should be targeted for management.

Acknowledgements BvW, TAZ, OLFW thank the DSI-NRF Centre for Invasion Biology and the South African National Research Foundation (Grants 109467, 103602, 109015, 110507) for support. SK acknowledges support from the South African National Department of Environmental Affairs through its funding of the South African National Biodiversity, and the DSI-NRF Centre of Excellence for Invasion Biology. John Measey and John Wilson provided useful comments on an earlier draft of this chapter. The South African Department of Environment, Forestry, and Fisheries (DEFF) are thanked for funding the South African National Biodiversity Institute noting that this publication does not necessarily represent the views or opinions of DEFF or its employees.

References

Alpers DL, van Vuuren BJ, Arctander P, Robinson TJ (2004) Population genetics of the roan antelope (*Hippotragus equinus*) with suggestions for conservation. Mol Ecol 13:1771–1784. https://doi.org/10.1111/j.1365-294X.2004.02204.x

Anderson S, Ankor BL, Sutton PC (2017) Ecosystem service valuations of South Africa using a variety of land cover data sources and resolutions. Ecosyst Serv 27:173–178. https://doi.org/10.1016/j.ecoser.2017.06.001

Aplin K, Suzuki H, Chinen CH et al (2011) Multiple geographic origins of commensalism and complex dispersal history of black rats. PLoS One 6:1–20. https://doi.org/10.1371/journal.pone.0026357

Appleton CC, Forbes AT, Demetriades NT (2009) The occurrence, bionomics and potential impacts of the invasive freshwater snail *Tarebia granifera* (Lamarck, 1822) (Gastropoda: Thiaridae) in South Africa. Zool Med Leiden 83:525–536.

Bacher S, Blackburn TM, Essl F et al (2018) Socio-economic impact classification of alien taxa (SEICAT). Methods Ecol Evol 9:159–168

Bellard C, Cassey P, Blackburn TM (2016a) Alien species as a driver of recent extinctions. Biol Lett 12:20150623. https://doi.org/10.1098/rsbl.2015.0623

Bellard C, Genovesi P, Jeschke JM (2016b) Global patterns in threats to vertebrates by biological invasions. Proc R Soc B-Biol Sci 283:20152454. https://doi.org/10.1098/rspb.2015.2454

Berruti A (1986) The predatory impact of feral cats *Felis catus* and their control on Dassen Island. S Afr J Antarct Res 16:123–127

Bester MN, Bloomer JP, van Aarde RJ et al (2002) A review of the successful eradication of feral cats from sub-Antarctic Marion Island, southern Indian Ocean. S Afr J Wildl Res 32:65–73

Blackburn TM, Essl F, Evans T et al (2014) A unified classification of alien species based on the magnitude of their environmental impacts. PLoS Biol 12:e1001850. https://doi.org/10.1371/journal.pbio.1001850

Bond WJ, Slingsby P (1983) Seed dispersal by ants in shrublands of the Cape Province and its evolutionary implications. S Afr J Sci 79:231–233

Bond WJ, Slingsby P (1984) Collapse of an ant-plant mutalism: the Argentine Ant (*Iridomyrmex humilis*) and myrmecochorous Proteaceae. Ecology 65:1031–1037. https://doi.org/10.2307/1938311

Bownes SJ, McQuaid CD (2006) Will the invasive mussel *Mytilus galloprovincialis* lamark replace the indigenous *Perna perna* L. on the south coast of South Africa? J Exp Mar Biol Ecol 338:140–151. https://doi.org/10.1016/j.jembe.2006.07.006

Brooks ML, D'antonio CM, Richardson DM et al (2004) Effects of invasive alien plants on fire regimes. Bioscience 54:677–688. https://doi.org/10.1641/0006-3568(2004)054[0677:EOIAPO]2.0.CO;2

Campbell K, Donlan CJ (2005) Feral goat eradications on Islands. Conserv Biol 19:1362–1374. https://doi.org/10.1111/j.1523-1739.2005.00228.x

Canavan S, Kumschick S, Le Roux JJ et al (2019) Does origin determine environmental impacts? Not for bamboos. Plants People Planet 1:119–128. https://doi.org/10.1002/ppp3.5

Cerfonteyn M, Ryan PG (2016) Have burrowing petrels recovered on Marion Island two decades after cats were eradicated? Evidence from sub-Antarctic skua prey remains. Antarct Sci 28:51–57. https://doi.org/10.1017/S0954102015000474

Child MF, Roxburgh L, Do Linh San E et al (eds) (2015) The red list of mammals of South Africa, Swaziland and Lesotho. South African National Biodiversity Institute; Endangered Wildlife Trust, Pretoria; Lethabong

Christie JD, Edward J, Goolaman K et al (1981) Interactions between St. Lucian *Biomphalaria glabrata* and *Helisoma duryi*, a possible competitor snail, in a semi-natural habitat. Acta Trop 38:395–417

Colville JC, Potts AJ, Bradshaw PL et al (2014) Floristic and faunal Cape biochoria: do they exist? In: Allsopp N, Colville JF, Verboom GA (eds) Fynbos: ecology, evolution, and conservation of a megadiverse region. Oxford University Press, Oxford, pp 73–93. https://doi.org/10.1093/acprof:oso/9780199679584.003.0004

Courchamp F, Pascal M, Chapuis JL (2003) Mammal invaders on islands, impact, control and control impact. Biol Rev 78:347–383. https://doi.org/10.1017/S1464793102006061

Davies SJ, Clusella-Trullas S, Hui C, McGeoch MA (2013) Farm dams facilitate amphibian invasion. Austral Ecol 38:851–863. https://doi.org/10.1111/aec.12022

Davies SJ, Jordaan MS, Karsten M et al (2020) Experience and lessons from alien and invasive animal control projects in South Africa. In: van Wilgen BW, Measey J, Richardson DM, Wilson JR, Zengeya TA (eds) Biological invasions in South Africa. Springer, Berlin, pp 625–660. https://doi.org/10.1007/978-3-030-32394-3_22

De Graaf G (1981) The rodents of southern Africa: notes on their identification, distribution, ecology and taxonomy. Butterworth, Durban

De Lange WJ, van Wilgen BW (2010) An economic assessment of the contribution of biological control to the management of invasive alien plants and to the protection of ecosystem services in South Africa. Biol Invasions 12:4113–4124

de Villiers FA, de Kock M, de Measey GJ (2016) Controlling the African clawed frog *Xenopus laevis* to conserve the Cape platanna *Xenopus gilli* in South Africa. Conserv Evid 13:17

De Wit M, Crookes D, van Wilgen BW (2001) Conflicts of interest in environmental management: estimating the costs and benefits of a tree invasion. Biol Invasions 3:167–178. https://doi.org/10.1023/A:1014563702261

Dean WRJ, Anderson MD, Milton SJ, Anderson TA (2002) Avian assemblages in native *Acacia* and alien *Prosopis* drainage line woodland in the Kalahari, South Africa. J Arid Environ 51:1–19. https://doi.org/10.1006/jare.2001.0910

Dilley BJ, Schoombie S, Schoombie J, Ryan PG (2016) "Scalping" of albatross fledglings by introduced mice spreads rapidly at Marion Island. Antarct Sci 28:73–80. https://doi.org/10.1017/S0954102015000486

Ellender BR, Weyl OLF (2014) A review of current knowledge, risk and ecological impacts associated with non-native freshwater fish introductions in South Africa. Aquat Invasions 9:117–132. https://doi.org/10.3391/ai.2014.9.2.01

Ellender BR, Rivers-Moore NA, Coppinger CR et al (2016) Towards using thermal stress thresholds to predict salmonid invasion potential. Biol Invasions 18:513–3525. https://doi.org/10.1007/s10530-016-1244-9

Ellender BR, Wasserman RJ, Chakona A et al (2017) A review of the biology and status of Cape Fold Ecoregion freshwater fishes. Aquat Conserv 27:867–879

Ellender BR, Weyl OLF, Alexander ME et al (2018) Out of the pot and into the fire: explaining the vulnerability of an endangered small headwater stream fish to black-bass *Micropterus* spp. invasion. J Fish Biol 92:1035–1050. https://doi.org/10.1111/jfb.13562

Evans T, Kumschick S, Blackburn TM (2016) Application of the environmental impact classification for Alien Taxa (EICAT) to a global assessment of alien bird impacts. Divers Distrib 22:919–931

Firmat C, Alibert P, Losseau M et al (2013) Successive invasion-mediated interspecific hybridizations and population structure in the endangered cichlid *Oreochromis mossambicus*. PLoS One 8:e63880. https://doi.org/10.1371/journal.pone.0063880

Fitzgerald BM, Veitch CR (1985) The cats of Herekopare Island, New Zealand: their history, ecology and effects on birdlife. New Zeal J Zool 12:319–330. https://doi.org/10.1080/03014223.1985.10428285

Frest TJ, Sanders Rhodes R (1982) *Oxychilus draparnaldi* in Iowa. Nautilus 96:36–39

Furman B, Cauret C, Colby G et al (2017) Limited genomic consequences of hybridization between two African clawed frogs, *Xenopus gilli* and *X. laevis* (Anura: Pipidae). Sci Rep 7:1091. https://doi.org/10.1038/s41598-017-01104-9

Greve M, Mathakutha R, Steyn C, Chown SL (2017) Terrestrial invasions on sub-Antarctic Marion and Prince Edward Islands. Bothalia 47:a2143. https://doi.org/10.4102/abc.v47i2.2143

Greve M, von der Meden CEO, Janion-Scheepers C (2020) Biological invasions in South Africa's offshore sub-Antarctic territories. In: van Wilgen BW, Measey J, Richardson DM, Wilson JR, Zengeya TA (eds) Biological invasions in South Africa. Springer, Berlin, pp 205–226. https://doi.org/10.1007/978-3-030-32394-3_8

Griffiths CL, Robinson TB, Lange L, Mead A (2010) Marine biodiversity in South Africa: an evaluation of current states of knowledge. PLoS One 5:e12008. https://doi.org/10.1371/journal.pone.0012008

Guthrie G (2007) Impacts of the invasive reed *Arundo donax* on biodiversity at the community-ecosystem level. MSc Thesis. University of the Western Cape

Hagen BL, Kumschick S (2018) The relevance of using various scoring schemes revealed by an impact assessment of feral mammals. NeoBiota 38:37–75. https://doi.org/10.3897/neobiota.38.23509

Hart LA, Downs CT (2014) Public surveys of rose-ringed parakeets, *Psittacula krameri*, in the Durban Metropolitan area, South Africa. Afr Zool 49:283–289. https://doi.org/10.1080/15627020.2014.11407644

Hawkins CL, Bacher S, Essl F et al (2015) Framework and guidelines for implementing the proposed IUCN Environmental Impact Classification for Alien Taxa (EICAT). Divers Distrib 21:1360–1363. https://doi.org/10.1111/ddi.12379

Heelemann S, Krug CB, Esler KJ et al (2013) Soil seed banks of remnant and degraded Swartland Shale Renosterveld. Appl Veg Sci 16:585–597. https://doi.org/10.1111/avsc.12026

Henderson L, Wilson JRU (2017) Changes in the composition and distribution of alien plants in South Africa: an update from the Southern African Plant Invaders Atlas. Bothalia 47:a2172

Hertling UM, Lubke RA (1999) Indigenous and *Ammophila arenaria*-dominated dune vegetation on the South African Cape coast. Appl Veg Sci 2:157–168. https://doi.org/10.2307/1478979

Holmes PM, Richardson DM, Esler KJ et al (2005) A decision-making framework for restoring riparian zones degraded by invasive alien plants in South Africa. S Afr J Sci 101:553–564

Hulme PE (2014) Invasive species challenge the global response to emerging diseases. Trends Parasitol 30:267–270. https://doi.org/10.1016/j.pt.2014.03.005

IUCN (2018) The IUCN red list of threatened species. Version 2018-1. www.iucnredlist.org. Downloaded 01 Oct 2018

Jackson MC, Woodford DJ, Bellingan TA et al (2016) Trophic overlap between fish and riparian spiders: potential impacts of an invasive fish on terrestrial consumers. Ecol Evol 6:1745–1752. https://doi.org/10.1002/ece3.1893

Jassat W, Naicker N, Naidoo S, Mathee A (2013) Rodent control in urban communities in Johannesburg, South Africa: from research to action. Int J Environ Health Res 23:474–483

Jones MGW, Ryan PG (2010) Evidence of mouse attacks on albatross chicks on sub-Antarctic Marion Island. Antarct Sci 22:39–42. https://doi.org/10.1017/S0954102009990459

Julius RS, Schwan EV, Chimimba CT (2018) Molecular characterization of cosmopolitan and potentially co-invasive helminths of commensal, murid rodents in Gauteng Province, South Africa. Parasitol Res 117:1729. https://doi.org/10.1007/s00436-018-5852-4

Karatayev AY, Burlakova LE, Karatayev VA, Padilla DK (2009) Introduction, distribution, spread, and impacts of exotic freshwater gastropods in Texas. Hydrobiologia 619:181–194. https://doi.org/10.1007/s10750-008-9639-y

Karssing RJ, Rivers-Moore NA, Slater K (2012) Influence of waterfalls on patterns of association between trout and Natal cascade frog *Hadromophryne natalensis* tadpoles in two headwater streams in the Ukhahlamba Drakensberg Park World Heritage Site, South Africa. Afr J Aquat Sci 37:107–112. https://doi.org/10.2989/16085914.2012.666381

Kesner D, Kumschick S (2018) Gastropods alien to South Africa cause severe environmental harm in their global alien ranges across habitats. Ecol Evol 8:8273–8285. https://doi.org/10.1002/ece3.4385

Khosa D, Marr SM, Wasserman RJ et al (2019) An evaluation of the current extent and potential spread of Black Bass invasions in South Africa. Biol Invasions 21:1721. https://doi.org/10.1007/s10530-019-01930-0

Kimberg PK, Woodford DJ, Roux H, Weyl OLF (2014) Species-specific impact of introduced largemouth bass *Micropterus salmoides* in the Groot Marico Freshwater Ecosystem Priority Area, South Africa. Afr J Aquat Sci 39:451–458. https://doi.org/10.2989/16085914.2014.976169

Kumschick S, Vimercati G, de Villiers FA et al (2017) Impact assessment with different scoring tools: how well do alien amphibian assessments match? NeoBiota 33:53–66. https://doi.org/10.3897/neobiota.33.10376

Kumschick S, Wilson JR, Foxcroft LC (2018) Framework and guidelines for conducting risk analyses for alien species. Preprints. https://doi.org/10.20944/preprints201811.0551.v1

Kumschick S, Foxcroft LC, Wilson JR (2020) Analysing the risks posed by biological invasions to South Africa. In: van Wilgen BW, Measey J, Richardson DM, Wilson JR, Zengeya TA (eds) Biological invasions in South Africa. Springer, Berlin, pp 569–592. https://doi.org/10.1007/978-3-030-32394-3_20

Kurdila J (1998) The introduction of exotic species into the United States: there goes the neighborhood! BC Envtl Aff L Rev 16:95–119

Le Maitre DC, Versfeld DB, Chapman RA (2000) The impact of invading alien plants on surface water resources in South Africa: a preliminary assessment. Water SA 26:397–408

Le Maitre DC, Gush MB, Dzikiti S (2015) Impacts of invading alien plant species on water flows at stand and catchment scales. AoB Plants 7:plv043. https://doi.org/10.1093/aobpla/plv043

Le Roux J (2002) The biodiversity of South Africa 2002: indicators, trends and human impacts. World Wildlife Fund, Endangered Wildlife Trust and Nedbank Green. Struik Publishers, Cape Town

Long J (2003) Introduced mammals of the world: their history, distribution and influence. CSIRO, Melbourne. https://doi.org/10.1071/9780643090156

Mace GM, Collar NJ, Gaston KJ et al (2008) Quantification of extinction risk: IUCN's system for classifying threatened species. Conserv Biol 22:1424–1442. https://doi.org/10.1111/j.1523-1739.2008.01044.x

Marr SM, Ellender BR, Woodford DJ et al (2017) Evaluating invasion risk for freshwater fishes in South Africa. Bothalia 47:a2177. https://doi.org/10.4102/abc.v47i2.2177

Marris JWM (2000) The beetle (Coleoptera) fauna of the Antipodes Islands, with comments on the impact of mice; and annotated checklist of the insect and arachnid fauna. J R Soc N Z 30:169–195. https://doi.org/10.1080/03014223.2000.9517616

Mathee CA, Robinson TJ (1999) Mitochondrial DNA population structure of roan and sable antelope: implications for the translocation and conservation of the species. Mol Ecol 8:227–238. https://doi.org/10.1046/j.1365-294X.1999.00556.x

Means DB, Travis J (2007) Declines in ravine-inhabiting dusky salamanders of the southeastern US coastal plain. Naturalist 6:83–96. https://doi.org/10.1656/1528-7092(2007)6[83:DIRDSO]2.0.CO;2

Measey J, Davies SJ, Vimercati G et al (2017) Invasive amphibians in southern Africa: a review of invasion pathways. Bothalia 47:a2117. https://doi.org/10.4102/abc.v47i2.2117

Miranda NAF, Perissinotto R (2014) Effects of an alien invasive gastropod on native benthic assemblages in coastal lakes of the iSimangaliso Wetland Park, South Africa. Afr Invertebr 55:209–228.

Mittermeier RA, Robles-Gil P, Mittermeier CG (1997) Megadiversity: Earth's biologically wealthiest nations. CEMEX, Conservation International, Arlington

Mittermeier RA, Robles-Gil P, Hoffmann M et al (2004) Hotspots revisited: Earth's biologically richest and most endangered terrestrial ecoregions. CEMEX, Mexico City

Mo M (2015) The red-whiskered bulbul *Pycnonotus jocosus* in Australia – a global perspective, history of introduction, current status and potential impacts. Aust Zool 37:461–471. https://doi.org/10.7882/AZ.2015.014

Moolman HJ, Cowling RM (1994) The impact of elephant and goat grazing on the endemic flora of South Africa succulent thicket. Biol Conserv 68:53–61. https://doi.org/10.1016/0006-3207(94)90546-0

Mucina L, Rutherford MC (eds) (2006) The vegetation of South Africa. Lesotho and Swaziland, South African National Biodiversity Institute

Mugwedi LF (2012) Invasion ecology of *Glyceria maxima* in KZN Rivers and wetlands. MSc Thesis, University of the Witwatersrand

Myers N, Mittermeier RA, Mittermeier CG et al (2000) Biodiversity hotspots for conservation priorities. Nature 403:853–858. https://doi.org/10.1038/35002501

Nelufule T (2018) An assessment of alien terrestrial invertebrate species in the pet trade in South Africa. MSc Thesis, University of Pretoria

Nkuna VK (2018) Risk analysis of alien grasses occurring in South Africa. MSc Thesis, Stellenbosch University

Nkuna VK, Visser V, Wilson JRU, Kumschick S (2018) Global environmental and socio-economic impacts of selected alien grasses as a basis for ranking threats to South Africa. NeoBiota 41:19–65. https://doi.org/10.3897/neobiota.41.26599

Noss RF (1990) Indicators for monitoring biodiversity: a hierarchical approach. Conserv Biol 4:355–364. https://doi.org/10.1111/j.1523-1739.1990.tb00309.x

Odendaal LJ, Haupt TM, Griffiths CL (2008) The alien invasive land snail *Theba pisana* in the West Coast National Park: is there cause for concern? Koedoe 50:93–98. https://doi.org/10.4102/koedoe.v50i1.153

Picker MD (1985) Hybridization and habitat selection in *Xenopus gilli* and *Xenopus laevis* in the south-western Cape Province. Copeia 1985:574–580. https://doi.org/10.2307/1444746

Picker M, Griffiths C (2011) Alien & invasive animals: a South African perspective. Struik Nature, Cape Town

Potgieter L, Douwes E, Gaertner M et al (2020) Biological invasions in South Africa's urban ecosystems: patterns, processes, impacts and management. In: van Wilgen BW, Measey J, Richardson DM, Wilson JR, Zengeya TA (eds) Biological invasions in South Africa. Springer, Berlin, pp 273–310. https://doi.org/10.1007/978-3-030-32394-3_11

Richardson DM, van Wilgen BW (2004) Invasive alien plants in South Africa: how well do we understand the ecological impacts? S Afr J Sci 100:45–52

Richardson DM, Macdonald IAW, Forsyth GG (1989) Reductions in plant species richness under stands of alien trees and shrubs in the Fynbos biome. South Afr For J 149:1–8. https://doi.org/10.1080/00382167.1989.9628986

Rivers-Moore NA, Fowles B, Karssing RJ (2013) Impacts of trout on aquatic macroinvertebrates in three Drakensberg rivers in KwaZulu–Natal, South Africa. Afr J Aquat Sci 38:93–99. https://doi.org/10.2989/16085914.2012.750592

Robinson TB, Griffiths CL, McQuaid CD, Rius M (2005) Marine alien species of South Africa – status and impacts. Afr J Mar Sci 27:297–306. https://doi.org/10.2989/18142320509504088

Robinson TB, Branch GM, Griffiths CL et al (2007) Effects of the invasive mussel *Mytilus galloprovincialis* on rocky intertidal community structure in South Africa. Mar Ecol Prog Ser 340:163–171. https://doi.org/10.3354/meps340163

Robinson TB, Peters K, Brooker B (2020) Coastal invasions: the South African context. In: van Wilgen BW, Measey J, Richardson DM, Wilson JR, Zengeya TA (eds) Biological invasions in South Africa. Springer, Berlin, pp 227–246. https://doi.org/10.1007/978-3-030-32394-3_9

Rodda GH, Fritts TH (1992) The impact of the introduction of the colubrid snake *Boiga irregularis* on Guam's lizards. J Herpetol 26:166–174. https://doi.org/10.2307/1564858

Sadchatheeswaran S, Branch GM, Robinson TB (2015) Changes in habitat complexity resulting from sequential invasions of a rocky shore: implications for community structure. Biol Invasions 17:1799–1816. https://doi.org/10.1007/s10530-014-0837-4

SANBI (2019) Statistics: red list of South African plants version 2019.1. Downloaded Redlist. sanbi.org on 2019/01/22

Schachtschneider K, February EC (2013) Impact of *Prosopis* invasion on a keystone tree species in the Kalahari Desert. Plant Ecol 214:597–605. https://doi.org/10.1007/s11258-013-0192-z

Scholes RJ, Biggs R (2005) A biodiversity intactness index. Nature 434:45–49. https://doi.org/10.1038/nature03289

Schreiner C, Rödder D, Measey GJ (2013) Using models to test Poynton's predictions. African. J Herpetol 62:49–62. https://doi.org/10.1080/21564574.2013.794865

Shackleton RT, Le Maitre DC, Richardson DM, van Wilgen BW (2015) The impact of invasive alien *Prosopis* species (mesquite) on native plants in different environments in South Africa. S Afr J Bot 97:25–31. https://doi.org/10.1016/j.sajb.2014.12.008

Shackleton RT, Novoa A, Shackleton CM et al (2020) The social dimensions of biological invasions in South Africa. In: van Wilgen BW, Measey J, Richardson DM, Wilson JR, Zengeya TA (eds) Biological invasions in South Africa. Springer, Berlin, pp 697–726. https://doi.org/10.1007/978-3-030-32394-3_24

Shelton JM, Samways MJ, Day JA (2015a) Non-native rainbow trout change the structure of benthic communities in headwater streams of the Cape Floristic Region, South Africa. Hydrobiologia 745:1–15. https://doi.org/10.1007/s10750-014-2067-2

Shelton JM, Samways MJ, Day JA (2015b) Predatory impact of non-native rainbow trout on endemic fish populations in headwater streams in the Cape Floristic Region of South Africa. Biol Invasions 17:365–379. https://doi.org/10.1007/s10530-014-0735-9

Shelton JM, Weyl OLF, Esler KJ et al (2018) Temperature mediates the impact of non-native rainbow trout on native freshwater fishes in South Africa's Cape Fold Ecoregion. Biol Invasions 20:2927–2944. https://doi.org/10.1007/s10530-018-1747-7

Spear D, Chown SL (2009) Non-indigenous ungulates as a threat to biodiversity. J Zool 279:1–17. https://doi.org/10.1111/j.1469-7998.2009.00604.x

Steadman DW (1995) Prehistoric extinctions of Pacific Island birds' biodiversity meets zooarchaeology. Science 267:1123–1131. https://doi.org/10.1126/science.267.5201.1123

Steenkamp HE, Chown SL (1996) Influence of dense stands of an exotic tree *Prosopis glandulosa* Benson, on a savanna dung beetle (Coleoptera: Scarabeidae) assemblage in southern Africa. Biol Conserv 78:305–311. https://doi.org/10.1016/S0006-3207(96)00047-X

Stephens K, Measey J, Reynolds C et al (2020) Occurrence and extent of hybridization between the invasive Mallard Duck and native Yellow-billed Duck in South Africa. Biol Invasions 22:693–707. https://doi.org/10.1007/s10530-019-02122-6

Taylor MR, Peacock F, Wanless RM (eds) (2015) The Eskom red data book of birds of South Africa, Lesotho and Swaziland. BirdLife South Africa, Johannesburg

Tolley KA, Davies SJ, Chown SL (2008) Deconstructing a controversial local range expansion: conservation biogeography of the painted reed frog (*Hyperolius marmoratus*) in South Africa. Divers Distrib 14:400–411. https://doi.org/10.1111/j.1472-4642.2007.00428.x

Tracey JP, Lukins BS, Haselden C (2008) Hybridisation between mallard (*Anas platyrhynchos*) and grey duck (*A. superciliosa*) on Lord Howe Island and management options. Notornis 55:1–7

Treasure AM, Chown SL (2014) Antagonistic effects of biological invasion and temperature change on body size of island ectotherms. Divers Distrib 20:202–213. https://doi.org/10.1111/ddi.12153

Turpie JK, Forsythe KK, Letley GK et al (2017) Mapping and valuation of South Africa's ecosystem services: a local perspective. Ecosyst Serv 27:179–192. https://doi.org/10.1016/j.ecoser.2017.07.008

van der Walt JA, Weyl OLF, Woodford DJ, Radloff FGT (2016) Spatial extent and consequences of black bass (*Micropterus* spp.) invasion in a Cape Floristic Region river basin. Aquat Conserv 26:736–748. https://doi.org/10.1002/aqc.2589

van Rensburg PJJ (1983) The feeding ecology of a decreasing feral house cat, *Felis catus*, population at Marion Island. In: Siegfried WR, Condy PR, Laws RM (eds) Antarctic nutrient cycles and food webs. Springer, Berlin, pp 620–624. https://doi.org/10.1007/978-3-642-82275-9_84

van Wilgen BW, Richardson DM (1985) The effects of alien shrub invasions on vegetation structure and fire behaviour in South African fynbos shrublands: a simulation study. J Appl Ecol 22:955–966. https://doi.org/10.2307/2403243

van Wilgen BW, Wilson JR (eds) (2018) The status of biological invasions and their management in South Africa in 2017. South African National Biodiversity Institute, Kirstenbosch and DST-NRF Centre of Excellence for Invasion Biology, Stellenbosch

van Wilgen BW, Reyers B, Le Maitre DC et al (2008) A biome-scale assessment of the impact of invasive alien plants on ecosystem services in South Africa. J Environ Manag 89:336–349. https://doi.org/10.1016/j.jenvman.2007.06.015

van Wilgen BW, Measey J, Richardson DM et al (2020a) Biological invasions in South Africa: an overview. In: van Wilgen BW, Measey J, Richardson DM, Wilson JR, Zengeya TA (eds) Biological invasions in South Africa. Springer, Berlin, pp 3–30. https://doi.org/10.1007/978-3-030-32394-3_1

van Wilgen NJ, van Wilgen BW, Midgley GF (2020b) Biological invasions as a component of South Africa's global change research effort. In: van Wilgen BW, Measey J, Richardson DM, Wilson JR, Zengeya TA (eds) Biological invasions in South Africa. Springer, Berlin, pp 851–874. https://doi.org/10.1007/978-3-030-32394-3_29

van Wyk AM, Dalton DL, Kotzé A et al (2019) Assessing introgressive hybridization in roan antelope (*Hippotragus equinus*): lessons from South Africa. BioRxiv 569830. https://doi.org/10.1101/569830

Vimercati G, Hui C, Davies SJ, Measey GJ (2017) Integrating age structured and landscape resistance models to disentangle invasion dynamics of a pond-breeding anuran. Ecol Model 356:104–116. https://doi.org/10.1016/j.ecolmodel.2017.03.017

Visser V, Wilson JRU, Canavan K et al (2017) Grasses as invasive plants in South Africa revisited: patterns, pathways and management. Bothalia 47:a2169. https://doi.org/10.4102/abc.v47i2.2169

Vogt S, de Villiers FA, Ihlow F et al (2017) Competition and feeding ecology in two sympatric *Xenopus* species (Anura: Pipidae). PeerJ 5:e3130. https://doi.org/10.7717/peerj.3130

Watkins BP, Cooper J (1986) Introduction, present status and control of alien species at the Prince Edward Islands. S Afr J Antarct Res 16:86–94

Weyl PSR, de Moor FC, Hill MP, Weyl OLF (2010) The effect of largemouth bass *Micropterus salmoides* on aquatic macro-invertebrate communities in the Wit River, Eastern Cape, South Africa. Afr J Aquat Sci 35:273–281. https://doi.org/10.2989/16085914.2010.540776

Weyl OLF, Ellender BR, Ivey P et al (2017a) Africa: brown trout introductions, establishment, current status, impacts and conflicts. In: Lobón-Cerviá J, Sanz N (eds) Brown trout: biology, ecology and management. Wiley-Blackwell, Hoboken, pp 623–639. https://doi.org/10.1002/9781119268352.ch24

Weyl OLF, Schirrmann MK, Hargrove JS et al (2017b) Invasion status of Florida bass *Micropterus floridanus* (Lesueur, 1822) in South Africa. Afr J Aquat Sci 42:359–365

Wilson JRU, Faulkner KT, Rahlao SJ et al (2018) Indicators for monitoring biological invasions at a national level. J Appl Ecol 55:2612–2620. https://doi.org/10.1111/1365-2664.13251

Wilson JR, Foxcroft LC, Geerts S et al (2020) The role of environmental factors in promoting and limiting biological invasions in South Africa. In: van Wilgen BW, Measey J, Richardson DM, Wilson JR, Zengeya TA (eds) Biological invasions in South Africa. Springer, Berlin, pp 353–384. https://doi.org/10.1007/978-3-030-32394-3_13

Winter L, Wallace GE (2006) Impacts of feral and free-ranging cats on bird species of conservation concern. A report from the American Bird Conservancy. Accessed at: ww.abcbirds.org/newsandreports/NFWF.pdf

Yapi TS, O"Farrell PJ, Dziba LE et al (2018) Alien tree invasion into a South African montane grassland ecosystem: impact of *Acacia* species on rangeland condition and livestock carrying capacity. Int J Biodivers Sci Ecosyst Serv Manage 14:105–116. https://doi.org/10.1080/21513732.2018.1450291

Zengeya T, Ivey P, Woodford DJ et al (2017) Managing conflict-generating invasive species in South Africa: challenges and trade-offs. Bothalia 47:a2160. https://doi.org/10.4102/abc.v47i2.2160

Part V
Handling the Invasions

18

Developing and Enforcing Policies and Legislation

Peter Lukey and Jenny Hall

Abstract This Chapter describes and reviews the evolving biological invasion policy and legislation development and its implementation in South Africa over approximately the last 160 years. Despite the lack of formal, published government policy on biological invasions, there has been an almost continuous process of law-making over the years, with 50 pieces of being passed since the Xanthium Spinosum Act of 1861. The fundamental legal approach has changed little over this time, with a strong preference for what we have called the 'identify and direct' approach—a 'problem' species is identified and specific people are directed to deal with that species in a specified way. The concept of 'faultless liability' often associated with this approach has been equally resilient (e.g. a landowner is held responsible for clearing invasions on their land even if they were not responsible for introducing the species to the area in the first place). The review also suggests that, from a purely biological invasion management perspective, the South African 'job-provision' policy driver that has dominated biological invasion management activities since the new democratic dispensation in 1994 may have some perverse impacts in the absence of formal biological invasion policy. One of the key conclusions (with the proviso that biological invasions are indeed a significant threat to South African society, the economy, and the environment) is that a comprehensive evidence-based policy-making process should be instituted as a matter of urgency. It is also suggested that climate change concerns and interest in the global Sustainable Development Goals may provide the perfect 'policy-development window' for the development of formal policy on biological invasion in South Africa.

P. Lukey (✉)
Department of Environment, Forestry and Fisheries, Pretoria, South Africa
e-mail: plukey@environment.gov.za

J. Hall
University of Johannesburg, Johannesburg, South Africa

18.1 Introduction

This chapter uses an evidence-based policy-making lens to trace biological invasion-related policy and legislation in South Africa from the mid-1800s to the present. It does this in an attempt to assess how the governance of biological invasions has evolved over time in response to changes in science, politics, fashion, public opinion and sentiment. The chapter begins with an explanation of what is meant by evidence-based policy and legislation. It then moves to an overview of historical developments and trends. In this regard it is noted that the time span covered in the chapter straddles a range of historically important political changes. While it cannot, within the scope of the work, address approaches in African customary law to the management of invasive alien species, it does address the approach under colonialism; unified, but apartheid-driven, South Africa; and the new democratic dispensation. This historical analysis is used to inform some tentative and speculative thoughts about how the policy and legislative regime may, or could, evolve to meet the new and emerging policy drivers.

18.2 Background to Current Governance, Policy, and Legislation

Government policy can be defined as a description of the things that government hopes to achieve and the methods and principles it will use to achieve them (Education and Training Unit, n.d.). It therefore defines government's goals and can be used as a basis for decision-making by providing direction on key positions, especially where there are no clear right or wrong answers. As its name implies, policy provides clarity on the political position of government.

Although policy itself is not law, it may often identify the need for legislation to be developed or improved. The resulting legislation will set out the binding and enforceable rules which have been adopted by a law making body—normally, at the national level, Parliament or the Minister in the case of subordinate legislation such as Regulations and Notices. It prescribes what is, or is not, allowed and in the case of things that are conditionally allowed—the standards, procedures and, sometimes, principles that must be followed. If a law is not followed, those responsible for breaking it can incur consequences such as being prosecuted.

If policy sets out the goals and planned activities of government, the law empowers government to put the necessary institutional and legal frameworks in place to achieve the policy goals. Therefore, ideally, law must be guided by current government policy (Education and Training Unit, n.d.).

Currently, the first step in South African government policy-making is the political party conference where political parties discuss, debate and agree their positions and approach to specific issues. The second step is the attempt by the ruling party to make its policy the official government policy. To this end the executive branch of government (the President, Deputy-President and Cabinet) develop new policies and laws which the legislative branch (Parliament) may, or may not, approve. This is often a long and slow process of debate and negotiation between the ruling party, opposition parties, the general public, non-government organisations and special interest groups. The process is often informed by discussion documents known as Green Papers and usually culminates in the publication of the policy in the form of a so-called White Paper in the *Government Gazette*. Although this process is political in nature, it is important to note that policy-making is triggered by current affairs and debates.

In this regard, one of the most important recent developments in policy-making practise is the concept of evidence-based policy-making. According to Marais and Matebesi (2013), the concept has risen to prominence internationally and it emphasises the provision of quality services that are ideologically free, pragmatic, forward-looking, strategic, responsive, effective, efficient, and scientific.

The South African Government's Department of Planning Monitoring and Evaluation (DPME) in the Presidency is a vocal advocate of evidence-based policy-making and offers courses on the concept to senior government officials as key policy-shapers. DPME believes that evidence-based policy-making helps policy-makers and providers of services make better decisions, and achieve better outcomes, by drawing upon the best available evidence from research and evaluation and other sources (DPME and UCT 2014). This includes decisions about the nature, size and dynamics of the problem at hand; policy options that might be considered to address the problem; effective and ineffective interventions to solve the problem; the likely positive and negative consequences of the proposed policy option; the intended and unintended consequences of the proposed policy option; effective and ineffective modes of delivery and implementation; how long the policy will have to run before positive results will be achieved; the resources that will be required to implement the policy; the costs and benefits of the proposed policy, and on whom will these costs and benefits fall; and the sustainability of the policy from an economic, social, and environmental perspective. With this, DPME believes that evidence-based policy-making is about making decisions based on knowing, with an estimated degree of certainty, what will work to achieve which outcomes, for which groups of people, under what conditions, over what time span, and at what costs (DPME and UCT 2014).

According to Strydom et al. (2010) evidence is made up of a range of components—not only scientific—and it is never used in isolation. Scientific evidence typically includes research, surveys, quantitative/statistical data, qualitative data, and the analysis thereof. However, Strydom et al. (2010) note that evidence also includes economic, attitudinal, behavioural and anecdotal evidence; together with knowledge of experts, as well as lay persons, propaganda, judgements, insight/experience, history, analogies, local knowledge and culture.

With this, and despite reservations about how policy-makers often source information with a particular agenda in mind, Strydom et al. (2010) still believe that policies based on evidence are more likely to be better informed, more effective and cost-efficient than policies which are formulated through ordinary time- and politically-constrained processes without evidence input. They believe that policy which is based on evidence is also likely to give policy-makers confidence in the decisions that they take and that scientific evidence exposes policy-making to a wider range of validated concepts and experiences, enables policies to be formulated based on solid technical bases and can open up a range of policy options for policy-makers to consider.

DPME's evidence-based policy and implementation process recognises four stages: the first is the 'diagnostic' stage that is effectively a problem analysis to define what the actual problem is through cause-and-effect relationships. It also includes identifying the various options for addressing the problem. The second stage is the 'planning' stage which defines the theory of change and describes what must be done to bring about the desired change efficiently and effectively. This is then followed by the 'output' stage where the policy is implemented, monitored, reviewed and refined. The final, fourth stage is an 'outcome and impact' evaluation where the impact of the policy is evaluated (DPME and UCT 2014). Ideally the development of a policy goes through this cycle a few times and is constantly improved and refined.

With this as background, the following sections look at South Africa's biological invasion policies and legislation from the mid-1800s to the present through an evidence-based policy-making lens and attempt to identify what the key policy drivers have been and may be in the future (see supplementary material for a full list of all laws relevant to biological invasions in South Africa since 1861). Although biological invasions involve many groups of species, the following sections deal chiefly with invasive plants as this has been the principal focus historically in formulating biological invasion policy and legislation. The limited mention of other species is therefore a reflection of policy and legislative priorities as opposed to a deliberate omission.

18.3 The Early Days: 1860 to 1909—Colonialism and Weeds

The earliest example of biological invasion legislation in South Africa that we are aware of was passed in 1861. At the time, western-style law-making took place in four politically demarcated areas in South Africa, namely, the British Colony of the Cape of Good Hope; the British Colony of Natal; the Transvaal/South African Republic; and the Orange Free State. (The latter were also briefly British colonies from 1902 to 1910, during which time they were known as the Transvaal and Orange River Colonies).

On 14 August 1861, the Second Parliament of the Colony of the Cape of Good Hope promulgated the Xanthium Spinosum Act (Cape of Good Hope Parliament 1871). This statute was *"[a]n Act for promoting the Extirpation of the Burr Weed called Xanthium Spinosum"* (Act 22 of 1861, Fig. 18.1). It was specifically aimed at eradicating and destroying *"the noxious plant known as the Xanthium Spinosum, or*

Fig. 18.1 The Xanthium Spinosum Act (Act 22 of 1861)

No. 22—1861.] AN ACT [August 14, 1861.

For Promoting the Extirpation of the Burr Weed called Xanthium Spinosum.

WHEREAS the growth of the noxious plant known as the Xanthium Spinosum, or Burr Weed, has increased to an alarming extent in various parts of the Colony, and whereas the presence of the burr will be most detrimental to the value of the wool fleece, and highly prejudicial to the wool-growing interests of the Colony: Be it enacted by the Governor of the Cape of Good Hope, with the advice and consent of the Legislative Council and House of Assembly thereof, as follows:

I. It shall be the duty of the civil commissioner of each division, in his capacity of chairman of the divisional council of such division, to cause a public notice in the English and Dutch languages, warning all occupiers of landed property of the liability they will incur by neglecting to eradicate or burn any of the burr weed called Xanthium Spinosum which may have sprung up upon their lands or ground, to be posted at or near the court-house of every resident magistrate in such division, at or near all places of public worship within such division, at the place in any town or village within such division as may be used by the inhabitants for posting public notices and at all other places within such division at which the civil commissioner shall deem it desirable to post the same.

II. All field-cornets in any division and occupiers of landed property in such division, are hereby authorized and required to give or cause to be given to the divisional council of such division notice in writing that any occupier of landed property within such division has neglected, for a period of sixty days or upwards, reckoned from the day of the posting at the court-house aforesaid of the notice aforesaid, to eradicate and burn certain of the said weed growing upon the farm or ground by him occupied.

III. Every divisional council receiving any such notice as aforesaid shall by notice in writing call upon the occupier in regard to whose alleged neglect such notice shall have been given, to show cause why the said council should not at his expense take immediate measures for eradicating and burning the weed aforesaid so growing as aforesaid, and unless such occupier shall give security to the said council that he will within a reasonable time to be fixed by such council eradicate and burn all of the said weed so growing as aforesaid, then it shall be lawful for such council, and it is hereby required, to employ all labourers necessary for eradicating and burning the said weed so growing as aforesaid, and all the charges thereby incurred shall, from the said occupier, be recoverable with costs in the court of the resident magistrate by action at the suit of the secretary to the divisional council: Provided that as often as the said council shall grant time as aforesaid to any such occupier as aforesaid to eradicate and burn the said weed, the said council shall reserve to itself sufficient time before the next seeding time of the said weed, to enable the said council to cause the said weed to be eradicated and burnt in the event of the neglect of the occupier aforesaid so to do.

IV. It shall be lawful for the divisional council of any division and it is hereby required to employ, through the instrumentality of the field-cornets or otherwise, labourers to eradicate and burn the said weed wherever it may be found growing on public roads on Crown lands, or on public outspan-places within such division, and to pay the charge so incurred out of any divisional council funds at the time or thereafter in the hands of the said council.

V. It shall and may be lawful for the divisional council of any and every division and they are hereby empowered to levy a special rate upon the fixed property of the division in which any such council is situated, sufficient to defray the costs and charges incurred under and in consequence of the provisions of the fourth section of this Act: Provided, also, that if in one year the costs and charges thereby incurred in any division by the divisional council thereof for the eradication and destruction of the said weed shall exceed the sum of one hundred pounds sterling (£100), then one half of the expense incurred over and above that sum shall be paid from and out of the public revenues of the Colony, and shall be recoverable by the divisional council in the usual manner of recovering claims for expenses incurred for and on behalf of Government: And provided that the valuation from the time being of the fixed property of any division for road purposes shall be the valuation thereof for the purpose of the rate by this section authorized.

VI. Whenever the said weed shall be found growing on municipal lands the commissioners of the municipality within the limits of which it shall be found are hereby required to cause the said weed to be destroyed, and in default it shall be lawful for the divisional council of the division to cause the same to be destroyed as by the third section provided.

VII. All road inspectors and overseers of free road parties of convicts are hereby required to cause the working parties under their direction to eradicate and burn all plants of the said weed whenever they may be found growing within the limits of their respective works.

VIII. This Act shall continue and be in force until the 30th of June, 1863, and no longer.

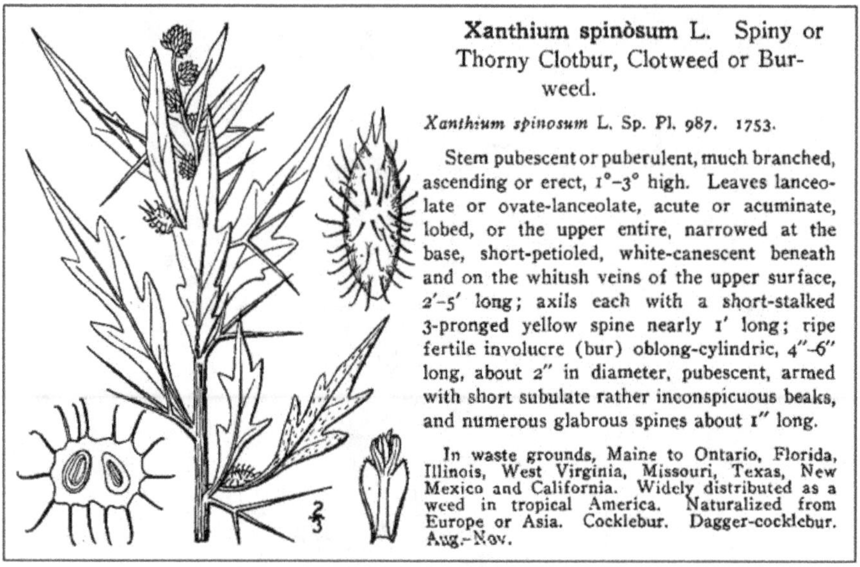

Xanthium spinòsum L. Spiny or Thorny Clotbur, Clotweed or Burweed.

Xanthium spinosum L. Sp. Pl. 987. 1753.

Stem pubescent or puberulent, much branched, ascending or erect, 1°–3° high. Leaves lanceolate or ovate-lanceolate, acute or acuminate, lobed, or the upper entire, narrowed at the base, short-petioled, white-canescent beneath and on the whitish veins of the upper surface, 2'–5' long; axils each with a short-stalked 3-pronged yellow spine nearly 1' long; ripe fertile involucre (bur) oblong-cylindric, 4''–6'' long, about 2'' in diameter, pubescent, armed with short subulate rather inconspicuous beaks, and numerous glabrous spines about 1'' long.

In waste grounds, Maine to Ontario, Florida, Illinois, West Virginia, Missouri, Texas, New Mexico and California. Widely distributed as a weed in tropical America. Naturalized from Europe or Asia. Cocklebur. Dagger-cocklebur. Aug.–Nov.

Fig. 18.2 A description of *Xanthium spinosum* (Cocklebur) from Britton and Brown's 1913 illustrated flora of the Northern United States, Canada and the British possessions (Britton and Brown 1913)

Burr Weed" which had spread *"to an alarming extent in various parts of the Colony". Xanthium spinosum* [commonly referred to as Spiny Cocklebur, or (more archaically) Burr Weed] is a South American annual herbaceous plant with many branches which grows up to 1.2 m high (Fig. 18.2). The fruit of the Spiny Cocklebur is an oval-shaped spiny bur about 10 mm long, green, with reddish, hooked spines that turn yellowish then brown. It is these burs that triggered the need for Act 22 of 1861 because they were *"most detrimental to the value of . . . wool fleece, and highly prejudicial to the wool growing interests of the Colony".*

Other than contaminating wool, the Spiny Cocklebur also competes with crop plants and native species along riverbanks and its spiny burs become entwined in the tails, manes and coats of domestic livestock, causing the animals much discomfort (Invasive Species South Africa, n.d.). The seedlings are particularly toxic to domestic livestock and it readily invades overgrazed pastures and spreads at the expense of the native species (Invasive Species South Africa, n.d.). It is also important to note that Spiny Cocklebur is not believed to have any social, economic or environmental benefits. It is clear from the preamble to the Act that wool-growing interests were the principal motivator for the statute.

The Xanthium Spinosum Act made it the duty of civil commissioners to publish notices in or near the court-house of every resident magistrate; near all places of public worship; at places used by the inhabitants for posting public notices and at all other places which the commissioner deemed desirable, *"warning all occupiers of landed property of the liability they will incur by neglecting to eradicate or burn any of the burr weed called Xanthium Spinosum which may have sprung up upon their lands or ground"*.

The occupiers of landed property were then given 60 days to comply before all field-cornets—civilian officials invested with the rank and responsibilities of a military officer and with judicial powers enabling them to act as a local administrator and magistrate—and other occupiers were required to report all cases of non-compliance to the Divisional Council. The Council, taking seeding cycles into account, had to then allow the occupier a reasonable chance to eradicate or burn the weed on their property, failing which the Council would do the job and claim the cost from the occupier—the so-called 'step in and pay the bill' principle. The same applied to errant municipalities where the weed was found growing on municipal land.

The law was not limited to private occupiers. It also required Council funds to be used to employ "*labourers to eradicate and burn the said weed wherever it may be found growing on public roads on Crown lands, or on public outspan-places*". Private landowners were indirectly responsible for part of these costs too because Council was allowed to raise these funds through property rates, although it could get half of the funds from "*the public revenues of the Colony*" if the eradication costs "*exceeded one hundred pounds sterling (£100)*". Moreover, "*all road inspectors and overseers of free road parties of convicts*" were required to ensure that "*the working parties under their direction*" eradicated and burned all Burr Weed plants "*whenever they may be found growing within the limits of their respective works*".

As if it were an experiment, the Act was only in force "*until the 30th of June, 1863, and no longer*" (own emphasis).

The basic structure of the Xanthium Spinosum Act, 1861 (Fig. 18.1) has remained remarkably resilient as an approach for biological invasion governance. First, it identifies the specific problem species. Secondly, it places a duty of care on private land-owners to eradicate the species on their land. Thirdly, it requires municipalities and other government custodians of state-owned land to eradicate the species on their land. The logic of this 'identify and direct' approach appears to be that if everyone who is directed to comply complies, then the identified problem will be completely solved. It may also be suggested that the Act was remarkably progressive in its funding model. However, if one accepts few, if any, people who were required to eradicate the weed were actually responsible for the weed's occurrence, it could also be seen as holding the 'victims' of the weed accountable for its control. This 'faultless liability' is an approach usually used with great circumspection in law because of the tension that it creates with the notion of fairness. It therefore suggests that the issue of managing the invasive species was considered to be sufficiently serious to merit the use of extreme regulatory measures.

It is noted, however, that the approach in the Act is entirely reactive as, although the Act regulates the impact of a particular problem species, it does not prevent the introduction or use of other alien species that may become problematic. In this regard, Bennett and van Sittert (2018) [citing Crosby (1986) and van Sittert (2000)] note that it was the farmers' purposeful introduction of alien agricultural plants, such as wheat, that led to the accidental introduction of *Xanthium spinosum* in South Africa in the first place.

The Xanthium Spinosum Act became the basis for biological invasion governance for years to come. It was extended through Act 27 of 1864 of the Cape of Good Hope and appears to have been largely replicated in the statutes of the Colony of Natal, specifically Law 20, 1861. The latter was repealed and re-enacted with amendments 13 years later by Law 38 of 1874 the purpose of which was "*To prevent the spread of the growth of the Xanthium Spinosum Burr Weed*". In this 1874 amendment, there is an admission that "*the growth of the noxious plant known as the Xanthium Spinosum Burr Weed has increased to an alarming extent in various parts of the Colony, and is prejudicial to the farming interests of the Colony*". Importantly, there is also the admission in the preamble that "*the Law hitherto in force has failed to check the spread thereof, and it has been necessary to make other and more stringent regulations to effect its extermination*". Despite these admissions, it was more the penalties that changed than the approach. In this regard, the more stringent regulations included occupiers of private land being liable to a recurring "*penalty for leaving Burr Weed upon their lands*". This penalty was no doubt why Burr Weed became known as *Boetebos* (a Dutch and Afrikaans term which can be literally translated as 'penalty bush'). In cases of non-compliance where the authority stepped in and claimed the bill the "*expense thus incurred [was] chargeable upon the land with 6 per cent interest*". The Transvaal Republic appears to have promulgated a similar law in 1872—the Xanthium Spinosum Law (Law 1 of 1872).

Up until this time, the emphasis had been overwhelmingly on weeds. However, 14 years later, the Government of the Transvaal Republic set the scene for the use of the 'identify and direct' approach to deal with a faunal species i.e. the Australian Bug (the Cottony Cushion Scale—*Icerya purchasi*) and other problem insects, when the Volksraad made the Australian Bug Resolution of 24 July 1888. The Resolution authorised the Executive Council to make Regulations and provisions in respect of the Australian Bug or other noxious insects. These Regulations and provisions included: the chopping down and destruction by the owner of trees and plants infected by the Australian Bug or other noxious insects; the power of the government to do the chopping or destroying at the cost of the owner and without any entitlement by the owner to claim compensation, if the owner is negligent; and the power to make additional Regulations. As *Icerya purchasi* is a pest to citrus crops, it is highly likely that it was influential citrus farmers who lobbied for this Resolution.

The focus, however, remained on weeds and by 1889, the Government of the Cape of Good Hope had expanded and mainstreamed their weed legislation into the Divisional Councils Act (Act 40 of 1889), Sub-Division VI, Part I, Extirpation of Xanthium Spinosum and other Noxious Weeds and Plants. These weed provisions, although similar to Act 22 of 1861, also included additional clauses such as allowing for the special exemption of river beds; the power to identify other "*noxious weeds and plants*" to be treated as if they were *Xanthium spinosum*, and the employment of weed inspectors with the power to access all land. Given that the Spiny Cocklebur competes with crop plants and native species along riverbanks, it is not clear why the

Governor on the request of the Council should exempt identified beds of public rivers from the requirement that every owner and every occupier of land must eradicate or destroy the weed.

Treating other noxious weeds and plants as if they were *Xanthium spinosum*—a weed with no benefits to the South African economy, society or environment—marks an interesting 'one-size-fits-all' policy development that, history has shown, appears to be at the root of much of the more recent public or special interest push-back against biological invasion legislation and its implementation (see the trout issue in Weyl et al. 2020, Chap. 6, as well as the controversy around the Himalayan Tahr in Sect. 18.7.4 below).

The Transvaal Republic made further advances in its regulation of weeds in 1897 with its Eradication of Burweed Law (Law 4 of 1897) which dealt with "*the control of certain weeds—the Xanthium Spinosum and the Scotch Thistle—which are increasing very considerably in various parts of the South African Republic, and which render the wool valueless where they grow*". Although the Cape's Divisional Councils Act (Act 40 of 1889) described above opened the door for the regulation of weeds other than *Xanthium spinosum*, this 1897 law appears to mark the start of the 'listing' of species. In this regard, despite Scotch Thistle (*Onopordum acanthium*) being a plant which can form dense, impenetrable stands that compete with field crops and forage plants, the protection of wool farmers still appears to be the overriding motivation for this law. As with the Cape legislation, Law 4 of 1897 makes it "*the duty of all burghers* (citizens) *to eradicate the* [weeds] *on their properties*". Like the Natal Law, there is also a provision for penalties for non-compliance, which are equally applicable to "*officials charged with the supervision of Government grounds*", where, if a person is "*negligent in respect of*" the duty to eradicate the weeds they "*shall be liable to a fine not exceeding £1, or, in case of non-payment thereof, to imprisonment with or without hard labour for a period not exceeding three days for the first offence, or for a second offence to a fine not exceeding £5, or in case of non-payment thereof to imprisonment for a period not exceeding 14 days with or without hard labour, and for each successive offence to a fine not exceeding £20 in addition to imprisonment for a period not exceeding six weeks with or without hard labour*". Interestingly, in what appears to be an attempt to promote compliance through improved detection of non-compliance, the Transvaal Law noted that "*the Government may grant a part, but not more than a third of the fines which have been imposed, to the informant.*"

In 1901, Natal identified another *Xanthium* species for control through their Xanthium Strumarium Burr Weed Act (Act 20 of 1901) which required that "*the Xanthium Spinosum Law No. 38, 1874, shall apply as fully and effectually to the Xanthium Strumarium Burr Weed as though such weed had been originally included in the said Law*".

In the 3 years before the four colonies became a union in 1910, there appears to have been a flurry of 'noxious weed' law-making. The Cape Colony started with its Divisional Councils Act Amendment Act (Act 17 of 1907) when the Act conferred "*upon Divisional Councils greater or additional powers for the eradication or extirpation of noxious weeds or plants, and in that respect to amend the*

Divisional Council Act, 1889". Once again, many of the basic 'identify and direct' provisions of Act 22 of 1861 remained visible and although *Xanthium Spinosum* was specifically mentioned, it also allowed for further plants to be identified and controlled. Importantly, the Act also included a "*penalty for neglect*" where "*upon proof of his neglect such owner or occupier shall be deemed to be guilty of an offence under this Act, and shall upon conviction be subject to a fine not exceeding twenty pounds and such owner or occupier shall be liable to the same penalty for every week or part of a week during which he shall fail, after his first conviction, to comply...*" Another interesting addition that appears to counter the strange 1889 'river bed exemption' is the provision that "*in the eradication of noxious weeds it shall be an offence to place such weeds in any river or any defined water course. Any person contravening this section shall upon conviction be liable to a penalty not exceeding twenty pounds.*" By identifying a specific dispersal pathway, namely rivers and water courses, this provision may also represent a broadening of the approach to invasive species management. The Act was followed up with The Rural Council (Cape Division) Act (Act 33 of 1909) which empowered Rural Councils to "*make, alter, revoke or amend regulations ...for the more effectual eradication or destruction or prevention of the spread from adjoining divisions of Xanthium Spinosum or any other noxious weed or plant.*"

The Transvaal followed in 1909 with their Noxious Weeds Act (Act 12 of 1909) to "*make better provision for the Eradication of Noxious Weeds*". Act 12 of 1909 empowered the Governor to make regulations: a) imposing a duty on the occupier or owners of land, mining title holders and holders of grazing rights on Crown Land to clear and keep clear their land of noxious weeds; b) prescribing the manner in which noxious weeds had to be eradicated in respect of this duty; c) empowering any official of the Department of Agriculture, field cornets, or police officers to inspect land and issue directions by written notice to clear that land of any noxious weed; d) empowering these officials to eradicate weeds in cases of non-compliance; e) providing for the recovery of the cost of eradication of noxious weeds from the person who is in default and the mode of such recovery; f) preventing the introduction into the Colony or the sale of any plant, seed or grain, which is likely to propagate or spread the growth of noxious weeds; and g) for generally preventing the spread of noxious weeds in the Colony. Penalties for non-compliance were also updated—"*Penalties may be imposed for a breach of or failure to comply with any such regulation not exceeding a fine of fifty pounds or in default of payment imprisonment with or without hard labour for a period of six months.*" In addition to the provision relating to eradication method prescription, the acknowledgement of the role of 'prevention'—restricting import or sale—is an important regulatory development. As with the Cape Colony's Divisional Councils Act Amendment Act (Act 17 of 1907) regulation of a specific dispersal pathway, the regulation of introduction pathways, namely import and sale, by Transvaal's Noxious Weeds Act (Act 12 of 1909) represents a further broadening of the approach to invasive species management. Furthermore, given the traditional and ongoing tension between mining and agriculture which was driven by competing land use, the specific mention of 'mining title holders' in an Act that clearly protects farming interests may also be a significant policy development. However, this was only a framework Act which could have no real impact until Regulations were published. As an acknowledgement of this fact, the new law made it clear that Law 4 of 1897 was only repealed from the date of the publication of any Regulations in terms of the Act. The Orange River Colony also published their own Noxious Weeds Act (Act 23 of 1909).

In summary, during the 50 year period 1860–1909, policy and legislation focused on a limited number of (mostly plant) species. In addition, the primary driver in policy and legislation dealing with biological invasions was to protect the interests of farmers by placing a general 'duty of care' on all landowners and land users to eradicate identified plants. The underlying policy consideration was therefore the protection of a specific economic sector by making everyone responsible for managing the problem—often in a faultless liability manner. Acknowledged failures of the policy and legislation in meeting stated objectives were on the whole met with increasingly stringent penalties for non-compliance rather than a shift in the underlying approach. As will be seen, the identification of species and direction of response—the 'identify and direct' approach—continues to remain a core component of biological invasion legislation, if not policy, today.

18.4 Dealing with Union: 1910 to 1934—Regulatory Hiatus

The Union of South Africa came into being on 31 May 1910 with the unification of the four British colonies of the Cape, Natal, Transvaal (formerly the Boer South African Republic) and Orange River (formerly the Boer republic of the Orange Free State). These former colonies became the Union's provinces known as the Cape, Natal, Transvaal and Orange Free State provinces respectively. At its establishment, the Union of South Africa was a self-governing autonomous dominion of the British Empire until independence from Britain through the 1926 Balfour Declaration and the 1931 Statute of Westminster. The Union became the Republic of South Africa on 31 May 1961 with the enactment of a new constitution.

On the biological invasion front, there appears to have been a regulatory hiatus following union in 1910. Apart from the new Transvaal Province's 1912 Ordinance—*"An Ordinance to Consolidate and Amend the Law relating to Municipal Government in this Province and the establishment of Health Committees therein, and to provide for matters incidental thereto"* (Ord. No. 9 of 1912)—which empowered Councils to make weed-related by-laws, the first 24 years following union was relatively quiet on the weed front. However, in addition to the usual suspect, *Xanthium spinosum*, Ord. 9 of 1912 also specifically identified *Cannabis indica* (Marijuana) and *Tagetes minuta* (Khakibos) as noxious weeds. This is significant because, unlike *Xanthium spinosum*, these plants provide benefits to some people—Marijuana for its narcotic and associated cultural uses, and, as hemp, for its use as a textile, for paper and as rope; and Khakibos for its essential oil (currently grown commercially in South Africa, France and North America for the same purpose). Furthermore, this could be the first time that weed regulations were being considered for use in addressing a problem that was not specifically related to the plant's impact on agriculture. For example, although the politics and economics of the early Marijuana industry is beyond the scope of this Chapter, it is probably safe to say that *Cannabis indica* was considered to be far more of a social problem than an agricultural economy problem.

According to Bennett and van Sittert (2018), from 1910 to 1913, the Union delegated weed control to the provinces through the Financial Relations Act (Act 10 of 1913). Bennett and van Sittert (2018) also note that from 1913 to 1937 provincial administrators were made responsible for controlling weeds through Part 5 of the Union of South Africa Act of 1909. Although each province still used their colonial laws to regulate weed control, Bennett and van Sittert (2018) describe an attempt in 1916 by the South African Farmers Union to get the Minister of Agriculture to rationalise and harmonise these provincial laws and the Provincial Administration Commission's failed suggestion that the Cape's system be applied nationally.

During this quiet period the Orange Free State amended its Noxious Weeds Act (Act 23 of 1909) with its 1920 Noxious Weeds Amendment Ordinance (Ordinance 6 of 1920); principally to replace the word 'Governor' with 'Administrator'. However, an interesting new provision was also slipped into this 'editorial' amendment that empowered the administrator to make grants, or to grant other assistance, monetary or otherwise, for "*the effectual eradication of noxious weeds*".

Further legislation, like Transvaal's 1928 Local Government (Noxious Weeds) Amendment (Ordinance 8 of 1928), concentrated largely on administrative amendments with few significant changes in policy or approach.

Apart from starting to include contested species—species that were considered problematic by some, but beneficial by others—and the potential government funding of eradication programmes, this period also appears to have introduced very few changes in policy or legislative approach. Notwithstanding this, it is possible that this period may have seen the introduction of a new policy-driver that was distinctly South African, namely, nationalism.

In this regard, the closing years of the nineteenth century saw botanists beginning to advocate for the recognition and protection of the Western Cape's unique native fynbos flora which was under threat from extensive transformations of the landscape due to farming, forestry, urban sprawl and wild flower-picking (Pooley 2010). Although this lobby was ignored by a largely indifferent public, Pooley (2010) believes that this may have changed in the context of post-South African war efforts to build unity among the English and Afrikaner populations, and where invasive alien plants presented a physical and symbolic opportunity for the botanist's advocacy. Indeed, Pooley proposes that the then new discipline of ecology suggested metaphors of integration (and segregation) that both politicians and natural scientists *"could exploit to their mutual benefit"* (Pooley 2010).

Pooley (2010) believes that botany, patriotism and the politics of national unity were closely bound up and that *"floral nativism... provided both a sense of identity for an emerging White settler nationalism and a justification for evicting the underclass from the commons"* (Pooley 2010, citing van Sittert 2003).

Bennett (2014) notes that there is a developed literature focusing on how fears of invasive alien plants also expressed South African nationalism in its apartheid and post-apartheid forms. In this regard there is a suggestion that a discourse around the 'danger' of invasive alien species in South Africa gained momentum in the late 1950s and early 1960s (Bennett 2014, citing Carruthers 2011) and that *"nationalism... provided justification for eradicating these ...species"*. Bennett (2014) notes that Peretti (2010) takes this argument further by linking South African interest in biological invasions in the 1980s with ideologies of apartheid that were *"concerned with separating the pure from impure"* (see also van Wilgen 2020, Sect. 2.14).

18.5 Intentionally Introduced Invasive Species 1860–1935 (Post Union But Pre-independence)

18.5.1 Alien Trees as a Solution to Problems

With reference to the most recent review of biological invasions, included in the 25 invasive species with 'severe impact', van Wilgen and Wilson's status report (2018) identifies 1 freshwater fish (4%), 4 freshwater invertebrates (16%), 17 terrestrial and freshwater plants (68%) and 3 terrestrial invertebrates (12%). Although the fact that plants make up the majority of species with severe impacts appears to align with the historical weed focus which was followed up until 1934, what does not align is the fact that included in the 17 terrestrial and freshwater plants are 11 tree species (44% of total species with severe impact). The reason for this is simply that before 1935 the policy was that trees, especially invasive trees, were regarded as being a solution rather than a problem.

According to Kruger and Bennett (2013, citing Barton 2002, Beinart 2003, Grove 1989, and Bennett 2010), the second half of the nineteenth century saw a number of forestry enthusiasts in Southern Africa propagating the idea that tree-planting would increase the amount of rain in dry areas and would improve streamflow. These claims motivated white settlers in all four colonies to deliberately introduce alien trees and to engage in tree-planting efforts to increase rainfall and soil fertility, slow erosion, stop desertification and relieve or cure tropical diseases, such as malaria (Kruger and Bennett 2013, citing Bennett 2010). Despite reservations based on observations to the contrary, the belief that alien forests stopped erosion and conserved water became and remained the official South African forest management policy until the mid-1930s (Kruger and Bennett 2013, citing Beinart 2003). This policy was reinforced by the findings of the Union-wide 1923 Drought Investigation Commission which supported the idea that forests had a positive climatic and hydrological influence (Kruger and Bennett 2013 citing Beinart 2003).

However, by the early 1930s this dogma was being formally challenged. For example, in 1932 the Department of Forestry directly contradicted the belief that trees improve streamflow when it adopted a policy of keeping a 20-m buffer zone between a stream and a plantation with the aim of mitigating the effects of their plantations on streamflow (Kruger and Bennett 2013, citing Malherbe 1968). According to Kruger and Bennett (2013), public and scientific criticism of afforestation—mainly that it decreased streamflow in higher rainfall zones—reached a crescendo prior to what they refer to as a seminal year in the history of South African forest hydrology research—1935, when South Africa hosted the Fourth Empire Forestry Conference which "*brought together foresters from around the British Empire to discuss current findings, to talk about problems and to coordinate empire-wide policies.*"

Kruger and Bennett (2013) believed that the 1935 Empire Forestry Conference crystallised political support for the emerging major, long-term research programme that would, among others, scientifically debunk the afforestation myths. Furthermore, in an early demand for evidence-based policy-making, the Conference provided a platform for politicians and critics of forestry to call for policies that were evidence-based, practical and under constant review. With regards to evidence, foresters agreed that in balancing theory and experience, experience should trump theory until the theory was proved in practice—an unusual step at a time when scientific theory generally enjoyed priority.

With reference to the concept of 'faultless liability' discussed above, it is important to note that in the 1880s and 1890s, in support of the 'forestry has a positive climatic and hydrological influence' policy, the Cape Colony's newly-founded Department of Agriculture distributed seeds and plants freely or at little cost (Kruger and Bennett 2013). Furthermore, from an invasive species management policy and legislation perspective, unlike accidentally-introduced weeds like *Xanthium spinosum*, many trees are examples of species which were intentionally introduced with very different pathways of entry. Consequently one would expect different regulatory approaches to the management of accidently and intentionally introduced invasive species. However, as discussed in Sect. 18.8.2 below, this is not necessarily the case.

18.5.2 The Jointed Cactus and Weeds Acts

The promulgation of the Jointed Cactus Eradication Act (Act 52 of 1934) marked a subtle change in the legislative approach. Although the Act was still largely based on the 'identify and direct' approach, it also contained a provision that was sensitive to the issue of 'faultless liability'. Other than simply putting the burden of eradication on the occupier or owner of land, section 3 of the Act provides for the eradication of *Opuntia aurantiaca* on private land by a designated official who is allowed to "*. . .take with him upon such land the labour, animals, vehicles, instruments, appliances, drugs or any other thing which in his opinion is necessary or required for the purpose of eradicating such cactus.*" After this initial clearing at no cost to the occupier or land owner, a clearance certificate was issued (section 7) together with a continuance order requiring the occupier or landowner to keep the land clear (section 8). Even when applying the traditional identify and direct approach, the Act therefore also provided for government assistance to the occupier and landowner through, what could be regarded as, an early form of our South Africa's current public works programmes.

Another important development was the inclusion of information and reporting provisions in terms of which government had to be informed of the presence of jointed cactus (section 1) and could collect such information (section 2) as it required.

These two important developments allowed for a more strategic approach to eradication based on reported or observed infestation data which could be used to inform government co-ordinated eradication campaigns. According to Bennett and van Sittert (2018), jointed cactus had become the most serious weed in the country by the 1920s, and government's response to the jointed cactus infestation in the Eastern Cape following the promulgation of the Jointed Cactus Eradication Act proved to be the one exception to the general rule of weak state enforcement of weed legislation. Using the powers provided under the Act, officials took a strong interventionist approach using teams of labourers to mechanically clear private farms. The Department of Agriculture then implemented a program of biological control using imported cochineal beetles and the moth *Cactoblastis cactorum* (Bennett and van Sittert 2018).

Three years after the Jointed Cactus Eradication Act (Act 52 of 1934), the Weeds Act (Act 42 of 1937) was promulgated. This Act repealed and replaced all of the existing colonial and post-colonial weeds legislation. Although the Weeds Act was largely modelled on the Jointed Cactus Eradication Act, it appears to have de-emphasised and diluted provisions for state-sponsored interventions on private land, effectively returning to a reliance on the concept of 'faultless liability'.

18.6 An Expanding Agenda (1935–1993)

The Weeds Act (Act 42 of 1937) remained the principal biological invasion legislation until it was replaced by the Conservation of Agricultural Resources Act (Act 43 of 1983) (CARA) some 46 years later. From the 1970s onwards, legislation started reflecting an expanding agenda. As discussed previously, until this time the main policy driver for passing legislation to manage invasive species was economic and specifically the economic interests of the agricultural sector. In other words, if invasive species were identified as having a negative impact on an economic sector or activity they needed to be controlled. Although this approach continued into the next phase of legislation, as is illustrated by the example of CARA (see Sect. 18.6.2 below), changes were afoot.

18.6.1 Expanding the Policy Drivers to Include Water and the Environment (1970–)

According to Kruger and Bennett (2013), the first scientific policy analysis regarding water and forests in South Africa, published in 1949, used scientific evidence and theoretical principles to challenge the myths of the hydrological benefits of forests, clarify policy options and introduce the concept that catchment management should be part of a broader system of ecosystem management. Subsequent reports of interdepartmental committees on the conservation of mountain catchments (1961) and afforestation and water supplies (1968) and those of a Commission of Inquiry into Water Matters (1970) provided the evidence that informed South African forestry and water policies for the next 30 years (Kruger and Bennett 2013).

The promulgation of the Mountain Catchment Areas Act (Act 63 of 1970) arguably signalled the first attempt at expanding economic policy drivers regarding biological invasion management to include environmental ones. Given the research and review processes described above, this Act was clearly a product of 'evidence-based policy-making'.

The Mountain Catchment Areas Act is primarily environmental in nature, as its purpose is to provide for the conservation, use, management and control of land situated in declared mountain catchment areas. Further evidence for the idea that the Act is underpinned by environmental concerns is that catchments which are declared in terms of the Act are now recognised as protected areas in terms of section 9 of the National Environmental Management: Protected Areas Act (Act 57 of 2003). It also has an underlying purpose of conserving water resources through the conservation of soil and vegetation. Section 3 of the Act empowers the Minister to issue directions, which may be applicable to catchment areas as well as areas within 5 km of a catchment area, relating to *"the destruction of vegetation which is, in the opinion of the Minister, intruding vegetation"*.

However, although these provisions represent a broadening in existing policy approaches, the power does not seem to have been utilised in practice. Furthermore, these environmental concerns did not extend to other protected areas such as national parks in any meaningful way. For example, the National Parks Act (Act 57 of 1976) focuses very much on protecting resources in the parks from immediate physical human activities such as hunting and picking plants (cf, for example, section 21). A possible exception to this in respect of alien animals is the provision in section 21 (g) which says that no person may *"introduce any animal or permit any domestic animal to stray into or enter a park"*.

18.6.2 CARA and the Return to Form (1983–)

Notwithstanding the broadening of policy drivers introduced by the Mountain Catchment Areas Act, the 1980s largely returned to form from a regulatory approach perspective, when the Conservation of Agricultural Resources Act (Act 43 of 1983) (CARA) was promulgated. CARA introduced a range of regulatory mechanisms that are aimed at protecting resources which the agricultural sector relies on. One of these is the power of the Minister to declare any plant to be a weed or invader plant, either nationwide or in specific areas only (section 2(3)). Unlike other provisions in the Act which expressly do not apply in urban areas, a more progressive approach is taken to weeds and invader plants in section 2(2) which provides that the *"provisions of this Act relating to weeds and invader plants shall also apply to land which is situated within an urban area"*. From a policy perspective, this was an interesting development because it extends the reach of regulatory control to prevent the spread of all the listed species, irrespective of where they are found.

The first set of Regulations aimed at giving effect to the weed and invader plant provisions in the Act were passed in 1984 (GNR 1048 *GG* 9238, 25 May 1984). These Regulations included a list of approximately 50 species which were categorised as being either 'weeds' or 'invader plants'. Regulations 15 and 16, which concern problem plants, were amended in March 2001. In this last 'problem plant' amendment to the CARA Regulations, the list of plants was expanded to include many more species (198) and it also categorised them as falling within one of three categories (GNR 280 *GG* 22166, 30 March 2001). The categorisation of the species shows the underlying policy issues which government has to wrestle with. Whilst category 1 plants were to be very tightly controlled and no person would be allowed to sell, advertise, exhibit, exchange or dispose of them, categories 2 and 3 showed some more flexibility because of the utility or value of the species. In this regard, category 2 species are known to be problematic, but because of their commercial value or other uses they are allowed in demarcated sites. Category 3 species include ornamental plants that can be kept, but not propagated.

Interestingly, with regards to category 2 species, a statement made by the then Minister of Agriculture and Land Affairs, Thoko Didiza, reflected the tension between wanting to exploit, or wanting to destroy invasive alien species when she said "*The government is fully supportive of the commercial ventures based on these species, recognizing as it does the important contribution that they make to the South African economy and the welfare of its people. ... However, this welfare is under threat from the significant impacts of alien invasive plants, and we must face up to, and deal with this problem*" (Hanks 2001).

18.6.3 Moving Beyond Plants (1983–)

Another way in which the underlying policy approach was expanded was a new or increased focus on regulating animals, marine species, and other organisms which cause disease. In this regard, the Agricultural Pests Act (Act 36 of 1983) and the Animal Diseases Act (Act 35 of 1984) both provided for the control of the importation of 'exotic animals' or parasites and diseases. Regulations passed in 1998 in terms of the Marine Living Resources Act (Act 18 of 1998)—which has the conservation of the marine ecosystem as one of its express objectives—require applications for mariculture activities to contain information regarding how the introduction of alien commensals, parasites and pathogens will be avoided and what measures will be taken to avoid the establishment of alien species in the wild (Regulation 61 of GNR 111 GG 19205, 2 September 1998). In 2013, a proposed amendment to the Act was published which allowed for more direct control over the release of alien species through a new section 43A (GN 434 GG 36413, 25 April 2013). The Bill, however, has not been passed to date.

18.6.4 1991: Rethinking Policy

Apart from these legislative developments in the late 1980s and early 1990s, the apartheid government of the day was also reassessing its environmental policy. One of the major initiatives during this period was the President's Council's investigations and recommendations on environmental policy. The mandate of the Council included *'a particular reference to the ecological, economic, social and legal implications thereof'.* In 1991 the President's Council released its report (The President's Council 1991) which made proposals for a national environmental management policy.

Amongst the many issues that the Council considered, it is clear that it was alive to the issue of invasive alien species; it contained just over a page dedicated to the issue, as well as references in other areas of the report. Although some of the Council's proposals appeared quite proactive, others seemed somewhat resigned, pragmatic and/or highly subjective. For example, it made comments such as "*trees improve any landscape*". In a seemingly progressive stance, and as seen from the extract below, the Council appeared willing to expand the traditional approach of managing invasive alien species for purely economic reasons by also including environmental considerations: "*In addition to the threat to agriculture, the rapid spread of invasive alien plants such as the jointed cactus, Australian acacias, hakea, lantana and many others are also a threat to the native flora and fauna and to natural habitats. Alien vegetation affects the productivity of river systems and wetlands. The banks of many of our rivers, particularly the upper reaches, are completely overgrown with Australian acacias (wattles). From here they spread up the mountain slopes. The open water of many rivers, dams and vleis is being choked by one or more of South American weeds e.g. the water hyacinth, and the water fern (Salvinia). Little has been done to control these invasive plants, because they affect only the natural productivity of the waters and their recreational potential, which as yet do not enjoy a high priority rating.*" (The President's Council 1991).

As a policy suggestion, the Council's opinion was that the strategy to address these negative impacts should be to "*clear sparsely infested areas*" and to "*contain the dense stands to prevent them from spreading until more effective control measures have been developed.*" The report also signals the need for a pragmatic approach. For example, it sounds a warning to 'purists' that "*. . .a fact that must be* *accepted by conservationists is that not all alien plants are undesirable. Many varieties of trees introduced over the past 300 years have been a decided asset and present no threat to our indigenous fauna or flora*". This remark is coupled with another that reveals some insight to the prevailing sentimentality for certain alien species "*Trees improve any landscape and play a vital role in urban areas. They soften the sharp edges of our cities and attract birds and other wildlife, bringing nature into the heart of the city. . . . Collectively, the gardens of suburban areas will play an increasingly important role in future in the total conservation effort, as natural habitat dwindles. While at best suburban gardens and parks are an artificial habitat, they can contribute to ensuring the survival of many species which might otherwise disappear as natural habitat is destroyed.*"

This sentimentality towards trees reveals a dualistic approach to the management of alien species that persists today. On the one hand nationalist policies stimulate a form of xenophobia towards all alien species. On the other hand, and in parallel, there is a distinct nostalgia and cultural attachment to certain species. This may explain why, for example, several of the trees that are still protected in terms of section 12 of the National Forests Act (Act 84 of 1998) are alien trees. These include the *Araucaria heterophylla*—Norfolk Island Pine—planted in 1826 by the wife of the last Landdrost (magistrate) of Stellenbosch and a range of other trees associated with historical milestones.

Under the report's discussion on the conservation of freshwater aquatic environments and aquaculture in particular, the Council again noted the potential benefits of aquaculture but provided the following caution "*It is very important to recognise, however, that as aquaculture is often based on the utilisation of foreign species of freshwater fish, marron and shrimps, there is always the danger of these becoming established in the wild to the detriment of natural species and the environment. There is also the risk of the introduction of diseases which could be disastrous to native fauna.*" At the same time it also said that, "*The early endeavours in aquaculture were largely concerned with the establishment of trout in the rivers of the Cape, Natal and the Transvaal. By 1920 trout were well established and provided excellent sport, even attracting anglers from overseas. Trout are still a great recreational amenity in Natal and the Transvaal, but the Cape Department of Nature Conservation gradually phased out the breeding and distribution of trout by 1989. Trout were considered a threat to small indigenous fishes, besides which the hatcheries were required for the propagation of indigenous species. Because of the great recreational potential of trout angling, rivers which are suitable for trout should be conserved.*"

18.6.5 The Immediate Pre-democracy Status Quo

Although by 1993 biological invasion legislation had expanded beyond weeds to include alien trees and animals, it still largely followed a 'identify and direct' approach. However, the policy drivers had expanded from purely economic ones to include broader natural resource concerns, particularly water, and the environment. Despite this broadening of policy drivers, it is clear from the President's Council Report (The President's Council 1991) that there was a still a policy vacuum in respect of, for example: how considerations of cost and benefit should influence control approaches; how interventions should be prioritised; what control approaches are acceptable; what control strategies should be employed; or whether interventions should be *ad hoc* or part of a co-ordinated national campaign, programme or strategy.

18.7 Governance in the Democratic South Africa (1994–Present): Internalising a Rights-Based Culture

The regulatory initiatives discussed above were undertaken in isolation of the rapid developments in international environmental law that were taking place at the time because of South Africa's ostracism by the international community. With the change in political dispensation, 1994—the year when South Africa elected a democratic government and was reaccepted into the international community in general and the United Nations Environment Programme activities in particular—marked the start of a new wave in the regulation of invasive alien species. There were perhaps two things that provided the impetus for new policy and regulation during this time. These were South Africa's ratification of the 1992 Convention on Biological Diversity in 1995, and the adoption of a new constitutional order in 1996.

With regards to the Convention, when South Africa ratified it on 2 November 1995, it incurred an obligation to domesticate the Convention's requirements, including those in article 8 and subarticle 8(h) which compel South Africa to "*as far as possible and as appropriate*", "*prevent the introduction of, control or eradicate those alien species which threaten ecosystems, habitats or species*". Giving effect to this obligation was not without its challenges because government acknowledged that, "*South Africa did not actively participate in the Convention negotiations and has largely been isolated from discussions around its issues. This has resulted in a general lack of awareness and understanding of the complex of issues that it raises.*" (Department of Environmental Affairs and Tourism 1997).

18.7.1 The Constitution

Both the 1993 interim Constitution and the final Constitution (Constitution of the Republic of South Africa 1996) laid the basis for new policy and legislation. This is because in moving to a constitutional democracy, South Africa also moved to a rights-based culture. In this regard, the environmental right which is contained in section 24 of the Bill of Rights says that, "*Everyone has the right—*

(a) *to an environment that is not harmful to their health or well-being; and*
(b) *to have the environment protected, for the benefit of present and future generations, through reasonable legislative and other measures that—*
   (i) *prevent pollution and ecological degradation;*
   (ii) *promote conservation; and*
   (iii) *secure ecologically sustainable development and use of natural resources while promoting justifiable economic and social development."*

The obligations in paragraph (b) to take reasonable measures to prevent ecological degradation, promote conservation and to secure sustainable development in essence provide a constitutional imperative for government to take action—legislative and otherwise—in respect of invasive alien species in so far as they contribute to degradation or threaten conservation or sustainable development.

18.7.2 The Draft Biodiversity Policy

Following the Constitution, government embarked on an extensive environmental policy and law reform process to ensure alignment with the requirements of the Constitution. Although there were many environmental laws that dealt with biodiversity, much of the legislation was outdated in that it tended to focus on managing species diversity and did not reflect the integrated and holistic approach reflected in the Convention on Biological Diversity which recognises that ecosystems, genes, and landscapes need to be managed in addition to species, and that people's interactions with biodiversity must be considered. In addition, existing legislation tended to focus on managing species in protected areas only, rather than across the country as a whole.

The policy process involved establishing an overarching environmental policy which was undertaken through the so-called Consultative National Environmental Policy Process (CONNEPP), as well as a series of sectorally focused policies. One of these sector-based policies was the White Paper on the Conservation and Sustainable Use of South Africa's Biological Diversity (Department of Environmental Affairs and Tourism 1997). The issue of invasive alien species features prominently in several places in the policy. For example, in the discussion on sustainability and the need to minimise negative impacts on biodiversity, the policy notes that, "*In* ***aquatic areas****, catchment changes, together with alien plant and animal invasions, and domestic, agricultural and industrial pollution, are among some of the primary mechanisms for biodiversity loss.*" [Emphasis added].

Also important for the purposes of this discussion is Goal 1 of the policy that deals with the conservation of biodiversity. One of the policy objectives for achieving the goal relates directly to alien species. The stated objective is to "*prevent the introduction of potentially harmful alien species and control and eradicate alien species which threaten ecosystems, habitats or species*".

In addition, the background discussion on how the policy objective will be achieved contains a number of points that are worth noting. First, as might be expected from an environmental policy, it is clear that the historical economic policy drivers for managing alien species are acknowledged but, for the first time, the main driver of the policy is clearly environmental. This is illustrated by the unambiguous statement that *"[t]his policy focuses upon alien organisms which threaten ecosystems, habitats or species"*. Secondly, the policy makes it clear that government's approach should not entail a form of alien xenophobia because it says that alien species can be categorised as being *"(a) those that are problematic and harmful, in that they negatively impact on biodiversity; and (b) those that are benign and in many instances serve useful purposes"*. Thirdly, the historical emphasis on plants is balanced by an acknowledgement of the impacts of alien animals and marine species. In this regard, the policy notes, for example, that *"in the Cape Peninsula ... invasive alien plants are chiefly responsible for the highest concentration of threatened taxa in the world"*. However, it also notes that, *"Introduced animals have also reduced South Africa's biodiversity, a few examples being the Argentinian ant, the Himalayan thar, the European starling, the house sparrow and the black rat, and on South Africa's islands, house mice, rabbits, and feral domestic cats. Some of the most drastic impacts of invasive animal species have been recorded in South African rivers, where alien fish, and to a lesser extent invertebrate and reptile species, have altered habitats and successfully outcompeted native fauna. Up to 60% of the threatened endemic freshwater fish of South Africa may be threatened by introduced fish species such as trout, carp and bass. Similarly in the marine environment, the accidental introduction of alien species through ballast water or on ship hulls has resulted in a number of alien species occupying our shores and coastal waters, in some instances displacing local species."*

Finally, the policy acknowledges that previous policy and legislation had not achieved their objectives, largely due to their *ad hoc* and reactive nature. The policy therefore proposes a *"proactive, preventative and precautionary approach"* which *"will take into consideration the need to balance the risks associated with introducing and releasing alien organisms with the potential social, economic and environmental benefits derived therefrom"*. To achieve this, the policy sets out a 13 point action plan which can be summarised as follows—

1. Reforming the law and strengthening compliance and enforcement of the law;
2. Developing legislative requirements for the introduction of alien species that pose a risk;
3. Continuing existing and creating new eradication programmes;
4. Preventing unintentional introductions of species;
5. Developing a policy on the translocation of species within the country;
6. Promoting local, native species in rehabilitation and re-vegetation schemes;
7. Providing incentives to landowners to control or eradicate harmful alien species;
8. Strengthening and coordinating institutional arrangements to identify harmful invasive species proactively and to catalogue invasions;
9. Supporting the development of biological and other control methods;

10. Improving awareness about the impacts of alien species;
11. Improving education and awareness about the risks posed by the planting or illegal importation of alien species, and identifying actions which can be taken to avoid these risks;
12. Improving capacity amongst the implementing agencies to implement the policy measures; and
13. Negotiating with neighbouring countries to harmonise legislation and practice.

It is noted that despite the policy being published as a White Paper (the usual final formal form of government policy), the White Paper was labelled a *"Draft for Discussion"* and called for comments to be submitted to the department's Director-General. The document also included a detailed summary of the policy's development process and specifically noted that the only outstanding step in the formalisation of the policy was "*the adoption of the White Paper by Parliament as formal policy*". Although it is Cabinet, not Parliament, who would usually have approved the White Paper's publication in the Gazette as government policy, the reason why this policy was again published for comment and then apparently abandoned is unclear. There is some speculation that, as this policy was developed through a process that was separate to the highly regarded CONNEPP process, it was abandoned for fear of it being branded as being an illegitimate product of a process that did not fully reflect the democratic ethos and commitment to broad-based participatory policy development espoused by the new democratically elected government at the time. Nevertheless, the White Paper still appears to have provided the policy direction for some aspects of law reform, including the National Environmental Management: Biodiversity Act (Act 10 of 2004) which is discussed below.

18.7.3 The Biodiversity Act

The approach of developing both an overarching environmental policy as well as a series of sectorally-focused policies was mirrored in the law reform process that followed. The subsequent promulgation of the National Environmental Management Act (Act 107 of 1998) (NEMA) provided a general framework for managing environmental matters and contained several provisions which may, and in some cases must, be considered when working with other sectorally-focused legislation, including the regulation of invasive alien species. These provisions included a set of environmental principles and definitions as well as an extensive range of compliance and enforcement powers. Whilst NEMA provides an important framework piece of legislation, because of its overarching nature it does not deal with invasive alien species specifically.

The main Act which was passed to supplement NEMA's overarching approach in respect of biodiversity in general is the National Environmental Management: Biodiversity Act (Act 10 of 2004) (Biodiversity Act). Chapter 5 of the Act is dedicated to the regulation of species and organisms that pose potential threats to biodiversity and provides the framework for managing alien species and invasive species. It creates an overlapping regulatory regime with that contained in CARA (King et al. 2018). The purpose of the Chapter includes—

- preventing the unauthorised introduction and spread of alien species and invasive species to ecosystems and habitats where they do not naturally occur;
- managing and controlling alien species and invasive species to prevent or minimise harm to the environment and to biodiversity in particular; and
- eradicating alien species and invasive species from ecosystems and habitats where they may harm such ecosystems or habitats.

The Biodiversity Act differentiates between 'alien species' and 'invasive species'. An alien species is one that is not native, or which is native but is found outside its normal distribution range; whereas an invasive species is a species that "*(a) threatens ecosystems, habitats or other species or has demonstrable potential to threaten ecosystems, habitats or other species; and (b) may result in economic or environmental harm or harm to human health*" (section 1). This distinction caters for the dualistic approach mentioned in Sect. 18.6.4 above and allows government to adopt different approaches to regulating species that pose a risk or threat to the environment and those that do not.

Chapter 5 adopts a four-pronged approach to the management of these species insofar as the public is concerned in a system of permit requirements, prohibitions, exemptions and duties of care. There are strong enforcement mechanisms to secure compliance with these provisions, including administrative enforcement directives which can be issued to compel compliance with the two duties of care and criminal sanctions of up to ZAR 10 million and/or imprisonment of up to 10 years, in addition to any supplementary penalties that can be imposed in terms of NEMA. However, because the Biodiversity Act itself is of a framework nature, it requires further mechanisms such as Regulations and Notices to be passed in order to operationalise and 'give teeth' to the Chapter (cf sections 66, 67, 70 and 97). These are currently found in the *Alien and Invasive Species Regulations* (GNR 598, *GG* 37885, 1 August 2014) and the *Alien and Invasive Species Lists* (GN 599, *GG* 37885, 1 August 2014, as amended by GN 864, *GG* 40166, 29 July 2016). The intention of the Regulations is to set out the more detailed rules and procedures that must be followed in respect of alien and invasive species, whereas the Lists (Notices)—as the name implies—sets out lists of taxa that are regulated and indicates the form of control that will be applicable to the species. As of the 2016 revised lists, 556 taxa which are considered to be present in South Africa are regulated, and 563 taxa which are considered to be absent from South Africa are prohibited from being imported; see Kumschick et al. (2020), Chap. 20 for more details of how the lists were developed.

The key to understanding the regulatory approach is the concept of 'restricted activities' and its relationship with the Lists. There are 13 restricted activities which relate to alien and invasive species. Five of these are set out in section 1 of the Act and include the importation, possession, growing, breeding, moving or otherwise translocating, selling, buying, donating or in any way acquiring or disposing of a species. The Regulations add a further seven activities:

(a) spreading or allowing the spread of, any specimen of a listed invasive species;
(b) releasing any specimen of a listed invasive species;
(c) the transfer or release of a specimen of a listed invasive fresh-water species from one discrete catchment system in which it occurs, to another discrete catchment system in which it does not occur; or, from within a part of a discrete catchment system where it does occur to another part where it does not occur as a result of a natural or artificial barrier;
(d) discharging of or disposing into any waterway or the ocean, water from an aquarium, tank or other receptacle that has been used to keep a specimen of an alien species or a listed invasive freshwater species;
(e) catch and release of a specimen of a listed invasive fresh-water fish or listed invasive freshwater invertebrate species;
(f) the introduction of a specimen of an alien or listed invasive species to off-shore islands; or
(g) the release of a specimen of a listed invasive fresh-water fish species, or of a listed invasive fresh water invertebrate species into a discrete catchment system in which it already occurs (regulation 6).

Any person who wants to undertake a restricted activity involving a listed species is generally subject to some form of control. In the case of alien species a permit will be required to undertake the restricted activity (section 65) unless the species has been exempted (section 66 and Listing Notice 2), or the Minister has prohibited the activity in respect of the species completely (section 67 and Listing Notice 4). Invasive species are subject to more stringent forms of control owing to the impact and actual harm that they cause (King et al. 2018). However navigating the requirements in respect of invasive species is more complex because of the way in which the Lists are drafted. In summary, invasive species are categorised in the Lists as falling within category 1a, 1b, 2 or 3. The reasoning behind the categories can be gleaned from the Regulations (regulations 2–5) which make it clear that the different categories relate to the degree to which the invasive species is considered to be problematic and the corresponding level of control that is considered to be appropriate for managing the problem. Species that are the most problematic—those falling in categories 1a and 1b—are subject to tighter forms of control than those which present the least threat—category 3. In a slight over-simplification, the approach to control in respect of each of the categories is:

- category 1a: the species must immediately be combatted and eradicated by the person in control of the species [this has been interpreted by some to mean a nation-wide eradication attempt is mandated, e.g. Wilson et al. (2013)];
- category 1b: the species must be controlled by the person in control of the species;
- category 2: a permit must be obtained to undertake a restricted activity; and
- category 3: these species are either exempt which means that restricted activities can be undertaken without a permit; or they are totally prohibited.

Like the CARA list, the development of categories under the Biodiversity Act creates a conflict between economic and environmental concerns. In this regard, the late Minister of Environmental Affairs, Edna Molewa, said "... *the most difficult category is the* ***Category 2*** *species. These are species that have value, such as plantation trees and fish-farming species, and yet can invade with very negative consequences outside of where they are being utilized. The Department has taken an approach that seeks to optimize the economic benefits of these species, whilst minimizing the damage that they cause. Permits are granted for their utilization, but they must be controlled outside of what is allowed in terms of the permit.*" (www.invasives.co.za).

The regulatory approach which has been described above follows the 'identify and direct' method which is evident in much of the legislation discussed so far. However, the Biodiversity Act and the Regulations have additional nuances which make the regime more progressive than previous laws. The first is that the Act creates a duty of care in respect of alien species and a duty of care in respect of invasive species (sections 69 and 73). In the case of the invasive species duty of care permit holders are required to comply with permit conditions. In addition—and in an expansion of the regulatory net—every landowner who has invasive species on their land is obliged to (a) notify government, (b) take steps to control and eradicate the species and to prevent it from spreading, and (c) take steps to prevent or minimise harm to biodiversity (section 73(2)). The second nuance is that of state responsibility. In this regard, the Act instructs the Minister to "*ensure the coordination and implementation of programmes for the prevention, control or eradication of species*" and empowers the Minister to establish a body of public servants to co-ordinate and implement these programmes (section 75(4) and (5)). It also requires management authorities of the various protected areas recognised by, or established in terms of, the National Environmental Management: Protected Areas Act (Act 57 of 2003) to include invasive species control and eradication strategies in their management plans and other organs of state to prepare monitoring, control and eradication plans for land under their control (section 76).

In addition to this, the legislation also provides a more coherent basis for evidence-based improvements of the regime by providing for information collection and management and research (Chapters 4 and 5 of the Regulations).

The Biodiversity Act therefore provides a far more comprehensive approach to the management of invasive species, both in the tools that are provided and the range of species that fall within the scope of the Act. There is however room for improvement. In this regard, difficulties in reconciling what is fair to hold individuals accountable and liable for is evident in some of the wording which uses ambiguous terms such as 'appropriate measures' and a poorly defined offences provision in the Regulations. See also Kumschick et al. (2020), Chap. 20 for a discussion on potential improvements to the risk analysis and biosecurity issues.

18.7.4 Legal Challenges

Understanding the legal challenges involving legislation can potentially provide useful insights to the policy drivers which are present. Given the relative newness of the Alien & Invasive Species Regulations and Notices, it is not surprising that there has been limited litigation involving them. But as with much of the litigation involving biodiversity, the underlying drivers for the litigation can be a clash of cultural ethics, self-interest (economic) or a desire to protect the environment.

One case, *Kloof Conservancy v Government of the Republic of South Africa and Others* (D) Case No: 12667/2012, 22 October 2014, was aimed at forcing government to give effect to its regulatory obligations. In this regard, in terms of section 70 (1)(a) of the Biodiversity Act, the Minister was obliged to publish the national list of invasive species within 2 years of the section coming into effect. In December 2012—more than 6 years after the deadline for publishing the list—the Kloof Conservancy applied to court for an order compelling the Minister to publish the list as well as the Regulations. Before the matter reached trial the Minister published interim 2013 Regulations and Notices, presumably in an effort to defeat the application. The Kloof Conservancy, however, launched a second application asking for the 2013 Regulations and Notices to be reviewed and set aside as well other orders to be made. Between arguing the case and the judgment being made, the Minister passed the 2014 Regulations and 2014 Notice. This meant that much of the Kloof Conservancy's requests became moot because the Regulations and Notice had been passed.

Notwithstanding this, the court showed its displeasure with the almost 8-year delay in passing the Regulations and Notice by awarding costs against government on an attorney-and-client basis i.e. a punitive costs order. It also declared the failure to publish the Regulations and Lists timeously to be unconstitutional and unlawful. In an unusual approach the court also made orders about implementation. It ordered that all steps had to be taken to ensure that organs of state complied with their duties and directed the Minister to appoint enough environmental management inspectors to ensure compliance with government's duties in relation to invasive alien species in Kwa-Zulu Natal within 6 months of the judgment (Note: the Minister successfully appealed aspects of the judgment related to the deployment of environmental management inspectors—See *Minster of Water and Environmental Affairs v Kloof Conservancy* (106/2015) [2015] ZASCA 177 (27 November 2015).

Although not directly related to his findings, the judge also expressed some reservations about the 'legality' of aspects of the 2014 Regulations. These *obiter dictum* comments may encourage future litigation on the Regulations.

By contrast to the *Kloof* matter, another case was initiated in the early 2000's (before the *Alien and Invasive Species Regulations* were promulgated) with the aim of preventing government from exercising its powers. This case related to the sentimental and ethical considerations that can arise in the context of alien species, particular animal species. In this instance government took a policy decision to extirpate the Himalayan Tahr (*Hemitragus jemlahicus*) on Table Mountain, whose ancestors had been introduced to South Africa from India by Cecil John Rhodes. The decision was met with outrage by many people, some of whom formed the "Friends of the Tahr" group, to challenge the culling. "Friends of the Tahr" launched an application to have the decision to cull reviewed. It was either withdrawn or dismissed (Butcher 2004). The matter may be resurrected in the future as a media article reports that the group was looking to approach the court about the possible culling of three remaining tahrs which were sighted in 2017 (Chambers 2017).

A third case shows how alien species regulation can be used in private disputes. *Appelgryn N.O. and Another v Jankielsohn and Others* ((FS) Case No. 3809/2016), 19 January 2017 involved a dispute about the ownership of a number of Red Lechwe Antelope. In this instance, the court found that because the antelope are alien species which are regulated in terms of the Biodiversity Act, the neighbour's keeping or allowing the antelope to be on her land was a restricted activity which required a permit. In the absence of a permit, it held that she was acting illegally and therefore had no right (*locus standi*) to defend the application.

18.7.5 The "Working for" Programmes: The Only Coherent National Programme for Managing Invasive Species?

The discussion on the legislation in previous sections has indicated that, apart from imposing obligations on the public, a co-ordinated government response is required to manage invasive species. In 1995, drawing heavily on hydrological and ecological insights from tree invasions in the Cape, the Working for Water (WfW) Programme became the first coherent national programme for managing invasive species. The 1997 *White Paper on the Conservation and Sustainable Use of South Africa's Biological Diversity* (Department of Environmental Affairs and Tourism 1997) recognised the fledgling WfW Programme as "*an RDP* (Reconstruction and Development Programme) *project . . .to clear invasive alien vegetation as part of a water conservation campaign and job-creation scheme*" and included providing "*. . .ongoing support to existing programmes. . .*" as a component of 1 of its 13 interventions aimed at "*preventing the introduction of potentially harmful alien species and controlling and eradicating alien species which threaten ecosystems, habitats or species*" (the RDP was the South African government's first socio-economic policy framework following democracy in 1994). With this, given that another of the draft policy's interventions was to "*improve capacity amongst implementing agencies to regulate the introduction, control and eradication of alien organisms that threaten biodiversity*", it seems clear that the 'water conservation and job-creation RDP project' was never envisaged as being the only 'coherent national programme for managing invasive species'.

According to van Wilgen and Wannenburg (2016), in establishing WfW, the Minister of Water Affairs and Forestry insisted that its projects should be implemented at a national scale as "*the need to create employment was ubiquitous*" and the problem of invasive species was not only confined to the area receiving the most attention, namely the Cape Floristic Region. This nation-wide approach was seen as making both political and ecological sense. Since its establishment, WfW has been expanded into more "Working for" programmes including *inter alia* Working for Wetlands, Working for Land and Working on Fire, dealing with wetland and degraded land rehabilitation and wildfire management respectively.

From a policy perspective, it is important to note the change in emphasis between this 'job-provision/ invasion problem' prioritisation balance of WfW and the abandoned draft policy's third intervention which reads, "*Develop control and eradication programmes, and provide ongoing support to existing programmes, based on a priority-rating system and in relation to costs and resources. This will consider threats posed to biodiversity, as well as social, economic, and environmental costs and benefits derived from using and removing identified organisms. The planning of intensive mechanical clearing operations will take account of job creation schemes and will provide for regular follow-up.*" [Emphasis added].

Van Wilgen and Wannenburg (2016) note that WfW projects are often selected to meet one or the other goal, "*resulting in confusion about how to prioritise projects*" and that the compromises inherent to the 'job-provision/invasion problem' prioritisation balance often result in projects that are not optimally selected.

Furthermore, there are also other challenges associated with the "Working for" programmes being the only coherent national programmes for managing invasive species, not least of which is the concern that local management capacity may be being undermined. According to van Wilgen et al. (2016), funds may be being diverted from organisations that have been established with the express goal of promoting the collective management of resources locally (for example, conservancies, Fire Protection Associations and Catchment Management Agencies) and conservation management agencies (for example, CapeNature and South African National Parks) to the detriment of such organisations. These agencies are then unable to build and maintain the necessary local capacity to efficiently and effectively manage the problem which, according to van Wilgen et al. (2016), essentially renders these organisations bystanders to the "Working for" programmes.

Despite its shortfalls, WfW is broadly considered, both nationally and internationally, to be a resounding success. However, the "Working for" programmes have effectively become the only significant national biological invasion response programme. From a purely biological invasion policy perspective this is not optimal—rather than being the only programme, the "Working for" programmes should be unashamedly job-provision focussed key components of a broader national programme that is unambiguously focussed on efficient and effective biological invasion management as envisaged by the abandoned policy.

18.8 Discussion and Future Directions

The recent report on the status of biological invasions and their management in South Africa in 2017 (van Wilgen and Wilson 2018) contains a number of "*policy-relevant messages*". It makes it clear that although there is a long legislative history of interventions aimed at controlling biological invasions, the number of problem species, the extent of invasion and magnitude of negative impacts is still growing. Although the report does provide a critique of the 2014 Alien & Invasive Regulations and Lists, it does not specifically refer to policy or legislative failures as being a root cause for the apparent lack of overall success in dealing with biological invasions. However, the report does make it clear that "*the lack of adequate planning and monitoring of the outcomes of control measures has been identified as a major weakness in South Africa*" (van Wilgen and Wilson 2018).

Bennett and van Sittert's recent historical review (2018) also concludes that "*the history of invasive alien plant and weed management has been chequered in its environmental and social outcomes.*" They believe governance efforts have faltered because of the difficulty of engaging private land owners, competition, local viewpoints and limited support for technical interventions by scientists and managers. Furthermore, differing perceptions of invasive species have meant that "*some regions are better endowed with facilities and awareness than others to tackle the complex challenges associated with invasion*" (Bennett and van Sittert 2018). Although invasive species, especially trees, are widely recognised as a national problem by scientists and growing public numbers, there is no broad consensus on what species cause the most problems.

As discussed below, it is possible that the lack of a dedicated biological invasion policy lies at the heart of van Wilgen and Wilson's concerns around inadequate planning and monitoring as well as the issues highlighted by Bennett and van Sittert (2018) that are bedevilling good governance and efficient and effective biological invasion management.

18.8.1 The Policy Vacuum

Despite over 160 years of concern around biological invasions, South Africa still has no formal national policy on the issue. Without formal policy there is no accepted government position on questions like—

- What is the overall vision for biological invasion management in South Africa?
- How do South Africans decide which are 'acceptable' and 'unacceptable' invasive species and under what circumstances?
- What would South Africans regard as 'acceptable change' to the land or landscape brought about by biological invasions and under what circumstances?
- Should some highly-invaded landscapes be 'sacrificed' to the invading species and under what circumstances?
- How should the invasive species legacy be dealt with?
- How should South Africa prioritise biological invasion management interventions?
- How, and how often, should South Africa review biological invasion management strategies and priorities?
- What types of biological invasion management interventions are accepted, not accepted or preferred, and under what circumstances?
- Who should implement biological invasion management interventions and under what circumstances?
- How will we measure the success, or otherwise, of biological invasion management interventions?

- How should biological invasion management interventions be co-ordinated, if at all?
- Who should pay for biological invasion management interventions and under what circumstances?
- How should South Africa build, maintain and retain efficient and effective biological invasion management capacity?

Although the draft *White Paper on the Conservation and Sustainable Use of South Africa's Biological Diversity* started to provide some initial guidance in this regard, its status as a draft rather than final policy has ensured that it has largely descended into obscurity and, as a result, it is not referenced at all in important policy-dependant or policy-relevant publications like the 2014 draft *National Strategy for dealing with Biological Invasions in South Africa* (Department of Environmental Affairs 2014) or the report on the status of biological invasions and their management in South Africa in 2017 (van Wilgen and Wilson 2018).

Without guidance on the important questions listed above, it is possible that biological invasion-related legislation is not tailored to achieve a specific desired and strategic outcome, but largely continues to follow the historical 'identify and direct' approach. Without nationally accepted guidance on desired outcomes and indicators for their measurement (cf. Wilson et al. 2018), all biological invasion management strategies and plans are likely to be no more than lists of activities and good intentions. Without guidance on prioritisation, valuable and limited resources may not be being used optimally. Without guidance on what interventions are preferred it is possible that some interventions may be resulting in perverse outcomes.

It is also probably safe to conclude that the over 6-year delay in the publication of the Regulations and Notices referred to in Sect. 18.7.4 above was probably exacerbated by the lack of policy, i.e. the 6-year development process would have been greatly reduced if the development of the Regulations and Notices were directed by clear and unambiguous national policy.

Finally, without policy, the dubious 'nationalistic' or 'species xenophobia' justification for invasive alien species eradication remains unchallenged. For example, Pooley (2010) suggests that the very existence of WfW "*is founded on antagonism to harmful introduced species, and advocates routinely use emotive language referring to 'cleaning' areas of 'infestations' of 'alien' plants, which 'threaten to engulf and exterminate the unique indigenous fauna and flora'*". Alarmingly, Pooley (2010) notes that parallels have been drawn between the focus on invasive alien plants in the emotional public response to the devastating veld fires in the Western Cape of 2000 and xenophobia in contemporary South Africa.

18.8.2 Law, Compliance and Enforcement

The first biological invasion-related 'action point' of the draft *White Paper on the Conservation and Sustainable Use of South Africa's Biological Diversity* is that the law required reforming and that strengthened compliance and enforcement of the law was needed. With regard to the latter point, there is a general rule of thumb for legal compliance that goes—for a just law, 10% of the regulated community will always comply, 10% will never comply and the remaining 80% will comply if they believe that the non-compliant 10% will be brought to book. What this means is that if the regulated community broadly believes that there is a relatively high likelihood that they will be caught and punished if they break the law, then the vast majority of the community will comply. However, if the regulated community broadly believes that there is a relatively low chance that they will be caught and punished then the vast majority of the community will not comply. Although this rule of thumb may be an oversimplification of deterrence and compliance promotion, which is a complicated and not well-understood area that also involves, among others, awareness of the law and ability to comply, it is used here for the sake of argument.

In the current context, the 'regulated community' comprises of all South African land owners and users, and it is probably fair to say that, of the regulated community that actually know their invasive species management legal obligations, very few are likely to believe that they will be caught and punished for non-compliance. Hence, it is highly likely that the level of non-compliance with invasive species law in South Africa is extremely high and, as a result, invasive species legislation is unlikely to be having the desired effect or impact. With this, it could also be argued that very little has changed in the legislative approach since the 1874 Burr Weed Law that noted that "*the Law hitherto in force has failed to check the spread*" of the weed. Thus it appears that enforcement may be an extremely important issue when dealing with invasive policy and legislation.

With reference to the compliance rule of thumb above, this 'rule' only holds true if the law is regarded as a 'just law'—law that is fair and reasonable. In a Constitutional democracy like South Africa, 'unjust law' can be prone to legal and popular challenge, making it difficult or impossible to enforce. Given that the majority of the regulated community are seldom directly or knowingly responsible for introducing an invasive species onto their land, it is easily argued that they are, in fact, the 'victims' of invasive species (Fig. 18.3). Indeed, it would probably be considered grossly unfair and unreasonable to hold a landowner responsible for polluted ambient air over her land if the pollution is from, say, a neighbouring factory. Thus, there could be a case to be made that government may be loath to enforce 'faultless liability' where the 'culprit' may, in fact, be the 'victim' of poor invasive species management by others.

Furthermore, perhaps the 'environmental justice principle' contained in NEMA also provides a useful test of whether South African biological invasion-related law is, in fact, 'just law'. Section 2(4)(c) of NEMA requires that "[e]*nvironmental justice must be pursued so that adverse environmental impacts shall not be distributed in such a manner as to unfairly discriminate against any person, particularly vulnerable and disadvantaged persons.*" With this it may be possible that the few people who have knowingly or purposefully introduced invasive species are distributing the adverse environmental impacts of these species to others, including vulnerable and disadvantaged persons. Despite this, the law requires these 'innocent' people to address these adverse environmental impacts.

Fig. 18.3 Garcia Pass, Western Cape—Pine trees invading Cape Nature land in the foreground with the commercial forest source of the pines in the background. Photograph courtesy of Brian W. van Wilgen

Other than justice considerations, enforceability, capacity and political will all impact on the effectiveness of enforcement, where enforceability refers to the extent to which a law may be efficiently and effectively enforced using available resources; capacity refers to structures, financial and technical resources, systems, strategies, skills, incentives and networks required for efficient and effective enforcement; and political will refers to the motivation, willingness, enthusiasm and political support for efficient and effective enforcement.

Although a full analysis of these considerations is beyond the scope of this Chapter, it is likely that all of the above factors have an impact on how useful the current legal regime is in managing biological invasions.

This notwithstanding, and despite the apparent challenges associated with the 'identify and direct' approach to biological invasion law that were recognised as far back as 1874, this approach still appears to be the core of our legislation.

18.8.3 The Future

It seems clear from the recent report on the status of biological invasions in South Africa (van Wilgen and Wilson 2018) that, despite relatively large public investments in dealing with the biological invasion problem, we appear to be losing the battle (cf. Wilson et al. 2020, Chap. 31). Furthermore, the report also makes it clear that the battle is becoming harder as the number of species causing major impacts and the magnitude of the impacts themselves grow, as further species become invasive, and as others enter a phase of exponential spread. Furthermore, according to Hellmann et al. (2008) and King et al. (2018), it will take more research to understand how specific invasive species may behave under an altered climate and which species will become invasive in the future.

With this, it seems that the current policy vacuum must be addressed as a matter of urgency in order to establish a national consensus on how we as South Africans view biological invasions, what our desired future looks like in terms of biological invasions, and how we intend to realise that desired future. As biological invasions affect all South Africans, this policy development process must be evidence-based and should follow the co-creation approach described by von der Heyden et al. (2016). In this regard, not only should the policy deal with the questions listed in Sect. 18.8.1 above, it should also deal with issues like the commercial exploitation of alien species, and how NEMA's polluter-pays principle may be applied in cases where such exploitation results in invasion of neighbouring land and rivers. Furthermore, a good-practise participatory policy development process is an excellent way of ventilating and addressing Bennett and van Sittert's (2018) issues of private land owner concerns and interests, faultless liability, conflicting interests and competition, local viewpoints, support for technical interventions by scientists and managers, differing perceptions of invasive species and the means of identifying and prioritising what species cause the most problems.

Perhaps growing concerns around climate change and interest in the global Sustainable Development Goals may provide the perfect 'policy-development window' for the development of formal policy on biological invasion in South Africa.

Acknowledgements We acknowledge the clear direction and support from Professor Brian van Wilgen of the Centre for Invasion Biology, Stellenbosch University, and Ms Marié Theron, Faculty Librarian: Science, Library and Information Service, Stellenbosch University for the links to the invaluable historical legislation databases. We would also like to thank SANBI's Sebataolo Rahlao for his comments and suggestions, as well as those of the peer reviewers.

References

Barton GA (2002) Empire forestry and the origins of environmentalism. University of Cambridge Press, Cambridge. https://doi.org/10.1017/CBO9780511493621

Beinart W (2003) The rise of conservation in South Africa: settlers, livestock, and the environment 1770–1950. Oxford University Press, Oxford

Bennett BM (2010) The El Dorado of forestry: the Eucalyptus in India, South Africa, and Thailand, 1850–2000. Int Rev Soc Hist Suppl 55:27–50. https://doi.org/10.1017/S0020859010000489

Bennett BM (2014) Model invasions and the development of national concerns over invasive introduced trees: insights from South African history. Biol Invasions 16:499–512. https://doi.org/10.1007/s10530-013-0601-1

Bennett BM, van Sittert L (2018) Historicising perceptions and the national management frame-

work for invasive alien. J Environ Manage 229:174–181. https://doi.org/10.1016/j.jenvman.2018.07.029
Britton N, Brown A (1913) An illustrated flora of the Northern United States, Canada and the British possessions, 2nd edn. Charles Scribner's Sons, New York. https://doi.org/10.5962/bhl.title.55504
Butcher T (2004) Anger as shaggy 'aliens' of Table Mountain face mass slaughter. The Telegraph, 22 May
Cape of Good Hope Parliament (1871) Statutes of the Cape of Good Hope, passed by the Second Parliament, during the sessions 1859–1863. Saul Solomon & Co., Cape Town
Carruthers J (2011) Trouble in the garden: South African botanical politics ca. 1870–1950. S Afr J Bot 77:258–267. https://doi.org/10.1016/j.sajb.2011.02.005
Chambers D (2017) Alive and on a high: three tahrs pictured on Table Mountain. Times Live, 13 July 2017
Crosby AW (1986) Ecological imperialism: the biological expansion of Europe, 900–1900. Cambridge University Press, Cambridge
Department of Environmental Affairs (2014) A national strategy for dealing with biological invasions in South Africa, Draft dated 25 March 2014. Department of Environmental Affairs, Pretoria
Department of Environmental Affairs and Tourism (1997) White Paper on the Conservation and Sustainable Use of South Africa's Biological Diversity, May 1997, Draft for discussion (Notice No. 1095 of 1997, Government Gazette No. 18163 of 28 July 1997). Government Printer, Pretoria
DPME, UCT (2014) Overview Paper: What is evidence-based policy-making and implementation? Department of Planning Monitoring and Evaluation. October. http://www.dpme.gov.za/keyfocusareas/evaluationsSite/Evaluations/What%20is%20EBPM%2014%2010%2013_mp.pdf. Accessed 23 July 2018
Education and Training Unit (n.d.) The policy and law making process. Education and Training Unit for Democracy and Development (ETU). http://www.etu.org.za/toolbox/docs/govern/policy.html. Accessed 5 June 2018
Grove R (1989) Scottish missionaries, evangelical discourses and the origins of conservation thinking in southern Africa 1820–1900. J S Afr Stud 15:163–187. https://doi.org/10.1080/03057078908708196
Hanks K (2001) A legal solution? The Conservation of Agricultural Resources Act explained. Department of Water Affairs and Forestry. www.dwaf.gov.za/wfw/Docs/Articles/CARA.doc. Accessed 22 Aug 2018
Hellmann JJ, Byers JE, Bierwagen BG, Dukes JS (2008) Five potential consequences of climate change for invasive species. Conserv Biol 22:534–543. https://doi.org/10.1111/j.1523-1739.2008.00951.x
Invasive Species South Africa (n.d.) Spiny cocklebur. http://www.invasives.org.za/legislation/item/314-spiny-cockleburxanthium-spinosum. Accessed 24 July 2018
King ND, Strydom HA, Retief FP (2018) Environmental management in South Africa, 3rd edn. Juta, Cape Town
Kruger FJ, Bennett BM (2013) Wood and water: an historical assessment of South Africa's past and present forestry policies as they relate to water conservation. Trans Roy Soc S Afr 68:163–174. https://doi.org/10.1080/0035919X.2013.833144
Kumschick S, Foxcroft LC, Wilson JR (2020) Analysing the risks posed by biological invasions to South Africa. In: van Wilgen BW, Measey J, Richardson DM, Wilson JR, Zengeya TA (eds) Biological invasions in South Africa. Springer, Berlin, pp 569–592. https://doi.org/10.1007/978-3-030-32394-3_20
Malherbe HL (1968) Report of the Interdepartmental Committee of investigation into afforestation and water supplies in South Africa. Government Printer, Pretoria
Marais L, Matebesi Z (2013) Evidence-based policy development in South Africa: the case of provincial growth and development strategies. Urb For 24:357–371. https://doi.org/10.1007/s12132-012-9179-4
Peretti JH (2010) Nativism and nature: rethinking biological nativism. In: Johnson S

(ed) Bio-invaders. White Horse Press, Winwick, Cambridgeshire, pp 28–36
Pooley S (2010) Pressed flowers: notions of indigenous and alien vegetation in South Africa's Western Cape, c. 1902–1945. J S Afr Stud 36:599–618. https://doi.org/10.1080/03057070.2010.507565
Republic of South Africa (1996) Constitution of the Republic of South Africa (Act No. 108 of 1996). Pretoria, Government Printer
Strydom WF, Funke N, Nienaber S et al (2010) Evidence-based policymaking: a review. S Afr J Sci 106(5/6), 8 pp. https://doi.org/10.4102/sajs.v106i5/6.249
The President's Council (1991) Report of the Three Committees of the President's Council on a National Environmental Management System. Government Printer, Pretoria
van Sittert L (2000) 'The seed blows about in every breeze': noxious weed eradication in the Cape Colony, 1860–1909. J S Afr Stud 26:655–674. https://doi.org/10.1080/713683604
van Sittert L (2003) Making the Cape floral kingdom: the discovery and defence of indigenous flora at the Cape ca. 1890–1939. Landscape Res 28(28):113–129. https://doi.org/10.1080/01426390306529
van Wilgen BW (2020) A brief, selective history of researchers and research initiatives related to biological invasions in South Africa. In: van Wilgen BW, Measey J, Richardson DM, Wilson JR, Zengeya TA (eds) Biological invasions in South Africa. Springer, Berlin, pp 31–64. https://doi.org/10.1007/978-3-030-32394-3_2
van Wilgen BW, Wannenburgh A (2016) Co-facilitating invasive species control, water conservation and poverty relief: achievements and challenges in South Africa's Working for Water programme. Curr Opin Environ Sustain 19:7–17. https://doi.org/10.1016/j.cosust.2015.08.012
van Wilgen BW, Wilson JR (2018) The status of biological invasions and their management in South Africa in 2017. South African National Biodiversity Institute, Kirstenbosch and DST-NRF Centre of Excellence for Invasion Biology, Stellenbosch
van Wilgen BW, Carruthers J, Cowling RM et al (2016) Ecological research and conservation management in the Cape Floristic Region between 1945 and 2015: history, current understanding and future challenges. Trans Roy Soc S Afr 71:207–303. https://doi.org/10.1080/0035919X.2016.1225607
von der Heyden S, Lukey P, Celliers L et al (2016) Science to policy – reflections on the South African reality. S Afr J Sci 112:1–6. https://doi.org/10.17159/sajs.2016/a0183
Weyl OLF, Ellender B, Wassermann RJ et al (2020) Alien freshwater fauna in South Africa. In: van Wilgen BW, Measey J, Richardson DM, Wilson JR, Zengeya TA (eds) Biological invasions in South Africa. Springer, Berlin, pp 151–182. https://doi.org/10.1007/978-3-030-32394-3_6
Wilson JRU, Ivey P, Manyama P et al (2013) A new national unit for invasive species detection, assessment and eradication planning. S Afr J Sci 109: Art. #0111, 0113 pp. https://doi.org/10.1590/sajs.2013/20120111
Wilson JRU, Faulkner KT, Rahlao SJ et al (2018) Indicators for monitoring biological invasions at a national level. J Appl Ecol 55:2612–2620. https://doi.org/10.1111/1365-2664.13251
Wilson JR, Foxcroft LC, Geerts S et al (2020) The role of environmental factors in promoting and limiting biological invasions in South Africa. In: van Wilgen BW, Measey J, Richardson DM, Wilson JR, Zengeya TA (eds) Biological invasions in South Africa. Springer, Berlin, pp 353–384. https://doi.org/10.1007/978-3-030-32394-3_13

19

Invasive Alien Plants and their Biological Control over the Years

Martin P. Hill, Vincent C. Moran, John H. Hoffmann, Stefan Neser, Helmuth G. Zimmermann, David O. Simelane, Hildegard Klein, Costas Zachariades, Alan R. Wood, Marcus J. Byrne, Iain D. Paterson, Grant D. Martin and Julie A. Coetzee

Abstract Invasive alien plant species negatively affect agricultural production, degrade conservation areas, reduce water supplies, and increase the intensity of wild fires. Since 1913, biological control agents i.e. plant-feeding insects, mites, and fungal pathogens, have been deployed in South Africa to supplement other management practices (herbicides and mechanical controls) used against these invasive plant species. We do not describe the biological control agent species

M. P. Hill (✉) · I. D. Paterson · G. D. Martin
Centre for Biological Control, Department of Zoology and Entomology, Rhodes University, Makhanda, South Africa
e-mail: m.hill@ru.ac.za

V. C. Moran · J. H. Hoffmann
Department of Biological Sciences, University of Cape Town, Cape Town, South Africa

S. Neser · H. G. Zimmermann · D. O. Simelane · H. Klein
Agricultural Research Council-Plant Health and Protection, Pretoria, South Africa

C. Zachariades
Agricultural Research Council-Plant Health and Protection, Cedara, South Africa

A. R. Wood
Agricultural Research Council-Plant Health and Protection, Stellenbosch, South Africa

M. J. Byrne
Centre for Invasion Biology, School of Animal, Plant and Environmental Sciences, University of the Witwatersrand, Johannesburg, South Africa

Centre for Biological Control, Rhodes University, Makhanda, South Africa

J. A. Coetzee
Centre for Biological Control, Department of Botany, Rhodes University, Makhanda, South Africa

that have been used, or what they do, or how they damage the target plant species. We focus instead on evaluations (informed opinions, evidence and quantifications) of what has been achieved in South Africa by using biological control to suppress populations and impacts of invasive plant species. Satisfactory long-term evaluations of outcomes are difficult and expensive, but many have been done, providing ample evidence that biological control is often highly successful. We use case studies from South Africa to support this assertion and to make the point that successes may be largely forgotten in a relatively short time. Biological control of invasive plants in South Africa is demonstrably cost effective and has become generally accepted as a preferred management strategy. However, it is not a panacea, and we discuss several issues which complicate our understanding of its effectiveness and which raise research and implementation challenges. Further studies are needed on the economic and social benefits of biological control of invasive plants to inform and involve the wider South African community.

19.1 Introduction

Biological control of invasive alien plants relies on suitably host-specific natural enemies (agents), mainly plant-feeding insects and mites, and fungal pathogens. The agent species are sourced from their respective countries of origin and released into the invaded range to suppress the aggressiveness of the targeted plant species. The rapid and spectacularly successful outcome of biological control against cactus species, particularly in Australia in the 1930s (Dodd 1940; Mann 1970), set an early precedent suggesting, deceptively, that biological control and the evaluation of its success is easy. This also led, ironically, to the misguided perception that other biological control projects targeting invasive plant species that had less-dramatic outcomes were relative failures. Over the years, unequivocal successes have made up a diminishing proportion of the escalating number of projects that have been initiated, and an increasing number of projects have resulted in significant levels of control, but which have been less obvious and harder to gauge (Raghu and Walton 2007).

Biological control, as a management tool, and as a supplement to physical destruction or removal of the problem plants, and to herbicidal applications, has been used in South Africa since 1913 and has accumulated an impressive number of documented successes (Moran et al. 2005, 2013; Klein 2011; Zachariades et al. 2017). Initially (from 1913 to the 1930s) invasive plants in agriculture, particularly invasive cacti, were the targets of biological control introductions (Zimmermann et al. 2009). Between the 1950s and 1980s, biological control projects were initiated on several species of plants that were invasive in the Cape Floristic Region, a 'biodiversity hotspot', including *Hypericum perforatum* (Clusiaceae; St. John's Wort; Gordon et al. 1986; Gordon and Kluge 1991), *Hakea sericea* (Proteaceae; Silky Hakea; Esler et al. 2010; Gordon and Fourie 2011), several Australian *Acacia* species (Fabaceae; wattles) (Impson et al. 2011), and *Leptospermum laevigatum* (Myrtaceae; Australian Myrtle; Gordon 2011). This era also saw projects initiated against floating, invasive aquatic plants (Coetzee et al. 2011a) and against the various hybrids in the *Lantana camara* L. hybrid-complex (Verbenaceae; Lantana; Urban et al. 2011). More recently, herbaceous plants, vines and various subtropical shrubs have been subjected to biological control (Moran et al. 2013).

For over a century, South Africa has been one of the five main countries conducting research on and implementation of biological control of invasive alien plants (McFadyen 1998). Since 1995, with major increases in funding from state sources (Zimmermann et al. 2004) and the consequent involvement of many more researchers, students, support staff (see also van Wilgen 2020, Chap. 2, Sect. 2.2), and implementers, there have been an increasing number of innovative and successful projects, allowing South Africa to play a leadership role in the use of biological control in the management of invasive alien plants (Moran and Hoffmann 2015; Schwarzländer et al. 2018).

All the projects and programmes on biological control of invasive plants in South Africa have been thoroughly documented in three sets of reviews (Hoffmann 1991; Olckers and Hill 1999; Moran et al. 2011), in Muniappan et al. (2009), and in a recent overview of the entire South African biological control effort (Zachariades et al. 2017). These compilations have been comprehensive in recording which agents have been sourced, tested, mass-reared, released, established and redistributed against each of the targeted, invasive plant species, and in recording their progress and the extent of damage caused on the target host-plants. However, these accounts have been less clear in summarising the overall achievements (i.e. outcomes) of biological control programmes against invasive alien plant species and consequently, that is the primary purpose of this account.

19.2 Evaluating the Effectiveness of Biological Control of Invasive Alien Plants

Across the world, many different approaches have been used to evaluate and express the outcomes of biological control projects against invasive plant species (e.g. Syrett et al. 2000; Morin et al. 2009). It has long been understood that the objective of biological control "... *is not the eradication* [of the target plant species] *but the reduction of their densities to non-economic levels*" (Huffaker 1964). In some instances, the success of biological control is self-evident and before-and-after photographic records (see, for example, Figs. 19.1, 19.2 and 19.3) suffice to make the general point. Evaluations of biological control accomplishments against invasive alien plant species, during the earlier eras, were mostly descriptive and subjective. Thus, an apparent understanding of the outcomes of biological control was, and often still is, based on 'proxy-studies' of the progress of the released biological control agents themselves. These studies aligned with the expertise of the scientists involved, who were overwhelmingly entomologists or, less often, pathologists, not plant ecologists. Many of these studies were appropriate for short-term funding cycles and provided excellent subject-matter for post-graduate student projects.

Fig. 19.1 Photographs taken at paired sites near Makhanda, Eastern Cape, before and after successful biological control, on hillsides covered with dense infestations of *Acacia longifolia* (Long-Leaved Wattle) trees in 1978 (**a**), subsequently replaced by native vegetation in 2013 (**b**). Photographs courtesy of JH Hoffmann

Fig. 19.2 Photographs taken at paired sites on Sunset Dam, Kruger National Park, Mpumalanga, before (**a**) and after (**b**) successful biological control of *Pistia stratiotes* (Water Lettuce). Photographs courtesy of LC Foxcroft

Fig. 19.3 Photographs taken at paired sites near Upington, Northern Cape, of flourishing populations of *Cylindropuntia fulgida* (Boxing-Glove Cactus) in 2012 (**a**) replaced by dead and dying stumps in 2015 (**b**) after biological control. Photographs courtesy of Travor Xivuri

But Crawley (1989), among others, has emphasised that measures of agent numbers and the levels of damage that they inflict, while useful determinants of progress, are not relevant to the population dynamics of the target plants themselves, i.e. to the outcomes of biological control. This plant-population perspective has been recognised by practitioners and has helped to change the emphasis in assessing the accomplishments of biological control of invasive alien plants.

The outcomes of biological control of invasive alien plants can be measured in many ways, including, for example, reductions in density of the invasive species, the extent of their distribution, changes in biomass, their longevity, and the numbers of seeds, or other propagules, produced and accumulated. Furthermore, as Hoffmann et al. (2019) have noted: "*Each of these measures must be considered in a context of scale (e.g. in a defined locality such as a reserve or forest, in a functional habitat such as a river catchment or mountain range, or at a landscape level that might include a biome, region or country).*" In the latest world catalogue of biological control of "*weeds*" (Winston et al. 2014), the "*scale of impact*" of biological control is categorised by subjective estimates as: "*localized, regional, widespread throughout* [the] *range* [of the target plant], *and unknown*", and descriptive notes are provided to support these opinions for each of 1555 intentional releases of "*weed*" biological control agents, worldwide. Thus, the impacts of biological control are invariably complex and often site-specific. For example, on *Chromolaena odorata* (Asteraceae; Triffid Weed), in South Africa, agent-impacts are sporadic and highly localised (Zachariades et al. 2011), and the situation is dynamic and not yet understood (Zachariades et al. 2016). Similarly, on the various taxa in the *L. camara* hybrid complex, agents undergo 'boom and bust' cycles, dependent on varietal types of the plants, and on altitudinal and other effects (Urban et al. 2011).

For all of these reasons, accurate quantitative measures of plant- and seed-population dynamics, and accompanying experimental evaluations of outcomes in biological control, are expensive and difficult, and require years or decades of study. The key here, with more recent South African programmes since the 1970s, has been sustained funding for biological control of invasive plants and the career-long-continuity of many of the research- and support-personnel involved. This has encouraged long-term research programmes (albeit never enough) and a disproportionately high number of carefully-quantified and unambiguous evaluations. For example: the 25-year evaluation of biological control of *Opuntia stricta* (Cactaceae; Australian Pest Pear; see Box 19.1) (Hoffmann et al. 1998; Paterson et al. 2011); 16 years of assessment on the biological control of *Sesbania punicea* (Fabaceae; Red Sesbania) at 22 sites (Hoffmann and Moran 1998); ongoing observations over 28 years on a fungal-pathogen biological control agent used against *Acacia saligna* (Fabaceae; Port Jackson Willow) (Wood and Morris 2007); continuing country-wide surveys over the past decade on several species of invasive aquatic plants (Coetzee et al. 2011a); and finally, the assessments of the degree of biological control against *L. camara*, ongoing since the 1970s (Cilliers 1987; Cilliers and Neser 1991; Urban et al. 2011). In all of these cases, and for the other cases cited throughout the text, conclusions about the status of the individual biological control projects, which have not been recently or fully published, have been brought up to date (to 2019) from the personal knowledge of one or more of the contributing authors.

Box 19.1 ***Opuntia stricta*** **(Cactaceae) (Australian Pest Pear)**
Invasion by *Opuntia stricta* (Australian Pest Pear) in the Kruger National Park was rapid, with about 60,000 ha being infested by the turn of the century, mainly through seeds being spread by baboons and elephants. An intensive herbicide-based control programme failed to achieve its objectives and eventually biological control was adopted as a management process despite the park management's reluctance to have additional non-native species (i.e. biological control agents) brought into a conservation area. Simulation models had correctly predicted that the release of the well-known cactus moth, cactoblastis, in 1988, would result in diminished invasion. But success was mainly due to the release of a cochineal insect species (open circles), in 1997, that kills the pads/cladodes (closed circles) and the plants. Long-term surveys over 20 years showed that there was a rapid decline in the density of the cactus plants and that the weed has now been permanently controlled well below an acceptable threshold by biological control (Figure updated from Impson et al. 2011).

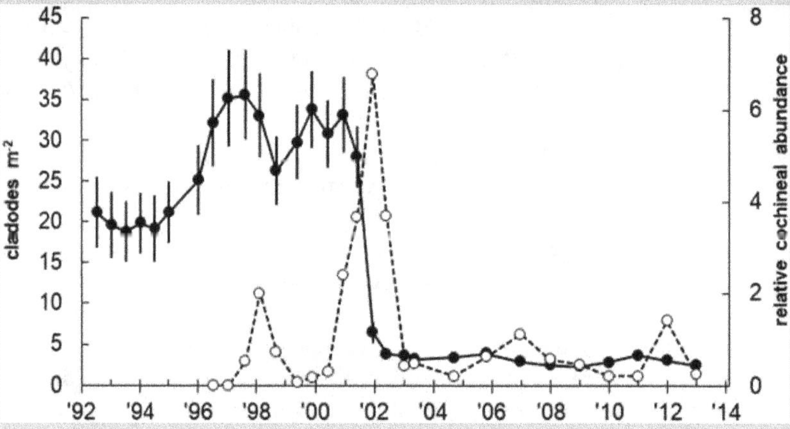

A system has been used in South Africa to describe overall success, i.e. outcomes of biological control against invasive alien plants, at the plant population (demographic) level, as 'complete', 'substantial', 'negligible', or 'unknown', and which relies on long-term formal evaluations, or on expert-opinion on broad categorisations of the reduced dependency on alternative control methods (chemical or mechanical), as a result of the introduction of biological control agents. This classification system was initiated by Hoffmann (1991). It has subsequently been refined and elaborated by Hoffmann (1995), Anonymous (1999), Hoffmann and Moran (2008), Klein (2011) and Zachariades et al. (2017), and has been widely adopted internationally (McFadyen 1998). Using this 'now-traditional' descriptive system as a basis, several assessments of the outcomes of biological control of invasive plants in South Africa at the level of a plant population, have been made, including:

1. Hoffmann (1991) who noted that of the 36 invasive plant species subjected to biological control in South Africa at that time, 14% were under 'complete' control, and 22% were under 'substantial' control.
2. Later, Klein (2011) recorded that of 48 invasive plant species on which agents had become established in South Africa, 21% were under 'complete' control and 38% under 'substantial' control.
3. Zachariades et al. (2017) recorded the following: "*Thirteen (72%) of the 18 worst taxa* [of invasive plants in South Africa] *from the work of van Wilgen et al.* (2012) *have biological control agents released on them already, and seven (39%) of these are under either complete or substantial biological control. When considering the more extensive list of species presented in van Wilgen et al.* (2008), *a list of 56 most damaging species to ecosystem functioning, 19 (34%) have active biological control and 13% are considered to be under substantial control.*"
4. From the most-recent lists compiled by Zachariades (2018a), counting only those instances in which the agents have been established for more than 10 years (i.e. allowing enough time to draw firm conclusions about outcomes), which coincidently also involves a total of 48 targeted, invasive plant species, 31% were recorded as under 'complete' control and 40% under 'substantial' control.
5. Lastly, taking a different approach, Henderson and Wilson (2017) found that there was an approximately 50% increase in the broad-scale documented range of alien plants in South Africa between 2000 and 2016. But these authors also concluded that "*some* [invasive plant] *species which have been the subjects of successful biological control programmes have shown very little expansion in their distribution*" and "*in general successful biological control seems to be associated with a reduction in the rate of spread*".

A feature of all these summaries is that they are based on the foundation of the 'traditional' system for describing and recording biological control achievements against invasive alien plants. Until now this has been the only way of doing so, but the descriptors and their definitions are vague, often subjective or ambiguous, and convey little or no information about details or context. An elaboration of the present system has therefore been proposed by Hoffmann et al. (2019) which is based on a conceptual framework (see Box 19.2) that categorises the outcomes of biological control, from barely perceptible (category C−), to spectacularly successful (categorised as A+) where, over decades, the agents have had a considerable impact on one or more of four 'invasion parameters' of their target hosts, namely, plant population density, distribution (area), biomass and number of propagules (seeds), for different regions, habitats or circumstances (see Box 19.3). Other parameters, such as the longevity of invasive plant populations, could be accommodated if required. This system has now been adopted in South Africa, and when implemented it will allow for more nuanced, detailed, and accurate analyses and descriptions of the outcomes of biological control of invasive alien plants.

Box 19.2 Envisaging the Outcomes of Weed Biological Control

The sigmoid curve in the figure below (reproduced with permission from Hoffmann et al. 2019) represents the hypothetical extent and progress of an invasion by an alien plant species, commonly measured as either area invaded, density, biomass, seed production or its equivalent. Given time (usually at least 10–20 years) after the initiation of biological control, damage by the biological control agents can reduce target-plant populations and their invasiveness: (1) if the effects are only slight, the trajectory of the curve will be moderated to a greater or lesser degree within the bounds of 'control category C'; or (2) as in about 40% of the cases in South Africa, biological control can 'reverse', i.e. reduce target-plant populations and the extent of the invasion, depressing thc trajcctory of the curve into 'control category B'; or (3) as in about 30% of the cases in South Africa, can spectacularly reduce weed populations and the extent of the invasion to levels below a tolerable threshold, moving the curve into 'control category A'. Conceptualising biological control in this way allows for categorisations and subsequent recording of outcomes, from C− (imperceptible levels of control) to A+ (the most favourable outcome), for different regions, habitats, and circumstances. Some examples that illustrate the application of this framework for viewing outcomes in weed biological control are given in Box 19.3.

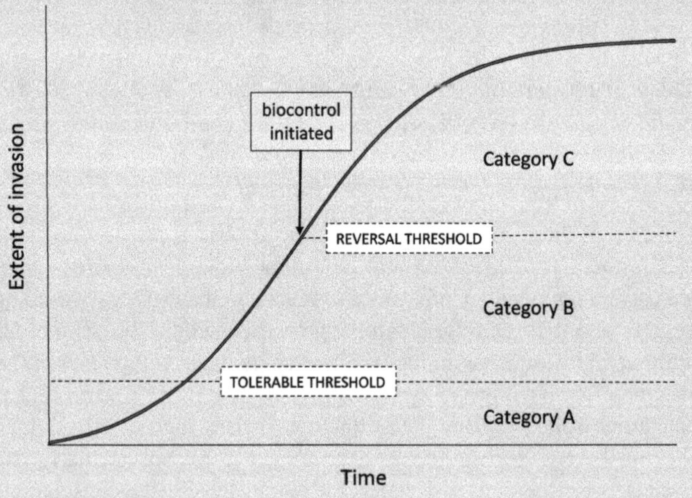

Box 19.3 A System for Recording the Outcomes of Weed Biological Control

Since the early 1990s, 'complete', 'substantial', or 'negligible' have been accepted descriptions of achievements in weed biological control at the plant population level (i.e. outcomes). These terms are too coarse and therefore can be misleading. For example, a biological control programme might result in the near-elimination ('complete' control), of the target weed in one area or situation, but in 'negligible' control elsewhere, an outcome that would 'average out' as 'substantial' control. This is clearly unsatisfactory. Recently a more elaborate and robust system has been developed for recording outcomes that considers four different 'invasion parameters': density, area, biomass and number of propagules (seeds), for different regions and habitats (Hoffmann et al. 2019). The examples, in the table below, of three well-known South African weed species, illustrate this system. The most positive 'outcomes' are recorded as A+ through to C− as the least favourable, based on categorisations as defined by Hoffmann et al. (2019) (and see Box 19.2). The categorisations that are dependent on expert opinion and observations are in light type, with those supported by quantified evidence in bold type. (1) *Opuntia ficus-indica* (Mission Prickly Pear), is not under satisfactory biological control in the coastal regions of the Eastern Cape, but is under excellent biological control in drylands, in all other regions, where populations are maintained well below an acceptable threshold. The biological control programme against *O. ficus-indica* has been ongoing for 86 years since the first agent was released against this plant in South Africa. (2) Invasions of *Acacia saligna* (Port Jackson Willow), in terms of density, biomass and numbers of seeds have been considerably moderated by biological control, but the distribution of the weed has remained static and is characterised by the appearance, over a wide area, of dense mats of seedlings from accumulated seed-banks after a fire or soil-disturbance. (3) *Eichhornia crassipes* (Water Hyacinth), is under far better biological control in the warmer, coastal areas and in clear waters, than it is in the highlands, under eutrophic conditions.

Weed species	Regions	Habitats	Biological control outcomes, over the years				
			Density	Area	Biomass	Seeds	Years
(1) *Opuntia ficus-indica* (L.) Mill. (Cactaceae) Mission Prickly Pear (Zimmermann and Moran 1991)	E. Cape All other	Coastal Drylands	B− A+	C− A+	C+ A+	C− A+	86

(continued)

(2) *Acacia saligna* (Labill.) H.L.Wendl. (Fabaceae) Port Jackson Willow (Wood and Morris 2007)	All	All	**B**	**C**	**B+**	**B**	32
(3) *Eichhornia crassipes* (C.Mart.) Solms (Pontederiaceae) Water Hyacinth (Coetzee et al. 2011a)	Highlands Highlands	Eutrophic Clear water	**B–** **B**	**B–** **B+**	**B–** **B+**	B– B+	45

19.3 What Has Been Forgotten over the Passage of Time?

One of the idiosyncrasies of biological control of invasive plant species is that society tends to forget the extent and/or intensity of the specific invasions that existed before biological control was implemented and fails to recall that there was ever a problem in the first place. Examples to support this statement and to elaborate on some of the successes achieved, are briefly described below:

1. At the beginning of the last century, *Opuntia monacantha* (Cactaceae; Cochineal Prickly Pear) was a prominent invasive species along the coastal regions from Cape Town to northern KwaZulu-Natal. Following biological control, populations declined so greatly that *O. monacantha* has become an insignificant and forgotten component of the South African flora. This situation has persisted for over a 100 years since biological control was initiated in 1913 (Lounsbury 1915; Moran et al. 2013).
2. *Opuntia ficus-indica* (Cactaceae; Mission Prickly Pear) is still considered problematic in some parts of the country. The densities of this plant have however been dramatically reduced over most of its distribution (see Box 19.3), as have its previously devastating socio-economic impacts (Annecke and Moran 1978; Zimmermann and Moran 1991; van Sittert 2002). Prickly-pear has beneficial properties and is used for its fruit and as fodder (Beinart and Wotshela 2011). Currently biological control has suppressed this species to a level at which the benefits of the plant to society now outweigh the costs of the invasion (Shackleton et al. 2007; Zachariades et al. 2017; Zimmermann 2017).
3. In many areas in South Africa, *Opuntia aurantiaca* (Cactaceae; Jointed Cactus) is under excellent and permanent biological control. But in the Eastern Cape it is still problematic because the species has not been reduced below an acceptable threshold (Moran and Zimmermann 1991), as evidenced by the frequency of complaints from landowners. As has been the case with previous examples, the extent of the invasion prior to biological control has been largely forgotten. The density of *O. aurantiaca* has also been greatly reduced even in the areas where the species is at its worst today (Moran and Annecke 1979; Moran and Zimmermann 1991).

4. *Hypericum perforatum* (Clusiaceae; St. John's Wort) became increasingly abundant and problematic in the Western Cape in the mid-1900s, but is now a rarity. Biological control agents that had succeeded in Australia and North America were released in South Africa in the 1960s and 1970s and stopped the invasion of *H. perforatum* at an early stage (Gordon et al. 1986; Gordon and Kluge 1991).
5. The problem of 'short institutional memories' is well illustrated by the example of *Sesbania punicea* (Fabaceae; Red Sesbania). This plant has been conclusively shown to have been rendered insignificant as an invader through biological control in South Africa (Hoffmann and Moran 1998). Densities of *S. punicea* declined by more than 95% as biological control took effect; woodlands of *S. punicea* with closed canopies of 5 m tall trees with up to 300 mm basal stem-diameters were reduced to patches of isolated plants that seldom survived long enough to develop beyond small bushy shrubs that are typically under 2 m tall, with spindly stems. This situation has persisted for the last 40 years, but *S. punicea* is still reported as being troublesome by landowners or land-managers who are making their own assessments on the extent of the problem for a suspected invasive species that they are encountering for the first time.
6. A variant on the theme discussed in (5) applies to *Acacia saligna* (Fabaceae; Port Jackson Willow). Dense forests of 5–10 m tall trees with closed canopies and barren understories once covered large areas in the Western Cape. Though still common and plentiful in this province, long-term studies have shown that it is a hugely-reduced problem since biological control was initiated in the late 1980s (Wood and Morris 2007) (see Box 19.3). Observations, since 1991, at 15 sites, some subjected to wild fires, have shown considerable site-specific variations but an overall decrease of more than 60% in tree densities. Invasive stands now consist predominantly of scattered trees and shrubs, all of which are infected by the pathogenic biological control agent and few will live for more than 5 years. The canopy seldom closes, and therefore other plant species can coexist with *A. saligna*. Furthermore, *A. saligna* phyllode biomass and seed outputs have markedly declined (Wood and Morris 2007). Given time, populations will become even less dense and abundant and other control measures will be needed only if landowners must clear their land for immediate use. Yet concerns about *A. saligna* as a highly-problematic invasive species persist.
7. The case study of *Acacia longifolia* (Fabaceae; Long-Leaved Wattle) is also informative, although there have been no formal quantifications on the changed status of *A. longifolia* invasions. Before any biological control agents were released, *A. longifolia* was an abundant and widespread species in the Western Cape. Dennill and Donnelly (1991) had shown that the biological control agents on *A. longifolia* were prolific and damaging, but there was no attempt to determine how this damage affected the population dynamics of the host plant. Currently, *A. longifolia* is seldom named, or ranks low on lists of the most-problematic invasive species (Zachariades et al. 2017). It is reasonable to assume that this reversal, i.e. the fact that the plant has all but disappeared from

the landscape, is principally due to biological control, albeit in conjunction with manual clearing operations, fires and other natural mortalities affecting the seeds. Over 30 years, biological control, and multiple fire cycles, have reversed the aggressiveness of *A. longifolia* which, in most situations, is now generally considered to have been rendered insignificant as an invasive species (Moran and Hoffmann 2012) (see Fig. 19.1).

8. The small serotinous tree, *Hakea sericea* (Proteaceae; Silky Hakea) is still a problematic invader in the Fynbos Biome, but 50 years ago it was far worse. Extensive clearing and burning operations were pursued in the 1980s and the distribution of dense infestations of hakea declined very significantly (Gordon and Fourie 2011) from about 530,000 ha in 1979 (Fugler 1979) to less than 190,000 ha in 2001 (Grand et al. 2007). The point that may have been overlooked is that these gains were made possible because management practices were backed up by increasing levels of biological control brought about by two species of seed-feeding biological control agents that were first released in South Africa in 1970. Esler et al. (2010), using models developed by Le Maitre et al. (2008), concluded that: "*In the virtual absence of mechanical control programs after the mid-1980s, the presence of seed-attacking insects* [as biological control agents] *provides an explanation for the failure of the previously invading species* [i.e. *H. sericea*] *to expand its range*"; and that "... *biological control was largely responsible for the failure of the species to re-colonize cleared sites, or to spread to new areas following unplanned wildfires.*"
9. As a result of biological control, and other management efforts, that date back to the mid-1970s, *Eichhornia crassipes* (Pontederiaceae; Water Hyacinth) is also far less of a problem than it used to be (Coetzee et al. 2011a) (see Box 19.3). In some areas and situations, biological control has completely supressed the species (Hill and Coetzee 2017a). In other areas biological control has significantly reduced plant populations and their impact, so that alternative management methods, such as herbicide applications and manual removal, are required, but far less frequently (Hill and Coetzee 2017b). Biological control of water hyacinth has been hindered by eutrophication of waters that promotes plant growth and by colder climates that slow the development of the insect biological control agents (Hill and Olckers 2001).
10. Other previously-problematic plant species have become uncommon, restricted, or rare under biological control and are now seldom mentioned in discussions on invasive species. Examples include: *Ageratina riparia* (Asteraceae; Mistflower) (Heystek et al. 2011), *Pistia stratiotes* (Araceae; Water Lettuce) (Coetzee et al. 2011a) (see Fig. 19.2), *Azolla filiculoides* (Azollaceae; Red Water Fern) (McConnachie et al. 2004; Hill et al. 2008; Coetzee et al. 2011a), and *Myriophyllum aquaticum* (Haloragaceae; Parrot's Feather) (see Box 19.4) (Cilliers 1999; Coetzee et al. 2011a). This is also the certain prognosis for *Cylindropuntia fulgida* var. *fulgida* (Cactaceae; Chain-Fruit Cholla), and *Cylindropuntia fulgida* var. *mamillata* (Cactaceae; Boxing-Glove Cactus),

both of which are currently being suppressed, over a wide area, to very low levels by biological control (Zimmermann 2017) (see Fig. 19.3).

Box 19.4 *Myriophyllum aquaticum* (Haloragaceae) (Parrot's Feather)
Myriophyllum aquaticum (Parrot's Feather) was regarded as one of South Africa's worst aquatic weeds, found in every province, where it caused a reduction in stream flow, blocked water pumps and interfered with recreational activities. Biological control was implemented in 1994 and a survey conducted in 2008 showed that the weed had been eliminated (i.e. 'clear') at 66 (63%) of the 105 sites surveyed (see lower section of 'stack' diagram). Surveys from 1995 to 2017 showed a significant reduction in the density of the weed at the other 39 sites (see lower histogram: mean number of shoots/$m^2 \pm$ SE) to levels below a tolerable threshold (Coetzee et al. 2011a)

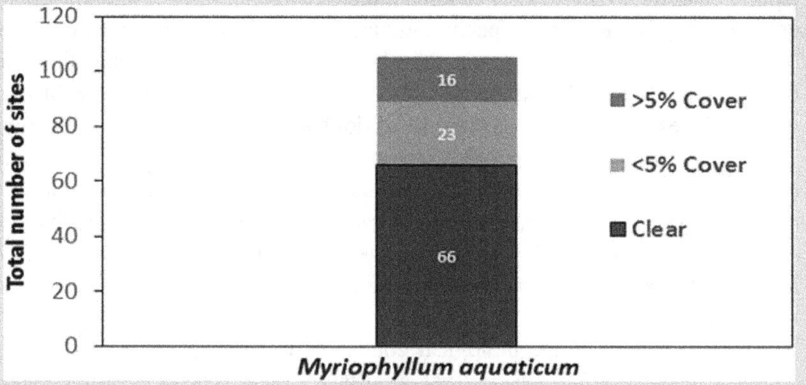

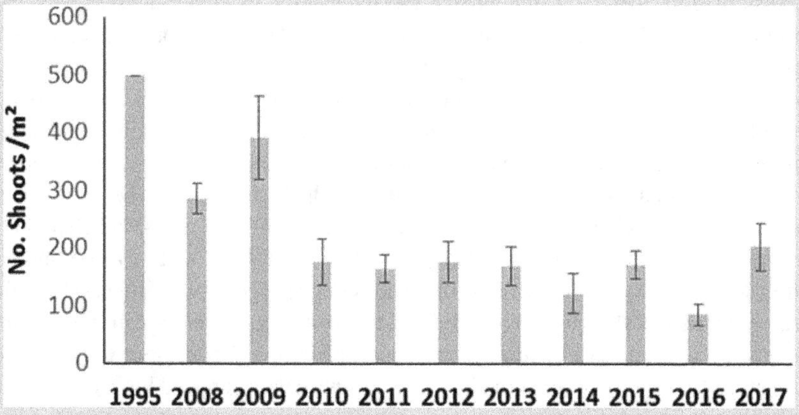

19.4 Additional Considerations, Caveats and Conclusions

Since 1995, and the advent of support from the Working for Water programme (Zimmermann et al. 2004), biological control of invasive plants in South Africa has gained enormous momentum, and many of the outcomes summarised in this chapter owe their successes to this. Much has been achieved, but some additional points and caveats need to be recorded:

1. Although in this account we have purposely not dealt with the details of the biological control agents that have been used against invasive plant species in South Africa since 1913, all the candidate agents have been subjected to internationally-accepted procedures for safety-testing (usually host-specificity trials): none of the eventual releases have resulted in unexpected consequences, or in damage to non-target plants. Over the years, in South Africa, 70 entities (species and their biotypes) of candidate biological control agents, under consideration in quarantine for prospective use against invasive alien plants, have been 'shelved', i.e. not released pending further developments or investigations, and 65 have been 'rejected' and the cultures destroyed, not usually because of issues with facilities or funding, but because of doubts about their safety (Zachariades 2018b). Klein (2011) has put this into perspective: "*It is of particular significance, in terms of safety practices and standards, that all the rejections were advocated and implemented by the researchers themselves, mostly following discussions within the South African weed biological control fraternity. No formal release applications from the researchers have been rejected by the regulatory authorities.*"
2. An integral feature of the biological control of invasive plants is dealing with conflicts of interest, i.e. where a potential target plant species is problematic in some circumstances or areas, but also has beneficial attributes. *Acacia mearnsii* De Wild. (Fabaceae; Black Wattle), for example, is a major problem in the Western and Eastern Cape provinces, especially in riparian areas, and was mooted for biological control in the early 1970s. This proposal met with strenuous objections lodged in 1977 by the wattle growers who, in other parts of the country, cultivate *A. mearnsii* and, to a lesser extent, two other closely-related wattle-species as the basis of lucrative timber production and tannin extraction industries. Previous initiatives and experience with other invasive *Acacia* species in South Africa led to the proposal that seed-destroying agents be released that reduce the reproductive capacity ('invasiveness') of their hosts, but that do not retard the vegetative growth. After protracted discussions and negotiations with the wattle-growers, the first biological control agent was released against *A. mearnsii* in 1993 (Impson et al. 2011). These precedents explain the frequent use of seed-destroying species in projects for biological control of invasive alien plants in South Africa. In some instances, as with the proposed biological control of two species of invasive *Pinus* species (Pinaceae; pine trees) in South Africa, attempts at biological control were abandoned, largely because of a failure to overcome conflicts of interest (Moran et al. 2000; Hoffmann et al. 2011).

3. More emphasis and research efforts are needed on detailed plant-population-level evaluations of the outcomes of biological control. But patience and persistence are the key: too many biological control efforts are perceived as failures because not enough time has elapsed, at least 10 years in many cases (McFadyen 1998), for the agents to have built up sufficiently in numbers and to have had an impact on target-plant populations. For example, an agent released against *Pereskia aculeata* (Cactaceae; Barbados Gooseberry), at a limited number of sites in South Africa, took 25 years, with numerous interventions to bolster its numbers, before it had a significant impact on its target host.
4. The economic costs and benefits of biological control of invasive plants and gains for agriculture, conservation, ecosystem services, and water supply in South Africa are discussed by van Wilgen et al. (2004), De Lange and van Wilgen (2010), van Wilgen and De Lange (2011), van Wilgen et al. (2012) and by Zachariades et al. (2017). These authors reach consensus that biological control of invasive plants is often more successful than other management practices, is prominently used against some of South Africa's worst invasive species, and is highly cost-effective. Certainly, the interactions of scientists with economists, over the last 25 years, has been crucial in convincing decision-makers and managers that biological control is essential in efforts to suppress problem plants.
5. Martin et al. (2018) have recently demonstrated the potential for extension of research and implementation practices in biological control of invasive plants to the wider community through employment opportunities and via innovative educational and outreach programmes. These precedents should be more widely applied in biological control and formally assessed for their socio-economic benefits.
6. Biological control is seldom a 'silver-bullet-solution': it is one of several options and strategies for the suppression of invasive alien plants. Early interventions using biological control, before the problem has been allowed to escalate, may maximise the suppression of plant invasions. South Africa has tried to adopt this as an approach for biological control (Olckers 2004) but has not always succeeded, largely because it is difficult to convince funders, landowners, and conservationists to commit resources to an invasive-plant problem that seems trivial and might never materialise.
7. Further, successful biological control is sometimes thwarted because other alien plants replace those that have been controlled. For example, when floating, aquatic invasive plant species are successfully controlled they may be replaced by submerged aquatic invaders (Coetzee et al. 2011b), or in the case of the suppression of invasive *Acacia* species, they are sometimes replaced by *Leptospermum laevigatum* (Australian Myrtle) which itself is refractory to biological control (Gordon 2011). Implementation and optimal integration of biological control with other forms of management, and close monitoring to document changes in populations of the invasive and the native flora, following biological control, remain an ongoing challenge and goal for researchers and managers.

We conclude that even small degrees of suppression of plant fitness, or reductions in the size of populations through biological control will be compounded over the years. In most cases, biological control will reduce the costs of other interventions and will make successful management of the problem more likely and easier. If there are any doubts about what has been achieved with biological control against invasive alien plants in South Africa over the years, the fundamental counterfactual question is: 'what would be the extent of the alien plant invasions if there had been no biological controls?'

Acknowledgements Our thanks and respects are extended to past and present invasive plant biological control practitioners in South Africa. Not all of them can be recorded as authors, but their work is the foundation for many of the cases and outcomes reported here. We appreciate the valuable inputs from the reviewers and the editors. We are also grateful to the Working for Water programme of the Department of Water Affairs and Forestry, subsequently the Department of Environmental Affairs, Natural Resources Management Programmes, for their generous financial support since 1995, which has put South Africa in the forefront of endeavours in biological control against invasive alien plants worldwide.

References

Annecke DP, Moran VC (1978) Critical reviews of biological pest control in South Africa. 2. The prickly pear, *Opuntia ficus-indica* (L.) Miller. J Entomol Soc South Afr 41:161–188

Anonymous (1999) Appendix. Releases of introduced biological control agents against weeds in South Africa. Afr Entomol Mem 1:175–182

Beinart W, Wotshela L (2011) Prickly pear, the social history of a plant in the Eastern Cape. Wits University Press, Johannesburg. https://doi.org/10.18772/22011115300

Cilliers CJ (1987) The evaluation of three insect natural enemies for the biological control of the weed *Lantana camara* L. J Entomol Soc South Afr 50:15–34

Cilliers CJ (1999) Biological control of parrot's feather, *Myriophyllum aquaticum* (Vell.) Verdc. (Haloragaceae), in South Africa. Afr Entomol Mem 1:113–118

Cilliers CJ, Neser S (1991) Biological control of *Lantana camara* (Verbenaceae) in South Africa. Agric Ecosyst Environ 37:57–75. https://doi.org/10.1016/0167-8809(91)90139-O

Coetzee JA, Hill MP, Byrne MJ et al (2011a) A review of the biological control programmes on *Eichhornia crassipes* (C.Mart.) Solms (Pontederiaceae), *Salvinia molesta* D.S.Mitch. (Salviniaceae), *Pistia stratiotes* L. (Araceae), *Myriophyllum aquaticum* (Vell.) Verdc. (Haloragaceae) and *Azolla filiculoides* (Lam.) (Azollaceae) in South Africa. Afr Entomol 19:451–468. https://doi.org/10.4001/003.019.0202

Coetzee JA, Bownes A, Martin GD (2011b) Prospects for the biological control of submerged macrophytes in South Africa. Afr Entomol 19:469–487. https://doi.org/10.4001/003.019.0203

Crawley MJ (1989) Plant life-history and the success of weed biological control projects. In: Delfosse ES (ed) Proceedings of the VII international symposium on biological control of weeds, Rome, 1988. CSIRO, Melbourne, pp 17–26

De Lange WJ, van Wilgen BW (2010) An economic assessment of the contribution of biological control to the management of invasive alien plants and to the protection of ecosystem services in South Africa. Biol Invasions 12:4113–4124. https://doi.org/10.1007/s10530-010-9811-y

Dennill GB, Donnelly D (1991) Biological control of *Acacia longifolia* and related weed species (Fabaceae) in South Africa. Agric Ecosyst Environ 37:115–135. https://doi.org/10.1016/0167-8809(91)90142-K

Dodd AP (1940) The biological campaign against prickly-pear. Commonwealth prickly pear board bulletin. Government Printer, Brisbane

Esler KJ, van Wilgen BW, te Roller KS et al (2010) A landscape-scale assessment of the long-term integrated control of an invasive shrub in South Africa. Biol Invasions 12:211–218. https://doi.org/10.1007/s10530-009-9443-2

Fugler SR (1979) Some aspects of the autecology of three *Hakea* species in the Cape Province, South Africa. MSc thesis, University of Cape Town

Gordon AJ (2011) Biological control endeavours against Australian myrtle, *Leptospermum laevigatum* (Gaertn.) F.Muell. (Myrtaceae), in South Africa. Afr Entomol 19:349–355. https://doi.org/10.4001/003.019.0206

Gordon AJ, Fourie A (2011) Biological control of *Hakea sericea* Schrad. & J.C.Wendl. and *Hakea gibbosa* (Sm.) Cav. (Proteaceae) in South Africa. Afr Entomol 19:303–314. https://doi.org/10.4001/003.019.0205

Gordon AJ, Kluge RL (1991) Biological control of St. John's Wort, *Hypericum perforatum* (Clusiaceae), in South Africa. Agric Ecosyst Environ 37:77–90

Gordon AJ, Kluge RL, Neser S (1986) Effect of the gall midge, *Zeuxidiplosis giardi* (Diptera: Cecidomyiidae), on seedlings of St. John's Wort, *Hypericum perforatum* (Clusiaceae). In: Delfosse ES (ed) Proceedings of the VI international symposium on biological control of weeds, Vancouver, 1984. Agriculture Canada, Ottawa, pp 743–748

Grand J, Cummings MP, Rebelo TG et al (2007) Biased data reduce efficiency and effectiveness of conservation reserve networks. Ecol Lett 10:364–374. https://doi.org/10.1111/j.1461-0248.2007.01025.x

Henderson L, Wilson JRU (2017) Changes in the composition and distribution of alien plants in South Africa: an update from the Southern African Plant Invaders Atlas. Bothalia 47(2):a2172. https://doi.org/10.4102/abc.v47i2.2172

Heystek F, Wood AR, Neser S et al (2011) Biological control of two *Ageratina* species (Asteraceae: Eupatorieae) in South Africa. Afr Entomol 19:208–216. https://doi.org/10.4001/003.019.0208

Hill MP, Coetzee J (2017a) The biological control of aquatic weeds in South Africa: current status and future challenges. Bothalia 47(2):a2152. https://doi.org/10.4102/abc.v47i2.2152

Hill MP, Coetzee JA (2017b) Chapter 7. Biological control of waterhyacinth restores ecosystem functioning to an impoundment in South Africa. In: van Driesche RG and Reardon RC (eds) Suppressing over-abundant invasive plants and insects in natural areas by use of their specialized natural enemies. The Forest Health Technology Enterprise Team (FHTET), FHTET-2017-02, Morgantown, WV, pp 48–52

Hill MP, Olckers T (2001) Biological control initiatives against water hyacinth in South Africa: constraining factors, success and new courses of action. In: Julien et al (eds) Biological and integrated control of water hyacinth, *Eichhornia crassipes*. Proceedings of the second global working group meeting for the biological and integrated control of water hyacinth, Beijing, 2000. ACIAR Proceedings, Australian Centre for International Research, Canberra 102: 33–38

Hill MP, McConnachie AJ, Byrne MJ (2008) *Azolla filiculoides* Lamarck (Pteridophyta: Azollaceae) control in South Africa: a 10-year review. In: Julien MH, Sforza R, Bon MC et al (eds) Proceedings of the XII international symposium on biological control of weeds, La Grande Motte, France, 2007. CAB International, Wallingford, pp 558–560. https://doi.org/10.1079/9781845935061.0558

Hoffmann JH (ed) (1991) Introduction. Biological control of weeds in South Africa. Agric Ecosyst Environ 37:1–3. https://doi.org/10.1016/0167-8809(91)90144-M

Hoffmann JH (1995) Biological control of weeds: the way forward, a South African perspective. British Crop Protection Council, Farnham (UK). Weeds in a changing world. Proc Brighton Crop Prot Conf 64:77–89

Hoffmann JH, Moran VC (1998) The population dynamics of an introduced tree, *Sesbania punicea*, in South Africa, in response to long-term damage caused by different combinations of three species of biological control agents. Oecologia 114:343–348. https://doi.org/10.1007/s004420050456

Hoffmann JH, Moran VC (2008) Assigning success in biological weed control: what do we really mean? In: Julien MH, Sforza R, Bon MC et al (eds) Proceedings of the XII international symposium on biological control of weeds, La Grande Motte, France, 2007. CAB International, Wallingford, pp 687–692. https://doi.org/10.1079/9781845935061.0687

Hoffmann JH, Moran VC, Zeller DA (1998) Long-term population studies and the development of an integrated management programme for control of *Opuntia stricta* in Kruger National Park. South Africa J Appl Ecol 35:156–160. https://doi.org/10.1046/j.1365-2664.1998.00283.x

Hoffmann JH, Moran VC, van Wilgen BW (2011) Prospects for the biological control of invasive *Pinus* species (Pinaceae) in South Africa. Afr Entomol 19:393–401. https://doi.org/10.4001/003.019.0209

Hoffmann JH, Moran VC, Hill MP (2019) Conceptualizing, categorizing and recording the outcomes of biological control of invasive plant species, at a population level. Biol Control. https://doi.org/10.1016/j.biocontrol.2019.02.005

Huffaker CB (1964) Fundamentals of biological weed control. In: DeBach P, Schlinger EI (eds) Biological control of insect pests and weeds. Chapman and Hall, London, pp 631–649

Impson FAC, Kleinjan CA, Hoffmann JH et al (2011) Biological control of Australian *Acacia* species and *Paraserianthes lophantha* (Willd.) Nielsen (Mimosaceae) in South Africa. Afr Entomol 19:186–207. https://doi.org/10.4001/003.019.0210

Klein H (2011) A catalogue of the insects, mites and pathogens that have been used or rejected, or are under consideration, for the biological control of invasive alien plants in South Africa. Afr Entomol 19:515–549. https://doi.org/10.4001/003.019.0214

Le Maitre DC, Krug RM, Hoffmann JH et al (2008) *Hakea sericea*: development of a model of the impacts of biological control on population dynamics and rates of spread of an invasive species. Ecol Model 212:342–358. https://doi.org/10.1016/j.ecolmodel.2007.11.011

Lounsbury CP (1915) Plant killing insects: the Indian cochineal. Agric J S Afr 1:537–543

Mann J (1970) Cacti naturalized in Australia and their control. Department of Lands, Brisbane

Martin GD, Hill MP, Coetzee JA et al (2018) Synergies between research organisations and the wider community in enhancing weed biological control in South Africa. BioControl 63:437–447. https://doi.org/10.1007/s10526-017-9846-4

McConnachie AJ, Hill MP, Byrne MJ (2004) Field assessment of a frond-feeding weevil, a successful biological control agent of red waterfern, *Azolla filiculoides,* in southern Africa. Biol Control 29:326–331. https://doi.org/10.1016/j.biocontrol.2003.08.010

McFadyen REC (1998) Biological control of weeds. Annu Rev Entomol 43:369–393. https://doi.org/10.1146/annurev.ento.43.1.369

Moran VC, Annecke DP (1979) Critical reviews of biological pest control in South Africa. 3. The jointed cactus, *Opuntia aurantiaca* Lindley. J Entomol Soc South Afr 42:299–329

Moran VC, Hoffmann JH (2012) Conservation of the fynbos biome in the Cape Floral Region: the role of biological control in the management of invasive alien trees. BioControl 57:139–149. https://doi.org/10.1007/s10526-011-9403-5

Moran VC, Hoffmann JH (2015) The fourteen international symposia on biological control of weeds, 1969–2014: delegates, demographics and inferences from the debate on non-target effects. Biol Control 87:23–31. https://doi.org/10.1016/j.biocontrol.2015.04.008

Moran VC, Zimmermann HG (1991) Biological control of jointed cactus, *Opuntia aurantiaca* (Cactaceae), in South Africa. Agric Ecosyst Environ 37:5–27. https://doi.org/10.1016/0167-8809(91)90136-L

Moran VC, Hoffmann JH, Donnelly D et al (2000) Biological control of alien, invasive pine trees (*Pinus* species) in South Africa. In: Spencer NR (ed) Proceedings of the X international symposium on biological control of weeds, Bozeman, 1999. Montana State University, Bozeman MT, pp 941–953

Moran VC, Hoffmann JH, Zimmermann HG (2005) Biological control of invasive alien plants in South Africa: necessity, circumspection, and success. Front Ecol Environ 3:71–77. https://doi.org/10.1890/1540-9295(2005)003[0071:BCOIAP]2.0.CO;2

Moran VC, Hoffmann JH, Hill MP (eds) (2011) Biological control of invasive alien plants in South Africa (1999–2010). Afr Entomol 19:177–549. https://doi.org/10.4001/003.019.0218

Moran VC, Hoffmann JH, Zimmermann HG (2013) 100 years of biological control of invasive alien plants in South Africa: history, practice and achievements. S Afr J Sci 109:5–10. https://doi.org/10.1590/sajs.2013/a0022

Morin L, Reid AM, Sims-Chilton NM et al (2009) Review of approaches to evaluate the effectiveness of weed biological control agents. Biol Control 51:1–15. https://doi.org/10.1016/j.biocontrol.2009.05.017

Muniappan R, Reddy GVP, Raman A (eds) (2009) Biological control of tropical weeds using arthropods. Cambridge University Press, Cambridge. https://doi.org/10.1017/CBO9780511576348

Olckers T (2004) Targeting emerging weeds for biological control in South Africa: the benefits of halting the spread of alien plants at an early stage of their invasion. S Afr J Sci 100:64–68

Olckers T, Hill MP (eds) (1999) Biological control of weeds in South Africa (1990–1998). Afr Entomol Mem 1:1–182

Paterson ID, Hoffmann JH, Klein H et al (2011) Biological control of Cactaceae in South Africa. Afr Entomol 19:230–246. https://doi.org/10.4001/003.019.0221

Raghu S, Walton C (2007) Understanding the ghost of *Cactoblastis* past: historical clarifications on a poster child of classical biological control. Bioscience 57:699–705. https://doi.org/10.1641/B570810

Schwarzländer M, Moran VC, Raghu S (2018) Constraints in weed biological control: contrasting responses by implementing nations. BioControl 63:313–317

Shackleton CM, McGarry D, Fourie S et al (2007) Assessing the effects of invasive alien species on rural livelihoods: case examples and a framework from South Africa. Hum Ecol 35:113–127. https://doi.org/10.1007/s10745-006-9095-0

Syrett P, Briese DT, Hoffmann JH (2000) Success in biological control of terrestrial weeds by arthropods. In: Gurr G, Wratten S (eds) Biological control: measures of success. Kluwer, Dordrecht, pp 189–230. https://doi.org/10.1007/978-94-011-4014-0_7

Urban AJ, Simelane DO, Retief E et al (2011) The invasive '*Lantana camara* L.' hybrid complex (Verbenaceae): a review of research into its identity and biological control in South Africa. Afr Entomol 19:315–348. https://doi.org/10.4001/003.019.0225

van Sittert L (2002) 'Our irrepressible fellow colonist': the biological invasion of prickly pear (*Opuntia ficus-indica*) in the Eastern Cape, c.1870–c.1910. In: Dovers S, Edgecombe R, Guest B (eds) South Africa's environmental history. Cases and comparisons. David Philip, Cape Town, pp 139–159. https://doi.org/10.1006/jhge.2001.0457

van Wilgen BW (2020) A brief, selective history of researchers and research initiatives related to biological invasions in South Africa. In: van Wilgen BW, Measey J, Richardson DM, Wilson JR, Zengeya TA (eds) Biological invasions in South Africa, Springer, Berlin, pp 31–64. https://doi.org/10.1007/978-3-030-32394-3_2

van Wilgen BW, De Lange WJ (2011) The costs and benefits of biological control of invasive alien plants in South Africa. Afr Entomol 19:504–514. https://doi.org/10.4001/003.019.0228

van Wilgen BW, de Wit MP, Anderson HJ et al (2004) Costs and benefits of biological control of invasive alien plants: case studies from South Africa. S Afr J Sci 100:113–122

van Wilgen BW, Reyers B, Le Maitre DC et al (2008) A biome-scale assessment of the impact of invasive alien plants on ecosystem services in South Africa. J Environ Manag 89:336–349. https://doi.org/10.1016/j.jenvman.2007.06.015

van Wilgen BW, Forsyth GG, Le Maitre DC et al (2012) An assessment of the effectiveness of a large, national-scale invasive alien plant control strategy in South Africa. Biol Conserv 148:28–38. https://doi.org/10.1016/j.biocon.2011.12.035

Winston RL, Schwarzländer M, Hinz HL et al (eds) (2014) Biological control of weeds: a world catalogue of agents and their target weeds, 5th ed. USDA Forest Service, Forest Health Technology Enterprise Team, Morgantown, WV. FHTET-2014-04

Wood AR, Morris MJ (2007) Impact of the gall-forming rust fungus *Uromycladium tepperianum* on the invasive tree *Acacia saligna* in South Africa: 15 years of monitoring. Biol Control 41:68–77. https://doi.org/10.1016/j.biocontrol.2006.12.018

Zachariades C (2018a) Biological control of invasive alien plants in South Africa: a list of all insects, mites and pathogens released as biological control agents from 1913–2018. http://www.arc.agric.za/arc-ppri/Documents/Table2-NaturalEnemiesReleased.pdf

Zachariades C (2018b) Biological control of invasive alien plants in South Africa: a list of all insects, mites and pathogens considered as biological control agents from 1913–2018. http://www.arc.agric.za/arc-ppri/Documents/Table1-NaturalEnemiesAll.pdf

Zachariades C, Strathie LW, Retief E et al (2011) Progress towards the biological control of *Chromolaena odorata* (L.) R.M.King & H.Rob. (Asteraceae) in South Africa. Afr Entomol 19:282–302. https://doi.org/10.4001/003.019.0229

Zachariades C, Uyi OO, Dube N et al (2016) Biological control of *Chromolaena odorata*: *Pareuchaetes insulata* spreads its wings. Proc South Afr Sugar Technol Assoc 89:291–306. https://www.researchgate.net/.../308721341

Zachariades C, Paterson ID, Strathie LW et al (2017) Assessing the status of biological control as a management tool for suppression of invasive alien plants in South Africa. Bothalia 47:2142–2161. https://doi.org/10.4102/abc.v47i2.2142

Zimmermann HG (2017) Global invasions of cacti: control, management and conflicts of interest. In: Inglese P, Mondragon C, Nefzaoui A et al (eds) Crop ecology, cultivation and uses of cactus pear. FAO, Rome, pp 172–185

Zimmermann HG, Moran VC (1991) Biological control of prickly pear, *Opuntia ficus-indica* (Cactaceae), in South Africa. Agric Ecosyst Environ 37:29–35. https://doi.org/10.1016/0167-8809(91)90137-M

Zimmermann HG, Moran VC, Hoffmann JH (2004) Biological control in the management of invasive alien plants in South Africa, and the role of the Working for Water Programme. S Afr J Sci 100:34–40

Zimmermann HG, Moran VC, Hoffmann JH (2009) Invasive cactus species (Cactaceae). In: Muniappan R, Reddy GVP, Raman A (eds) Biological control of tropical weeds using arthropods. Cambridge University Press, Cambridge, pp 108–129. https://doi.org/10.1017/CBO9780511576348.007

20

Biological Invasions and their Risk Analysis

Sabrina Kumschick, Llewellyn C. Foxcroft and John R. Wilson

Abstract Risk analysis is an important decision-support tool for the management of biological invasions. South Africa, as a signatory to international agreements, has enacted legislation requiring risk analyses to be conducted if trade is to be restricted or regulated and if alien species are to be introduced. In this chapter, we outline the various needs for risk analyses for biological invasions in South Africa, summarise the current status, and make recommendations for a way forward. In particular, we highlight the need to move away from approaches that are purely based on expert opinion or entirely reactive, and propose a new system and process which includes the use of a structured risk analysis framework with clear guidelines to avoid expert bias. We highlight the need to assess risk, consider risk management options (including benefits), and to develop clear recommendations. The proposed process also involves the review of recommendations by an independent panel. We further note that the effectiveness of such approaches will be defined by their transparency, their accuracy, how feasible they are to implement in practice, and the trust that people have in the system.

S. Kumschick (✉) · J. R. Wilson
Centre for Invasion Biology, Department of Botany and Zoology, Stellenbosch University, Stellenbosch, South Africa

South African National Biodiversity Institute, Kirstenbosch Research Centre, Cape Town, South Africa
e-mail: sabrina.kumschick@gmail.com

L. C. Foxcroft
Conservation Services, South African National Parks, Skukuza, South Africa

Centre for Invasion Biology, Department of Botany and Zoology, Stellenbosch University, Stellenbosch, South Africa

20.1 Risk Analysis for Biological Invasions

Frameworks for the management and regulation of alien species and associated processes have been developed all over the world to deal with undesirable consequences and to mitigate future impacts of species introduced outside of their native ranges. Risk analysis is a formal, evidence-based process to analyse the risk of a particular hazard to a certain area or situation. A hazard can be an event or phenomenon, like the increase of temperature globally or the pollution of a river with chemicals. A hazard can also be more specific, like the introduction of an alien species to a new area. This chapter reviews what has been done in South Africa in the context of international best practice, and outlines a proposed way forward.

In general terms, risk analysis is the combination of *hazard identification*, *risk assessment*, *risk management* and *risk communication* (Fig. 20.1). Biological invasions present various hazards that can be broadly grouped in terms of species, pathways, and areas. If the hazard is the alien species itself, risk assessment includes the likelihood of a particular species being introduced, establishing and spreading in an area, and the consequences (negative impacts) thereof; risk management focuses on management options available for the species, the ease of management more broadly, and people's perceptions and uses of the species that could lead to conflicts of interest around its control (Branquart et al. 2016). For pathway risk assessment, the likelihood and consequences of the pathway or vector bringing in harmful alien species is assessed (in this case, risk is often proportional to the number of harmful alien species introduced along a pathway, i.e. the colonisation pressure); and risk management focuses on interventions to make the pathways "cleaner" (e.g. the Ballast Water Convention; IMO 2004). Lastly, area-based risk assessments examine the suite of alien species in a certain area and the vulnerability of the area itself to

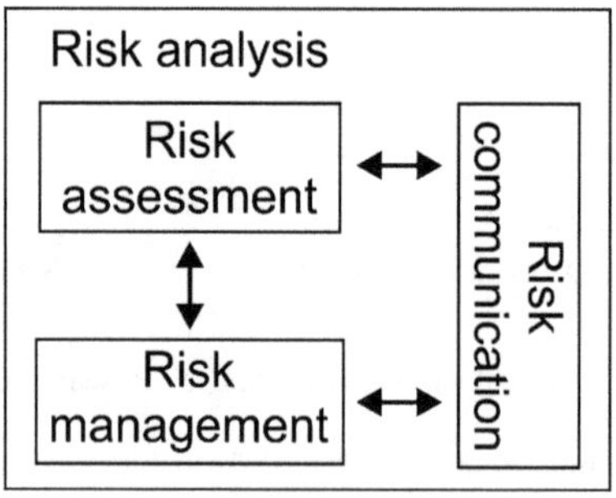

Fig. 20.1 The components of risk analysis. *Hazard identification* frames the problem that is being addressed (with respect to biological invasions this is often in terms of pathways, species, or areas—not shown in the figure). *Risk assessment* focuses on the likelihood and consequences of the hazard occurring. *Risk management* considers ways to mitigate the risks while maintaining any benefits. *Risk communication* aims to ensure effective communication and sharing of relevant information between stakeholders and assessors during the process of risk assessment and management, both to improve the information available for the risk analysis and to try to ensure that everyone ultimately agrees to abide by the outcome of the process

invasion, to distinguish high from low risk areas; and risk management efforts aim both to reduce the total level of invasion in the area and to limit the potential for all pathways to bring in new species. Since alien species generally affect a wide variety of stakeholders and are socially complex rather than clearly defined scientific problems (Woodford et al. 2016), communication with stakeholders is key throughout the process. Such communication needs to be a two-way process which aims to inform people about the outcomes of the analysis (*risk communication*), but also ensures stakeholders' knowledge, opinions, and concerns are considered during the risk assessment and risk management planning.

Besides the assessment of different hazards, risk analyses can also focus on different stages in the invasion process (Blackburn et al. 2011). One of the most prominent distinctions is between pre- and post-border analyses. It is often much more cost-effective to prevent an invasion from happening than to mitigate the harm or try to eradicate a species once it has established—pre-border risk analyses can help to make decisions about whether or not to allow the import of a new species, based on risk to the recipient environment. For species that are already present, we need ways to assess which to regulate and manage post-border. The main distinctions between pre- and post-border analyses are therefore: (1) the need for pre-border analyses to include the likelihood of arrival and introduction into the country; and (2) the need for post-border analyses to include risk management considerations focussing on reducing the negative impacts of invasions while maintaining benefits, and reaching agreement between stakeholders that might hold opposing entrenched positions.

Risk analysis methods ideally have a few key qualities that make them useful for helping to make management decisions including that they should be transparent, accurate, realistic, evidence-based, and free of biases as far as possible. Transparency ensures that decisions are traceable and the rationale behind them is clear. Accuracy is needed so that resources are not wasted on managing benign species, or so that potentially harmful species can be prevented from becoming problematic. Risk analysis also requires resources to implement, including financial and human capital (i.e. skilled practitioners). Given the inherent biases any assessor may have (e.g. Montibeller and Von Winterfeldt 2015) it is also important that such analyses are based on sound evidence. Furthermore, group assessments and independent review panels can minimise biases by a single assessor. Explicit guidance documents and training of assessors in the principles and methods of risk analysis can further improve the process and reduce assessor bias rooted in a different understanding of the questions asked. A group of European scientists developed a minimum set of 14 standards for risk assessment to aid their implementation for the EU Regulation on invasive alien species (1143/2014) (Roy et al. 2018), many of which overlap with what is discussed in this paragraph, and a few of which are specific to the requirements of EU regulation (like the assessment of impact on ecosystem services and effects of future climate change). But these standards are also useful in a South African context and have largely been followed in what is suggested here.

20.1.1 *Risk Assessment*

Many tools have been developed to assess the risk of biological invasions, but most focus on specific species as the hazard (Kumschick and Richardson 2013). Such tools typically cover a variety of aspects related to the risks posed by a species, which includes transport, establishment, abundance, spread, and impact (Leung et al. 2012). However, there are also many tools that cover or assess only a specific part of the risk of a species and which are often mistakenly put under the umbrella of "risk assessments". For example, much effort has been put into standardising the way we quantify impacts, and many recent developments were made around impact assessment tools (e.g. Blackburn et al. 2014; Bacher et al. 2018; Zengeya et al. 2020, Chap. 17). Similarly, species distribution models and climate matching tools are improving the way we can assess and map a species' climatically suitable area in a new range (see Wilson et al. 2020, Sect. 13.5, Chap. 13). Van Wilgen et al. (2009), for example, present climate suitability matches for amphibians and reptiles that can improve the way we assess risks for this group. In combination with other tools, climate suitability models and impact assessments can improve the assessment and analysis of risks of biological invasions in a transparent, standardised, and repeatable manner. However, by definition, a risk assessment needs to cover both the likelihood of the hazard occurring (i.e. the invasion process) as well as its consequences (i.e. the negative impacts).

Despite the aspects of risk to be covered being rather straightforward, there are many ways in which these can be assessed in practice. Several approaches to risk assessment have been suggested, including the trait-scoring, statistical, decision-tree, rapid screening, mechanistic, and detailed approach (Keller and Kumschick 2017). Each of these has its own benefits and weaknesses and is based on different premises:

Trait Scoring This assumes that species with specific traits have a higher chance of becoming invasive, and of having higher impacts. In other words, the more "invasive traits" a species has, the higher the risk it poses. This is the most common approach used to date, including in the Australian Weed Risk Assessment model (AWRA; Pheloung et al. 1999), which was developed as a pre-border tool to screen alien plants to be introduced to Australia. The AWRA has been used around the world and modified for different taxonomic groups including fish and freshwater invertebrates (Copp et al. 2005; Tricarico et al. 2010; see Appendix in Kumschick and Richardson 2013). This approach of assessing risk requires information about the taxon as alien and what makes some species more successful than others. It also assumes that the main contributor to invasion success is inherent to the introduced species' traits, and it is generally limited to a taxonomic group with comparable traits.

Statistical Approach The statistical approach is very similar to trait-scoring insofar as it uses traits of species to predict the potential for invasion and causing harm (Keller and Kumschick 2017). In the trait-scoring approach (seemingly) important traits are collated and rated by experts only. In the statistical approach the list of traits is refined using statistical or machine learning algorithms. These algorithms find patterns in the data and include only those traits that contribute to invasion success, as opposed to all traits considered important by experts. For example, for alien cacti it was found that native range size is a predictor of both impact and invasiveness in South Africa (Novoa et al. 2016a). The statistical approach can take into account interactions between traits and lead to much simpler tools than trait-scoring, but the underlying methods and mechanisms are not always easily understood and therefore less likely to be supported by managers and policy-makers (Keller and Kumschick 2017). Furthermore, due to the models including a very limited number of variables, no conclusion can be reached if data for any of the few selected variables is lacking.

Decision Trees These can be a subset of the statistical approach, but also be based on traits considered as important by experts. Similar issues therefore arise in this approach, which comprises of several questions of which each answer leads either to another question or a decision regarding a species' risk. If the answer to any question is not known, no decision can be reached, or tenuous assumptions are made. For that reason, this approach has not been used extensively. However, Tucker and Richardson (1995) suggested a decision tree approach to assess risks of tree invasions to the Fynbos Biome.

Rapid Screening The rapid screening approach can use elements of the previous three approaches, with the focus on a quick assessment of often a large number of species where little information is available. Rapid screening can be used as a stand-alone risk assessment, to prioritise species for more detailed risk analysis or to create watch lists. Faulkner et al. (2014) applied a rapid screening tool to almost 400 species to create a watch list of alien species for South Africa. Their tool included only three aspects: (1) the invasiveness of the species elsewhere, (2) climate match, and (3) tourism and trade data as a proxy for propagule pressure.

Mechanistic Approach The mechanistic approach is not based on traits per se, but it follows a species through the invasion process and assesses the likelihood that it will cross certain barriers (e.g. those proposed by based on Blackburn et al. 2011), and whether it has the potential to have an impact. For example, if a species is unlikely to find suitable climatic conditions and habitats in a new region, it would be unlikely to establish and subsequently become invasive, which can reduce its risk despite potential impacts. Such an approach is implemented in Belgium (D'hondt et al. 2015).

Detailed Approach The detailed approach, as the name suggests, requires a substantial amount of data on the alien species' ecology, biology and behaviour, as well as the recipient environment and the interaction between all these factors. It often requires additional research to fill knowledge gaps, which may include interviews with experts. Since it also includes management considerations and stakeholder perceptions, it resembles more closely a risk analysis approach. It has been used in Canada to assess risks of the highly contentious Asian carp species (Mandrak and Cudmore 2004).

Which approach is most useful depends on the circumstances and purpose. For example, rapid screening is quick and easy. However, results of such assessments provide limited knowledge on the mechanisms of invasion and expected behaviours of a species under specific circumstances as the underlying assumptions are often not explicit. This can be problematic for species with no previous history of introduction outside of their native range (no "invasion history"). Trade-offs between the optimal investment and the optimal outcome need to be made when selecting an approach for a risk assessment tool. Furthermore, the amount of knowledge available on a taxon can limit the use of certain approaches. The statistical approach requires the input of a substantial amount of data to train the model—without adequate knowledge about the taxon and its behaviour as an invader, such models cannot be developed. Furthermore, the final model requires data on only a few variables, which increases the likelihood of this information not being available for some species. This emphasises the need for a risk assessment tool to accommodate situations where information on a species to be assessed is scarce or lacking.

Many tools approach this by applying the precautionary principle—if no data are available on a species or a certain situation, preventing informed decision-making on the risks involved, the species is by default deemed to be a high risk (e.g. Nentwig et al. 2016). This principle is common practice and also anchored in international agreements. For example, in the management of alien species, the IUCN (2000) states that "unless there is a reasonable likelihood that an introduction will be harmless, it should be treated as likely to be harmful".

20.1.2 Risk Management

To reach sensible and effective management decisions, risk management considerations are fundamental. Ignoring management feasibility, benefits of the taxon, or potential conflicts between stakeholders, has been shown to lead to unsuccessful and wasteful management decisions (e.g. van Wilgen and Richardson 2012). The distinction between whether, as opposed to how, to regulate and manage a species relies on estimating the risks it poses to the recipient environment and economy. For taxa that are not yet present in an area, and for which decisions on importation are required, this can be a relatively straightforward process: if the species poses a high risk its importation should not be allowed, but if it poses a low risk the species could be considered safe for import (e.g. Keller and Kumschick 2017). However, decisions regarding taxa that are already present in an area, and in use for various purposes, cannot solely rely on estimates of the risks they pose, but also depend on management options available for the species. For example, many trees were introduced into South Africa for forestry. Even though many of them threaten native biodiversity, management decisions need to take into account the costs of management, the techniques and tools available, as well as the species life-history traits (Wilson et al. 2011; Richardson et al. 2015). Further, management does not happen in isolation from the rest of society; social perceptions and benefits need to be assessed and accounted for (e.g. van Wilgen and Richardson 2012; Zengeya et al. 2017; see Shackleton et al. 2020, Chap. 24). Unlike in the risk assessment where clear answers and probabilities are often provided to determine the level of risk, the inclusion of benefits is dependent on the agendas of various role players, priorities of decision makers, and the influence of key stakeholders (e.g. Kumschick et al. 2012; Woodford et al. 2017). To keep the process transparent, provision needs to be made to outline how the inclusion of benefits influences the management decisions, and which benefits are included.

Once a risk has been identified and assessed, and the species is regarded as of concern (i.e. high risk), one needs to consider what can be done to manage the risk (Fig. 20.2). For species already present in an area this will often require a detailed evaluation of management options, the development of management plans and

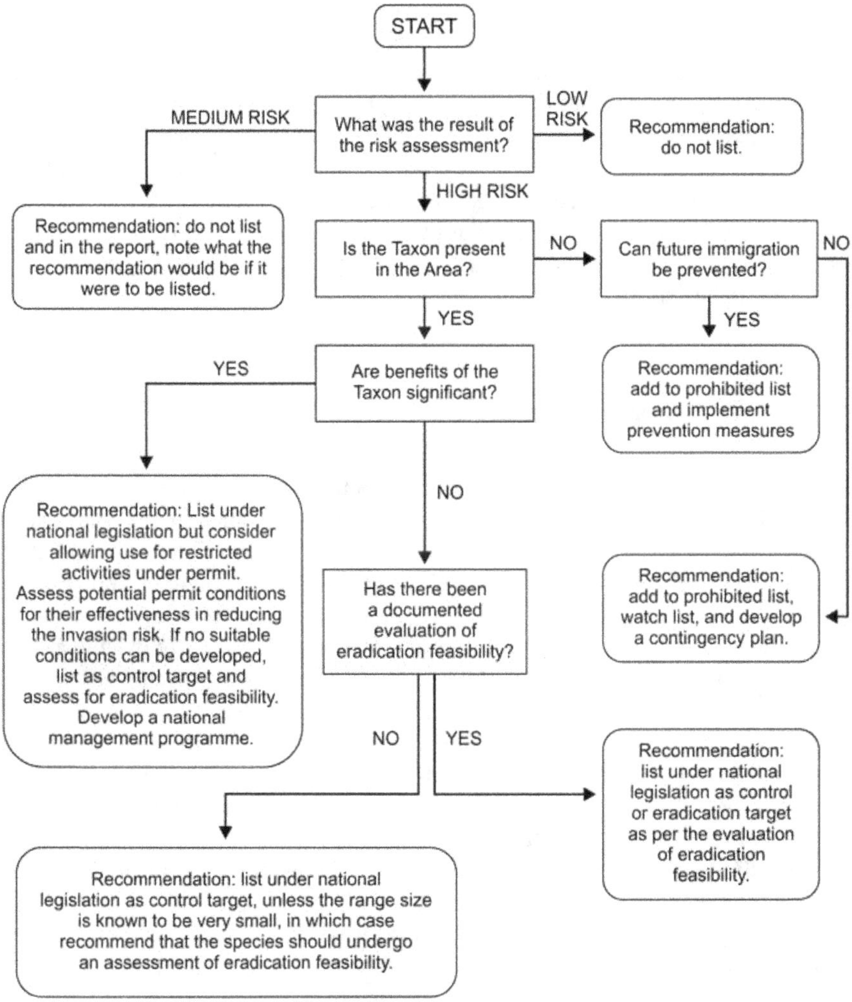

Fig. 20.2 The proposed process outlining the steps to develop recommendations for the listing of alien species under national legislation in South Africa. Reproduced from Kumschick et al. (2018) with permission

financial resources, and a process of prioritisation of potential interventions, including considerations related to the funds available to put the management plan into action (Wilson et al. 2017). Such detailed assessments are generally beyond the scope of risk analyses which are mostly desktop-based (but see example of biological control release applications below). However, some basic management considerations need to be taken into account that allow for a broad classification of how to treat a certain risk.

Risk management is generally more open-ended than the risk assessment, but needs to be documented in detail to assure that decisions are transparent and can be revisited when information becomes available. Risk management can include parameters like socio-economic and environmental benefits, the feasibility of stopping future immigration of a risky species, and basic considerations regarding eradication feasibility (e.g. Panetta and Timmins 2004; Wilson et al. 2017). Renteria et al. (2017), for example, suggested a rapid scheme for prioritisation of alien species in South Africa for eradication, using information on the species' distribution and "eradication feasibility syndromes". While such simple desktop studies are useful to flag species for further evaluation, they are not sufficient to determine whether a species is a suitable eradication target. In practice, eradication feasibility depends heavily on the biological (e.g. location of individuals, detectability, availability of effective control methods) and administrative context (e.g. funds available, and a dedicated and persistent leader and team).

20.1.3 Risk Communication

Once the level of risk has been determined and options for management and benefits evaluated, it is crucial to clearly communicate the outcomes of the analysis to stakeholders, including the general public, policy-makers and traders of the alien species. There are two important requirements for risk communication. Firstly, stakeholders must be engaged during the risk analysis, both for assessors to obtain information on the hazard, and to gain the support of stakeholders (e.g. Novoa et al. 2015, 2016b). There are often formal regulatory processes of stakeholder engagement. For example, before the promulgation of new regulations in South Africa, they are published for public comment, whereafter the comments needed to be addressed or acknowledged. In contentious cases, an independent scientific assessment might be needed (Scholes et al. 2017), but if conflicts are to be avoided, engagement should happen from the outset of the process.

Secondly, risk communication is important for providing stakeholders with sufficient information to understand the recommendations. Stakeholders need to be in a position to know under which circumstances decisions would be altered, for example, how new information or changing practices would influence risk. Therefore, communication needs to be simple enough to ensure understanding, but simultaneously enough information needs to be provided to underpin the decision. Decisions are often only successful and implementable if stakeholders understand the risks associated with the taxon. To gain the support from the general public and financial institutions (e.g. government departments), engagement and clear communication regarding risks is crucial. In practice, part of risk communication consists of clearly documenting the hazard, the circumstances of the assessment (including the area of assessment), and the results of the assessment, in an easily understandable manner. An easy-to-digest summary sheet including the main findings of each section (risk assessment and management), including short descriptions of the hazard (e.g. the alien species), pathways, impacts, management options, and benefits can serve this purpose.

20.2 Risk Analysis in South Africa

Various tools have been developed for use specifically in the South African context. Tucker and Richardson (1995) were amongst the pioneers in risk assessments for biological invasions. They developed a tool to assess the risks that woody plants pose to the Fynbos Biome, which requires information on the alien species, the region of origin, and various details specific to the Fynbos Biome, for example tolerance to fires. This approach is a combination of a trait-based scoring approach and a decision tree, and allowance is made for the lack of data on some traits or aspects. Due to the very specific focus of the scheme (the Fynbos Biome) it is most likely not applicable to a wider geographic range or different habitats, and it is limited to assessing woody plants. Similarly, other studies developing or applying risk assessments and analysis

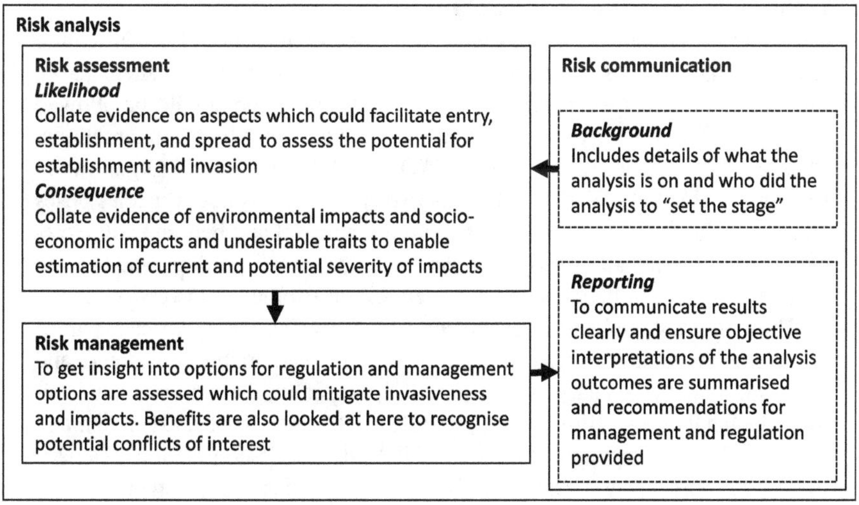

Fig. 20.3 Aspects of risk to be considered for the listing and regulation of alien species under national legislation in South Africa divided into risk assessment, risk management and risk communication based on a standard framework for risk analysis. The process suggested for risk management is shown in Fig. 20.2

in a South African context have been rather limited in scope. For example, van Wilgen et al. (2008) called for the use of risk assessments for the herpetofauna present in South Africa, mainly introduced through the pet trade. Alien fish have been extensively assessed for their risks in South Africa (Ellender and Weyl 2014; Marr et al. 2017), also necessitated by the conflicts between conservationists and recreational anglers (see Weyl et al. 2020, Chap. 6). An exception to this is the 'watch list' approach outlined above (Faulkner et al. 2014), which is applicable in any given region and can rapidly assess a wide variety of taxonomic groups. Most recently, we developed a risk analysis framework to underpin the regulation of alien taxa and aid management decisions (Kumschick et al. 2018). It is applicable to any taxonomic group and can be used pre- and post-border.

The need for risk analyses on biological invasions has been formally adopted in South African legislation (Box 20.1). The legislation outlines actions related to alien species that need risk analyses to be completed. This includes permits for the importation of new alien species into the country and to carry out restricted activities involving some listed species, the listing of alien species under national legislation (Fig. 20.3), and the introduction and subsequent release of alien species for biological control (see Hill et al. 2020, Chap. 19). For all these issues, processes are being instituted and frameworks developed in South Africa, which need to be followed to carry out the respective activities. In Table 20.1, we summarise the legislative needs for risk analysis in South Africa, how they are currently implemented, and provide recommendations for the future (for further details see the discussion below).

Box 20.1 Key International and South African Legislation Pertaining to the Risk Analysis of Biological Invasions

International agreements under the World Trade Organisation (WTO) require the assessment of risks before certain activities involving an alien taxon, especially trade, can be restricted, or before a new taxon should be allowed for import. These agreements recognise the standards set by the International Plant Protection Convention (IPPC; FAO 1996) and the World Organisation for Animal Health (OIE 2011). The assessment of the risk that a species poses allows for the distinction between potentially harmful and benign taxa. However, *risk analysis* additionally includes considerations regarding whether and how these risks can be managed (Convention on Biological Diversity 2002).

These international standards and agreements which South Africa is part of are complemented by legal obligations which need to be considered and followed where risk analysis is required. Specifically, the National Environmental Management: Biodiversity Act (NEM:BA, Act 10 of 2004) Alien and Invasive Species Regulations (in short, NEM:BA A&IS Regulations; Department of Environmental Affairs 2014) outline the necessary content of risk analyses for the application of permits, including for import and restricted activities.

Table 20.1 Risk analyses for alien species in South Africa as implemented and/or recommended based on international best practice

Type	Purpose of risk analysis	Process in South Africa	Recommended process
Import of alien species	To prevent the import of species with high potential of invasion and harm and allow import of beneficial species with low risk	From October 2014 as per the NEM:BA A&IS Regulations: detailed information required including on species' biology, ecology, behaviour and proposed activity, although there is no template for such information nor how such information is used to arrive at a risk analysis recommendation. The decision is made by the DEFF in consultation with other national and provincial departments. As of 2017, ASRARP has been acting as a scientific review body for such recommendations	Implementation of clear guidelines for risk analysis including standardised and transparent calculations of risk levels; review by a scientific body (e.g. ASRARP) set as a legal requirement. Decisions made by governmental committee chaired by DEFF.
Listing of species under national legislation	To underpin the regulation of alien species with scientific evidence in a transparent way, as required by international agreements	After 2004 a series of working groups met to provide expert opinion on the regulated lists. Recommendations and the rationale for decisions have not been clearly documented. Listing decisions are sent for public comment	Implementation of transparent decision-making process, including framework and guidelines as outlined in Figs. 20.2 and 20.3; review of analysis by ASRARP; decision for listing is made by government, and promulgated for public comment.
Permitting for restricted activities involving listed alien species (Category 2 under NEM:BA A&IS Regulations)	To ensure the safety of the proposed activity related to listed alien species and mitigate the risks posed	Risk analysis framework as outlined in the NEM:BA A&IS Regulations	Development and implementation of clear guidelines for risk analysis focussing on the nature of the proposed activity (risk management); development of permit conditions for all Category 2 listed species based on scientific evidence
Biological control	To ensure the safety of the release of biological control agents and limit non-target effects	Host specificity tests under quarantine; detailed release application reviewed by Biological control Release Committee	The current process has had an ideal safety record

Acronyms: *NEM:BA A&IS Regulations* National Environmental Management: Biodiversity Act 2004 (Act No. 10 of 2004) Alien and Invasive Species Regulations, 2014; *ASRARP* Alien Species Risk Analysis Review Panel; *DEFF* Department of Environment, Forestry, and Fisheries

20.2.1 Import of Alien Species

Several government departments administer import applications for alien species; on a national level those are the Department of Environment, Forestry, and Fisheries (DEFF) and the Department of Agriculture, Land Reform and Rural Development (DALRRD) mainly concerning new introductions into the country. For many other introductions of alien species and species new to a province, provincial departments are still largely responsible for the reviewing of import applications to introduce species from outside the country. They generally need to adhere to provincial legislation, but the processes followed vary significantly between provinces and taxa. Furthermore, they depend on the expertise available in the respective departments, and no consistent frameworks are followed to guide decision-making. In the following paragraphs, we mainly focus on the import of new alien species.

DALRRD (through South Africa's National Plant Protection Organisation, NPPOZA), is mandated to prevent the introduction, establishment and spread of alien pest or disease species on plants and plant products. DALRRD conducts pre-border risk analysis for agricultural species imported into South Africa. These analyses have a focus both on minimising the risks of introducing pathogens or pests on imported species and on the risks of the species themselves becoming invasive. The responsibility of DALRRD is to reduce risks of agriculture, but in the absence of other processes before the promulgation of the NEM:BA A&IS Regulations (see van Wilgen et al. 2020, Chap. 1, Box 1.1), DALRRD have issued permits for new introductions of non-agricultural alien species as well. This has not been sufficient to address all of the environmental risks associated with the introduction of new species, and the minimum standards for accepting introductions has historically been set differently for different organisms introduced for different purposes.

To address this, the NEM:BA A&IS Regulations were developed and DEFF mandated to issue permits for alien species not yet legally present in the country (i.e. new introductions) (Department of Environmental Affairs 2014). Every application for the import of a species not yet present in a country needs to be accompanied by a risk analysis according to international agreements (Box 20.1). The NEM:BA A&IS Regulations have since August 2014 outlined the information required for risk analyses accompanying import applications for the introduction of new species into South Africa (note these are currently termed "risk assessments" in the regulations, but are actually "risk analyses" as per the definitions in this chapter and in line with the international naming standards in Fig. 20.1). The requirements include information on the species to be imported, its taxonomy, biology and ecology, aspects linked to the likelihood of invasion like invasion history elsewhere, and traits linked to impact potential. This list includes several aspects that have been shown to be linked to invasion success and impact. Furthermore, it includes specific information regarding the planned activity in South Africa and measures proposed to manage the risks, which can be grouped under risk management. These risk analyses are the responsibility of applicants who wish to import new alien species, and the minimum requirements for assessors qualifications are specified in the NEM:BA A&IS Regulations, which makes the quality of analyses rather heterogeneous. Furthermore, while the regulations require detailed information on the biology and ecology of the species, the region and circumstances of its introduction and other relevant factors, they do not provide guidance on how to reach conclusions on the species' risk based on that information, or how to format the analysis (i.e. it is not actually a framework for risk analysis). Given the lack of clear guidelines on how to assess and weight each aspect, and how to reach a conclusion on the magnitude of risk, it is difficult to defend the accuracy and consistency when using this approach (Keller and Kumschick 2017). Further disadvantages of this approach are that it is time and resource-consuming.

To assist applicants for import permits to develop risk analyses and to render the decision-making process more transparent, a framework and guidelines for risk analysis have been developed—the Risk Analysis for Alien Taxa Framework (RAAT; Kumschick et al. 2018). The framework is based on international best-practice, and the guidelines cover most aspects related to the risk of alien species as outlined in the framework provided in the NEM:BA A&IS Regulations; information on the intended activities related to the permit application would need to be outlined separately. RAAT consist of 23–29 detailed questions. Answers to questions are given as scenarios to reduce assessor bias, and calculations provided to achieve risk levels (low, medium or high). The framework includes all aspects of risk as described in Fig. 20.3. In many developing countries like South Africa, decisions need to be made with limited funds and expertise available (see also Soliman et al. 2016 for southeast Asia). Even though RAAT can take several days to complete, it is less labour-intensive than collecting all the detailed information as outlined in the NEM:BA A&IS Regulations. Furthermore, conclusions regarding risk levels are less subjective as answer levels and calculations are outlined in detail. These guidelines are, however, not currently implemented for import applications (but see below for listing of species under the NEM:BA A&IS Regulations).

To support the DEFF in making decisions concerning the risks of biological invasions, the South African Alien Species Risk Analysis Review Panel (ASRARP) was formed. ASRARP is tasked with reviewing risk analyses attached to import applications to ensure they are scientifically robust and take into account the best available evidence, as well as risk analyses underpinning the listing of species under national legislation (see below and Fig. 20.4). ASRARP is an independent body (run through the South African National Biodiversity Institute) that incorporates scientists and taxon experts. It provides recommendations as to the validity and completeness of the information provided in the analysis, as well as recommendations regarding the outcomes of the application (i.e. whether to import and/or regulate a species). Recommendations are submitted to DEFF who then decide whether to grant the import permit (Fig. 20.4).

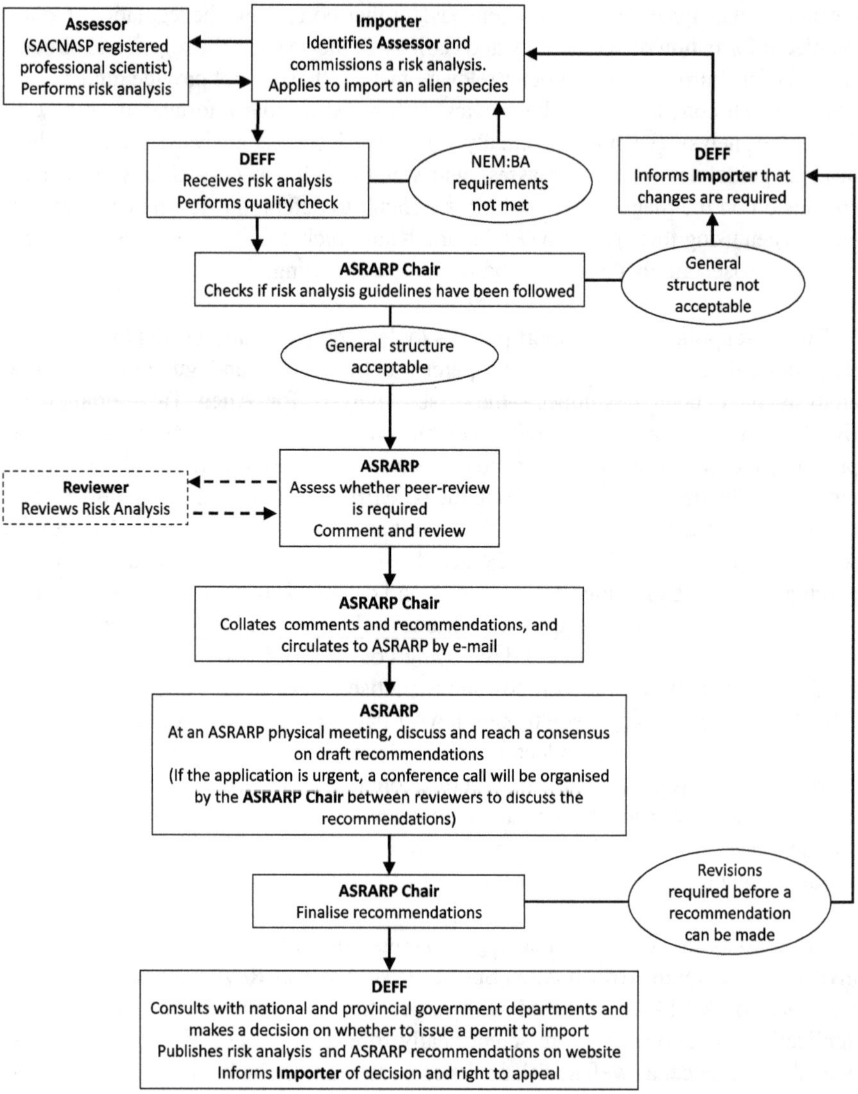

Fig. 20.4 Procedure for development of recommendations regarding the import of an alien species not currently legally in the country, specifically focusing on the review of applications by the Alien Species Risk Analysis Review Panel (ASRARP). Acronyms: *SACNASP* South African Council of Natural Science Practitioners, *DEFF* Department of Environment, Forestry and Fisheries, *NEM:BA* National Environmental Management: Biodiversity Act 2004 (Act No. 10 of 2004) Alien and Invasive Species Regulations, 2014

20.2.2 *Listing of Alien Species Under National Legislation*

There are two broad types of regulatory lists in South Africa, namely the prohibited list, which consists of species not yet present in the country, and a list of species which are in the country and need to be managed to reduce their impacts. The latter puts species in categories based on the benefits and the feasibility of control, namely: Category 1a—eradication targets; Category 1b—control targets needing a national management plan; Category 2—species requiring a permit for restricted activities (1b outside of permitted conditions—see next section for details); Category 3—species which can remain but need to be phased out (propagation to be prevented). The initial development of the lists of species regulated under national legislation included workshops with taxonomic experts to determine which species should be regulated and how (prohibited, or listed for control). Such workshops that include several experts can be very useful as they can reduce biases that could arise if a single expert was used (Sutherland and Burgman 2015). The inclusion of various stakeholders in these meetings was aimed at reducing conflicts and ensuring that stakeholders' voices are heard. However, in this case, decisions on which species and taxa to list were not based on a structured or transparent process, as no consistent framework for risk analysis was available. Each taxon-specific working group took a slightly different approach to develop recommendations on which species to regulate and how. The final decisions were, of course, taken by policy makers, but it is not clear why (and in some cases how) these decisions differed from the recommendations, as neither process was made publicly available.

Given the lack of transparency of the process leading to the regulation of certain alien taxa in South Africa, stakeholders have raised queries about the species that are regulated. As a result, various retrospective assessments and analyses have been carried out as emergency measures to respond to the criticisms, but none of those followed a framework specifically developed for South African conditions. A framework for decision-making related to the regulation of species needs to consider all aspects of risk and be explicit enough to transparently show the decision-making process and the evidence underpinning it (Roy et al. 2018). The Risk Analysis for Alien Taxa Framework—as described in the previous section—has been developed to this end (Kumschick et al. 2018). Besides the need to assess whether a species is of high risk, management feasibility and benefits of the species need to be assessed to underpin decisions regarding the listing category (Fig. 20.2). The benefits of using such a framework, as opposed to expert opinion alone, are that it leads to more transparent policy which is evidence-based and therefore more defendable. Furthermore, it provides stakeholders with reasons for management implications which should prevent queries from being raised.

20.2.3 *Permits for Restricted Activities*

Restricted activities, as defined in the NEM:BA A&IS Regulations, include, amongst others, possession, breeding, propagating or trading of listed alien species. Some alien species can be extremely harmful to the recipient environment, but at the same time have large benefits for some stakeholders. Such species are generally termed "conflict species" as they often lead to disputes around their management and control (Zengeya et al. 2017). On the one hand, once there is a large enough market for a species it is often difficult to enforce control. On the other hand, in many cases some of the impacts can be avoided and risks mitigated with risk management strategies. For example, keeping pets in escape-safe cages can ensure they do not get introduced into natural areas and cause harm to native species. Similarly, importing only males of certain species can (in most cases) prevent propagation and therefore the establishment of the species in the new area, even should some individuals escape. This issue is tackled with the requirement of permits for certain species (Category 2 listed species under the NEM:BA A&IS Regulations in South Africa; although permits to conduct research or biological control tests can be issued for work on taxa listed in other categories) and the development of permit conditions that mitigate the risks of the species. Each species regulated under national legislation would have gone through a thorough risk analysis process in order to be listed (see above). Therefore permit applications for restricted activities for some species mainly focus on risk management aspects including, for example, the nature and location of the planned activity and the number and sex of individuals to be used. This is similar to requirements for import permits for new alien species (see above). These considerations should be detailed enough to cover risk related to any of the activities involved, including for example transport from breeding facilities elsewhere. Permits already issued and refused for species listed as Category 2 under the NEM:BA A&IS Regulations in South Africa are listed in Table 20.2.

20.2.4 Non-regulated Alien Species

Various other activities would benefit from having frameworks and processes in place to prevent unintended negative consequences. This includes for example the commercialisation of alien species present in the country but not (yet) regulated under national legislation. At present, such species can be cultivated and traded without restrictions under the NEM:BA A&IS Regulations (although they are subject to other regulations like the Convention on International Trade in Endangered Species of Wild Fauna and Flora, CITES). Given that propagule pressure plays an important role in the establishment success of species, certain uses could enhance the chances of some species becoming invasive and causing harm (e.g. Kumschick et al. 2016). An example in South Africa to illustrate the potential issue is the "giant flag" project (www.giantflag.co.za), to build a South African flag consisting of colourful desert plants (including alien cacti). The alien cacti initially proposed to be used for this project were not considered as a particularly high risk in South Africa by the project co-ordinator, partly because of their main use as horticultural species and the low propagule pressure related to their use. However, the proposed project involves the planting of millions of cacti in their preferred habitat in South Africa, which changes the risk of these species becoming invasive due to the sheer number of individuals planted (Colautti et al. 2006).

Table 20.2 Number of permits issued for species listed as Category 2 under the NEM:BA A&IS Regulations in South Africa based on data up to December 2016 (van Wilgen and Wilson 2018)

Taxon	NEM:BA Category	Species	Number of permits granted	Number of permits refused
Terrestrial and freshwater plants (45 permits applied for)	1b	*Cryptostegia grandiflora* (Rubber Vine)	1	0
	2	*Pinus patula* (Patula Pine) and hybrids	3	0
		Acacia mearnsii (Black Wattle) and hybrids	6	0
		Acacia melanoxylon and hybrids, varieties and selections (Blackwood)	5	0
		Acacia dealbata (Green Wattle)	3	0
		Agave sisalana (Sisal)	1	0
		Nasturtium officinale (Watercress)	4	0
	Context specific	*Casuarina cunninghamiana* (Beefwood)	10	0
		Eucalyptus camaldulensis (River Red Gum) and hybrids	3	0
		Murraya paniculata (listed as *Murraya exotica* on the permit) (Orange Jessamine)	1	0
		Pinus pinaster (Cluster Pine) and hybrids	4	0
		Pinus radiata (Monterey Pine) and hybrids	4	0
Freshwater invertebrates (15 permits applied for)	2	*Cherax cainii* (Smooth Marron)	5	0
		Cherax tenuimanus (Hairy Marron)	8	2
Marine invertebrates (12 permits applied for)	1b	*Carcinus maenas* (Green Crab)	12	0
Freshwater fishes (121 permits applied for)	Context specific	*Ctenopharyngodon idella* (Grass Carp)	27	0
		Ctenopharynngodon idella (Tripod Grass Carp)	22	0
		Cyprinus carpio (Common Carp)	3	0
		Gambusia affinis (Mosquito Fish)	1	0
		Micropterus dolomieu (Smallmouth Bass)	1	0
		Micropterus salmoides (Largemouth Bass)	2	0
		Oreochromis niloticus (Nile Tilapia)	61	4
Reptiles (59 permits applied for)	2	*Basiliscus plumifrons* (Green Baslisk)	1	0
		Bitis gabonica rhinoceros (Gaboon Viper)	4	0
		Centrochelys sulcata (Spur-thighed Tortoise)	5	0
		Chelydra serpentina (Common Snapping Turtle)	2	0

(continued)

Table 20.2 (continued)

Taxon	NEM:BA Category	Species	Number of permits granted	Number of permits refused
		Gekko gecko (Tokay Gecko)	1	0
		Macrochelys temminckii (Alligator Snapping Turtle)	2	0
		Morelia spilotes (Carpet/Diamond Python)	14	0
		Python bivittatus (Burmese Python)	16	0
		Trachemys species (Slider turtles)	1	0
	Context specific	*Iguana iguana* (Green Iguana)	12	1
Birds	2	*Acridotheres fuscus* (Jungle Mynah)	2	0
(85 permits applied for)		*Psittacula krameri* (Rose-ringed Parakeet)	81	0
	Context specific	*Alectoris chukar* (Chukar Partridge)	2	0
Mammals	2	*Addax nasomaculatus* (Addax)	1	0
(318 permits applied for)		*Aepyceros melampus petersi* (Black-faced Impala)	1	0
		Ammotragus lervia (Barbary Sheep)	14	0
		Antilope cervicapra (Indian Blackbuck)	3	0
		Axis axis (Axis Deer/Chital)	7	0
		Axis porcinus (Hog Deer)	7	0
		Cervus elaphus (Red Deer)	3	0
		Dama dama (Fallow Deer)	71	0
		Hydrochaeris hydrochaeris (Capybara)	1	0
		Kobus ellipsiprymnus defassa (Defassa Waterbuck)	1	0
		Kobus leche kafuensis (Kafue Lechwe)	19	0
		Kobus leche leche (Red Lechwe)	165	1
		Oryx dammah (Oryx)	22	0
		Ovis aries musimon (Mouflon)	1	0
	Context specific	*Erythrocebus patas* (Pata's Monkey)	1	0

'Context specific' refers to species that are listed in several categories depending on the area in which they occur (regulation by area)

20.2.5 *Release Applications for Biological Control Agents*

A separate process to ensure that the introduction and release of classical biological agents for alien plants poses an acceptable environmental risk has been highly efficient (Moran et al. 2005). The levels of risk and protocols followed are based on international best practice, and largely determined by the research community itself. As a result, the process for assessing risks of the deliberate release of biological control agents against various invasive alien plants is very sophisticated and includes the collection of new data and conducting of experiments under quarantine over several years (Klein et al. 2011; for more information on biological control, see Hill et al. 2020, Chap. 19). These experiments include host-specificity tests to make sure the agent only feeds on or attacks the target species to be controlled and no other, closely related native plants and crops. All these experiments take place in secure quarantine facilities which reduces the risk of unintentional release of the species under investigation and allows for the screening of parasites and pathogens on the potential agent. Permit applications for the release of biological control agents, including an estimate of the likelihood and consequences of non-target effects and an assessment on the potential benefits of the introduction, are compiled by a dedicated team of scientists and submitted to the decision making body (in this case chaired by DALRRD). The applications including the risk assessments are then reviewed by an independent panel of experts similar to ASRARP, namely the National Biological Control Release Application Review Committee (Zachariades et al. 2017), and sent for review to international experts. The record of safety is enviable. Despite South Africa being among the top five countries globally with regard to the use of weed biological control, no direct significant non-target effects of weed biological control agents in South Africa have been detected over the past century (Moran et al. 2005).

20.3 Conclusions

South Africa's environment and biodiversity need to be protected from the negative impacts of biological invasions (see Le Maitre et al. 2020, Chap. 15; O'Connor and van Wilgen 2020, Chap. 16; Zengeya et al. 2020, Chap. 17). Risk analysis can aid towards reaching this goal insofar as it can help us distinguish the "good" from the "bad" alien species. Furthermore, there are clear economic benefits that come about by using risk analyses to underpin management decisions on alien species (e.g. Keller et al. 2007). This chapter provides guidelines on how to optimise the use of risk analysis to regulate imports and permitting of alien species and to underpin decisions regarding their management.

However, risk analyses have limitations. They can provide important information to prioritise species for management, but risk analyses are not, by themselves, prioritisation tools. Decisions as to which species to manage, and how, depend on the funds available, the costs of management, and many other factors that are not addressed in most risk analyses. Risk analyses are also not equivalent to, and cannot replace, management plans. While risk analyses compile much useful information for the management of a species, pathway, or area, additional information is needed and SMART (Specific, Measurable, Achievable, Realistic, and Timely) goals need to be set for management. Furthermore, as mentioned earlier, desktop-based risk analyses cannot replace detailed assessments of the feasibility of eradication (e.g. Kaplan et al. 2012; Jacobs et al. 2014), but they can provide useful information to aid prioritisation of potential eradication targets. South Africa has historically taken a pragmatic or reactive approach to the issue, but we believe that the solutions provided here offer an opportunity for more transparent and evidence-based decision-making.

Acknowledgements We thank Reuben Keller for his work on approaches to risk assessments, Kim Canavan for advice with biological control risk assessments, and Craig Whittington-Jones for a report on provincial processes related to import applications. We also thank the reviewers for helpful comments. The South African Department of Environment, Forestry, and Fisheries (DEFF) are thanked for funding the South African National Biodiversity Institute noting that this publication does not necessarily represent the views or opinions of DEFF or its employees.

References

Bacher S, Blackburn TM, Essl F et al (2018) Socio-economic impact classification of alien taxa (SEICAT) with data theft. Methods Ecol Evol 9:159–168. https://doi.org/10.1111/2041-210X.12844

Blackburn TM, Pyšek P, Bacher S et al (2011) A proposed unified framework for biological invasions. Trends Ecol Evol 26:333–339. https://doi.org/10.1016/j.tree.2011.03.023

Blackburn TM, Essl F, Evans T et al (2014) A unified classification of alien species based on the magnitude of their environmental impacts. PLoS Biol 12:e1001850. https://doi.org/10.1371/journal.pbio.1001850

Branquart E, D'hondt B, Vanderhoeven S et al (2016) From impact studies to management actions: practicing risk analysis of introduced trees. In: Krumm F, Vitkova L (eds) Introduced tree species in European forests: opportunities and challenges. European Forest Institute, Joensuu, pp 114–125

Colautti RI, Grigorovich IA, MacIsaac HJ (2006) Propagule pressure: a null model for biological invasions. Biol Invasions 8:1023–1037. https://doi.org/10.1007/s10530-005-3735-y

Convention on Biological Diversity (2002) COP 6 decision VI/23. The 6th conference of the parties of the convention on biological diversity decision VI/23, The Hague, 7–19 April 2002. https://www.cbd.int/decision/cop/?id=7197. Accessed 18 May 2018

Copp GH, Garthwaite R, Gozlan RE (2005) Risk identification and assessment of non-native freshwater fishes: concepts and perspectives on protocols for the UK. Cefas Science Technical Report No. 129. Cefas, Lowestoft. https://doi.org/10.1111/j.1439-0426.2005.00692.x

D'hondt B, Vanderhoeven S, Roelandt S et al (2015) Harmonia+ and Pandora+: risk screening tools for potentially invasive plants, animals, and their pathogens. Biol Invasions 17:1869–1883. https://doi.org/10.1007/s10530-015-0843-1

Department of Environmental Affairs (2014) National Environmental Management: Biodiversity Act 2004 (Act No. 10 of 2004) Alien and Invasive Species Regulations, 2014. Government Gazette 590(37885)

Ellender BR, Weyl OLF (2014) A review of current knowledge, risk and ecological impacts associated with non-native freshwater fish introductions in South Africa. Aquat Invasions 9:117–132. https://doi.org/10.3391/ai.2014.9.2.01

FAO (1996) Pest risk analysis for quarantine pests. International standards for phytosanitary measures publication no. FAO, Rome, p 11

Faulkner KT, Robertson MP, Rouget M et al (2014) A simple, rapid methodology for developing invasive species watchlists. Biol Conserv 179:25–32. https://doi.org/10.1016/j.biocon.2014.08.014

Hill MP, Moran VC, Hoffmann JH et al (2020) More than a century of biological control against invasive alien plants in South Africa: a synoptic view of what has been accomplished. In: van Wilgen BW, Measey J, Richardson DM, Wilson JR, Zengeya TA (eds) Biological invasions in South Africa. Springer, Berlin, pp 549–568. https://doi.org/10.1007/978-3-030-32394-3_19

International Maritime Organization (2004) International convention for the control and management of ships' ballast water and sediments, 2004. BWM/CONF/36. IMO, London

IUCN (2000) IUCN guidelines for the prevention of biodiversity loss caused by alien invasive species approved by 51st meeting of the IUCN Council, February 2000. http://www.issg.org/infpaper_invasive.pdf

Jacobs L, Richardson DM, Wilson JRU (2014) *Melaleuca parvistaminea* Byrnes (Myrtaceae) in South Africa: invasion risk and feasibility of eradication. S Afr J Bot 94:24–32. https://doi.org/10.1016/j.sajb.2014.05.002

Kaplan H, van Zyl HWF, Le Roux JJ et al (2012) Distribution and management of *Acacia implexa* in South Africa: is eradication an option? S Afr J Bot 83:23–35. https://doi.org/10.1016/j.sajb.2012.07.016

Keller RP, Kumschick S (2017) Promise and challenges of risk assessment as an approach for preventing the arrival of invasive species. Bothalia 47:131–138. https://doi.org/10.4102/abc.v47i2.2136

Keller RP, Lodge DM, Finnoff DC (2007) Risk assessment for invasive species produces net bioeconomic benefits. Proc Natl Acad Sci USA 104:203–207. https://doi.org/10.1073/pnas.0605787104

Klein H, Hill MP, Zachariades C et al (2011) Regulation and risk assessment for importations and releases of biological control agents against invasive alien plants in South Africa. Afr Entomol 19:488–497. https://doi.org/10.4001/003.019.0215

Kumschick S, Richardson DM (2013) Species-based risk assessments for biological invasions: advances and challenges. Divers Distrib 19:1095–1105. https://doi.org/10.1111/ddi.12110

Kumschick S, Bacher S, Dawson W et al (2012) A conceptual framework for prioritization of invasive alien species for management according to their impact. NeoBiota 15:69–100. https://doi.org/10.3897/neobiota.15.3323

Kumschick S, Devenish A, Kenis M et al (2016) Intentionally introduced terrestrial invertebrates: patterns, risks, and options for management. Biol Invasions 18:1077–1088. https://doi.org/10.1007/s10530-016-1086-5

Kumschick S, Wilson JR, Foxcroft LC (2018) Framework and guidelines for conducting risk analyses for alien species. Preprints. https://doi.org/10.20944/preprints201811.0551.v1

Le Maitre DC, Blignaut JN, Clulow A et al (2020) Impacts of plant invasions on terrestrial water flows in South Africa. In: van Wilgen BW, Measey J, Richardson DM, Wilson JR, Zengeya TA (eds) Biological invasions in South Africa. Springer, Berlin, pp 429–456. https://doi.org/10.1007/978-3-030-32394-3_15

Leung B, Roura-Pascual N, Bacher S et al (2012) TEASIng apart alien species risk assessments: a framework for best practices. Ecol Lett 15:1475–1493. https://doi.org/10.1111/ele.12003

Mandrak NE, Cudmore B (2004) Risk assessment for Asian carps in Canada. Department of Fisheries and Oceans Canada. Burlington, Ontario

Marr SM, Ellender BR, Woodford DJ et al (2017) Evaluating invasion risk for freshwater fishes in South Africa. Bothalia 47(2):150–159. https://doi.org/10.4102/abc.v47i2.2177

Montibeller G, Von Winterfeldt D (2015) Cognitive and motivational biases in decision and risk analysis. Risk Anal 35(7):1230–1251. https://doi.org/10.1111/risa.12360

Moran VC, Hoffmann JH, Zimmermann HG (2005) Biological control of invasive alien plants in South Africa: necessity, circumspection, and success. Front Ecol Environ 3:77–83. https://doi.org/10.2307/3868513

Nentwig W, Bacher S, Pyšek P, Vilà M, Kumschick S (2016) The generic impact scoring system (GISS): a standardized tool to quantify the impacts of alien species. Environ Monit Assess 188:315. https://doi.org/10.1007/s10661-016-5321-4

Novoa A, Kaplan H, Kumschick S et al (2015) Soft touch or heavy hand? Legislative approaches for preventing invasions: insights from Cactaceae in South Africa. Invasive Plant Sci Manag 8:307–316. https://doi.org/10.1614/IPSM-D-14-00073.1

Novoa A, Kumschick S, Richardson DM et al (2016a) Native range size and growth form in Cactaceae predicts invasiveness and impact. NeoBiota 30:75–90. https://doi.org/10.3897/neobiota.30.7253

Novoa A, Kaplan H, Wilson JRU et al (2016b) Resolving a prickly situation: involving stakeholders in invasive cactus management in South Africa. Environ Manag 57:998–1008. https://doi.org/10.1007/s00267-015-0645-3

O'Connor T, van Wilgen BW (2020) The impact of invasive alien plants on rangelands in South Africa. In: van Wilgen BW, Measey J, Richardson DM, Wilson JR, Zengeya TA (eds) Biological invasions in South Africa. Springer, Berlin, pp 457–486. https://doi.org/10.1007/978-3-030-32394-3_16

OIE (2011) Guidelines for assessing the risk of non-native animals becoming invasive. http://www.oie.int/fileadmin/Home/eng/Our_scientific_expertise/docs/pdf/OIEGuidelines_NonNativeAnimals_2012.pdf. Accessed 30 Aug 2017

Panetta FD, Timmins SM (2004) Evaluating the feasibility of eradication for terrestrial weed invasions. Plant Prot Q 19:5–11

Pheloung PC, Williams PA, Halloy SR (1999) A weed risk assessment model for use as a biosecurity tool evaluating plant introductions. J Environ Manag 57:239–251. https://doi.org/10.1006/jema.1999.0297

Renteria JL, Rouget M, Visser V (2017) Rapid prioritisation of alien plants for eradication based on climatic suitability and eradication feasibility. Austral Ecol 42:995–1005. https://doi.org/10.1111/aec.12528

Richardson DM, Le Roux JJ, Wilson JRU (2015) Australian acacias as invasive species: lessons to be learnt from regions with long planting histories. South For 77:31–39. https://doi.org/10.2989/20702620.2014.999305

Roy HE, Rabitsch W, Scalera R et al (2018) Developing a framework of minimum standards for the risk assessment of alien species. J Appl Ecol 55(2):526–538. https://doi.org/10.1111/1365-2664.13025

Scholes RJ, Schreiner GO, Snyman-van der Walt L (2017) Scientific assessments: matching the process to the problem. Bothalia 47:191–199. https://doi.org/10.4102/abc.v47i2.2144

Shackleton RT, Novoa A, Shackleton CM et al (2020) The social dimensions of biological invasions in South Africa In: van Wilgen BW, Measey J, Richardson DM, Wilson JR, Zengeya TA (eds) Biological invasions in South Africa. Springer, Berlin, pp 697–726. https://doi.org/10.1007/978-3-030-32394-3_24

Soliman T, MacLeod A, Mumford JD et al (2016) A regional decision support scheme for pest risk analysis in Southeast Asia. Risk Anal 36(5):904–913. https://doi.org/10.1111/risa.12477

Sutherland WJ, Burgman MA (2015) Use experts wisely. Nature 526(7573):317–318. https://doi.org/10.1038/526317a

Tricarico E, Vilizzi L, Gherardi F et al (2010) Calibration of FI-ISK, an invasiveness screening tool for nonnative freshwater invertebrates. Risk Anal 30:285–292. https://doi.org/10.1111/j.1539-6924.2009.01255.x

Tucker KC, Richardson DM (1995) An expert system for screening potentially invasive alien plants in South African Fynbos. J Environ Manag 44:309–338. https://doi.org/10.1016/S0301-4797(95)90347-X

van Wilgen BW, Richardson DM (2012) Three centuries of managing introduced conifers in South Africa: benefits, impacts, changing perceptions and conflict resolution. J Environ Manag 106:56–68. https://doi.org/10.1016/j.jenvman.2012.03.052

van Wilgen BW, Wilson JR (2018) The status of biological invasions and their management in South Africa in 2017. South African National Biodiversity Institute, Kirstenbosch and DST-NRF Centre of Excellence for Invasion Biology, Stellenbosch

van Wilgen NJ, Richardson DM, Baard EHW (2008) Alien reptiles and amphibians in South Africa: towards a pragmatic management strategy. S Afr J Sci 104:13–20

van Wilgen NJ, Roura-Pascual N, Richardson DM (2009) A quantitative climate-match score for risk-assessment screening of reptile and amphibian introductions. Environ Manag 44:590–607. https://doi.org/10.1007/s00267-009-9311-y

van Wilgen BW, Measey J, Richardson DM et al (2020) Biological invasions in South Africa: an overview. In: van Wilgen BW, Measey J, Richardson DM, Wilson JR, Zengeya TA (eds) Biological invasions in South Africa. Springer, Berlin, pp 3–30. https://doi.org/10.1007/978-3-030-32394-3_1

Weyl OLF, Ellender B, Wassermann RJ et al (2020) Alien freshwater fauna in South Africa. In: van Wilgen BW, Measey J, Richardson DM, Wilson JR, Zengeya TA (eds) Biological invasions in South Africa. Springer, Berlin, pp 151–182. https://doi.org/10.1007/978-3-030-32394-3_6

Wilson JRU, Gairifo C, Gibson MR et al (2011) Risk assessment, eradication, and biological control: global efforts to limit Australian acacia invasions. Divers Distrib 17:1030–1046. https://doi.org/10.1111/j.1472-4642.2011.00815.x

Wilson JR, Panetta FD, Lindgren C (2017) Detecting and responding to alien plant incursions. Cambridge University Press, Cambridge. https://doi.org/10.1017/CBO9781316155318

Wilson JR, Foxcroft LC, Geerts S et al (2020) The role of environmental factors in promoting and limiting biological invasions in South Africa. In: van Wilgen BW, Measey J, Richardson DM, Wilson JR, Zengeya TA (eds) Biological invasions in South Africa. Springer, Berlin, pp 353–384. https://doi.org/10.1007/978-3-030-32394-3_13

Woodford DJ, Richardson DM, MacIsaac HJ et al (2016) Confronting the wicked problem of managing biological invasions. NeoBiota 31:63–86. https://doi.org/10.3897/neobiota.31.10038

Woodford DJ, Ivey P, Jordaan MS et al (2017) Optimising invasive fish management in the context of invasive species legislation in South Africa. Bothalia 47:245–253. https://doi.org/10.4102/abc.v47i2.2138

Zachariades C, Paterson ID, Strathie LW et al (2017) Assessing the status of biological control as a management tool for suppression of invasive alien plants in South Africa. Bothalia 47:254–272. https://doi.org/10.4102/abc.v47i2.2142

Zengeya T, Ivey P, Woodford DJ et al (2017) Managing conflict-generating invasive species in South Africa: challenges and trade-offs. Bothalia 47:273–283. https://doi.org/10.4102/abc.v47i2.2160

Zengeya TA, Kumschick S, Weyl OLF et al (2020) An evaluation of the impacts of alien species on biodiversity in South Africa using different assessment methods. In: van Wilgen BW, Measey J, Richardson DM, Wilson JR, Zengeya TA (eds) Biological invasions in South Africa. Springer, Berlin, pp 487–512. https://doi.org/10.1007/978-3-030-32394-3_17

21

Alien Plant Control Projects and their Assessment

Brian W. van Wilgen, John R. Wilson, Andrew Wannenburgh and Llewellyn C. Foxcroft

Abstract Since 1995, the South African government has spent at least ZAR 15 billion (unadjusted for inflation; approximately USD1 billion) on alien plant control operations across South Africa. The amount spent per year has risen exponentially since 2010, and in 2019 annual spending is around ZAR 2 billion per year. Based on a small (but growing) number of case studies that have assessed management effectiveness, it is clear that the cover of invasive alien plants has been reduced in some localised areas, but continues to grow in others. A number of factors contribute to success, but the effort and resources required for successful control appear to be routinely under-estimated, with actual costs between 1.5 and 8.6 times higher than initial budget estimates. Currently, therefore, control measures (other than biological control) have largely failed to check invasions at a national scale, and there have been no documented eradications of plant invasions from continental South Africa. We argue that control can be considerably improved by effective prioritisation, goal-setting and planning; monitoring of outcomes rather than of inputs;

B. W. van Wilgen (✉)
Centre for Invasion Biology, Department of Botany and Zoology, Stellenbosch University, Stellenbosch, South Africa
e-mail: bvanwilgen@sun.ac.za

J. R. Wilson
Centre for Invasion Biology, Department of Botany and Zoology, Stellenbosch University, Stellenbosch, South Africa

South African National Biodiversity Institute, Kirstenbosch Research Centre, Cape Town, South Africa

A. Wannenburgh
Department of Environment, Forestry and Fisheries, Cape Town, South Africa

L. C. Foxcroft
Centre for Invasion Biology, Department of Botany and Zoology, Stellenbosch University, Stellenbosch, South Africa

Conservation Services, South African National Parks, Skukuza, South Africa ensuring that the existence of multiple goals does not lead to confusion over priorities; developing methods to reduce the under-estimation of the costs of control; adherence to best practices and standards; simplifying the currently complex contracting and employment models; and using a variety of methods to resolve or reduce conflicts over species that have commercial or other value, but cause significant environmental damage. Addressing these challenges will be difficult, but would be essential if plant invasions in South Africa are to be brought under control.

21.1 Introduction

Attempts to control invasive alien plants have a long history in South Africa. While we do not have detailed knowledge of early control efforts, regulations relating to invasive plant management date back to 1861 (see Lukey and Hall 2020, Chap. 18). The first biological control agents were introduced over 100 years ago (Moran et al. 2013). The government made attempts to control pines (*Pinus* species), gums (*Eucalyptus* species) and hakeas (*Hakea* species) in grassland and fynbos areas near Makhanda (Grahamstown) as early as the 1930s (Macdonald 2004). In 1943, operations were introduced to control invasive alien pines, wattles (Australian *Acacia* species) and gums in the Cape of Good Hope Nature Reserve on the Cape Peninsula (Macdonald et al. 1989). Attempts to control invasive alien plants in the Kruger National Park began in the 1950s (Foxcroft and Freitag-Ronaldson 2007). In 1968, legislation was enacted and an eradication programme initiated against *Solanum elaeagnifolium* (Satansbos), although eradication was never achieved (Wilson et al. 2013). In 1976, the Department of Forestry scaled up its efforts to control invasive alien plants in the mountain catchment areas in the Western Cape (Fugler 1983; Fenn 1980), but after a decade the programme fell behind schedule, and essentially came to a halt when the responsibility for managing catchment areas was transferred to the provinces in the late 1980s (van Wilgen and Wannenburgh 2016). In 1995, efforts to control invasive alien plants across the whole country were started afresh under the auspices of the Working for Water (WfW) Programme. This public works programme has the dual goals of controlling invasive alien plants while at the same time creating employment and development opportunities for disadvantaged people in rural areas (van Wilgen and Wannenburgh 2016). This chapter reviews the extent to which projects dealing with terrestrial plant invaders have been implemented across the country, and their costs and effectiveness. Chapter 4 (Hill et al. 2020a) discusses progress and challenges relating to the management of aquatic plant invaders.

21.2 Alien Plant Control Projects

Many conservation agencies at national and provincial level, private landowners, volunteer "hack" groups, and NGOs have implemented alien plant control projects (Fenn 1980; Attwell 1985; Macdonald et al. 1985; van Wilgen et al. 2017; van Rensburg et al. 2017). However, monitoring data for these efforts were either not available to us, or were never collected in the first place. This section therefore provides a brief summary of the extent of alien plant control projects funded by the WfW programme between 1995 and 2017, both because data are available, and because WfW has provided the bulk of funding for alien plant control projects over the past two decades.

Since 1995, WfW has spent ZAR 15 billion (unadjusted for inflation) on alien plant control operations across South Africa. The amount spent per year has risen exponentially since 2010, reaching around ZAR 2 billion per year in 2017 (Fig. 21.1a). During this time, WfW has cleared an average of about 200,000 condensed ha (Fig. 21.1c) per year, and conducted follow-up operations on about 600,000 ha per year (cleared areas are subjected to an average of three follow-up operations over time). The apparent decrease in area treated since 2014 is due to a relaxation of the requirement to record areas treated. This essentially means that recent figures are underestimates. In terms of the species targeted, wattles received more than three times the funding (ZAR 3.5 billion) than any other taxon (Table 21.1). The other groups on which large amounts have been spent include *Lantana camara* (Lantana), trees in the genera *Prosopis* and *Eucalyptus*, and *Chromolaena odorata* (Triffid Weed) (Table 21.1). WfW has dual goals, which require it to create employment and to clear invasive alien plants (van Wilgen and Wannenburgh 2016); the programme has created between 2000 and 23,000 full-time equivalent jobs per year.

Alien plant control operations funded by WfW are carried out by implementing agents who often operate on adjacent land parcels owned or managed by different agencies. Plant invasions, however, do not respect jurisdictional boundaries. To gain control over invasions in any given area, it would therefore be necessary for neighbouring landowners to collaborate closely with each other, and to co-ordinate control efforts, which brings additional challenges. Box 21.1 profiles case studies where alien plants are managed in large areas across several jurisdictions to illustrate challenges and achievements.

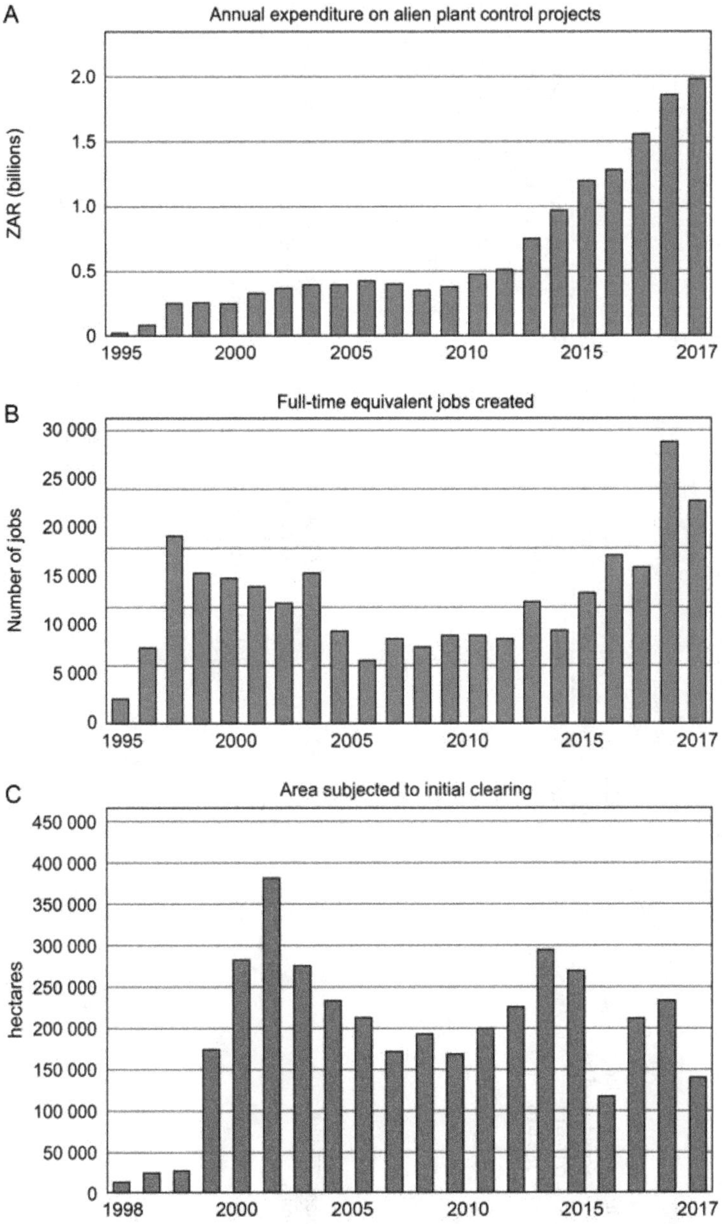

Fig. 21.1 The contribution of the Working for Water programme to alien plant control in South African in terms of: (**a**) the amount of money invested (ZAR unadjusted for inflation); (**b**) the number of full-time equivalent jobs created. This number only includes jobs created for previously unemployed people as part of poverty alleviation efforts; (**c**) the area treated per annum in condensed ha (see Table 21.1 for a definition of condensed ha). Note that the apparent decrease in area treated between 2013 and 2014 is due to a relaxation of the requirement to record areas treated. Data sourced from Working for Water Information Management System

Table 21.1 The top ten alien plant taxa targeted for control in South Africa since 1995, with estimates of area invaded, area treated, and rates of spread between 2000 and 2018

Taxon	Growth form	Estimated area invaded (condensed ha)[a]	Area subjected to initial clearing (condensed ha)	Area subjected to follow-up clearing (condensed ha)	QDGCs occupied (2000)	QDGCs occupied up to 2016 (% increase)[b]	Cost (millions of ZAR in 2018 value)	Notes
Acacia species (Australian wattles)	Trees and shrubs	719,912[c] 582,016[d]	73,516	159,999	428[1] 158[2] 166[3] 256[4] 101[5]	463 (8%) 168 (6%) 175 (5%) 302 (18%) 126 (25%)	3450	Widely planted and highly invasive group of trees and shrubs. Control is problematic as all species resprout when cut, and germinate *en masse* from vast reserves of soil-stored, long-lived seed banks. Species with established biological control agents have spread markedly less than those without biological control agents
Cactaceae (cacti)	Succulent shrubs	77,430[d]	14,211	32,353	0[6] 131[7] 10[8] 861[9] 25[10]	83 (NA) 208 (59%) 65 (550%) 1002 (16%) 99 (296%)	330	Cacti are among the most widespread and dominant groups of invasive plants in South Africa, with 35 species listed as invaders. Fifteen species are under effective biological control, but other species have recently started to spread rapidly
Campuloclinium macrocephalum (Pompom Weed)	Perennial herb	No estimate available	3506	597	14	108 (671%)	47	This species invades grasslands, and has recently spread rapidly
Chromolaena odorata (Triffid Weed)	Shrub	43,227[c]	9802	27,469	93	119 (28%)	722	A weed of the eastern coastal belt and escarpment

(continued)

Table 21.1 (continued)

Taxon	Growth form	Estimated area invaded (condensed ha)[a]	Area subjected to initial clearing (condensed ha)	Area subjected to follow-up clearing (condensed ha)	QDGCs occupied (2000)	QDGCs occupied up to 2016 (% increase)[b]	Cost (millions of ZAR in 2018 value)	Notes
Eucalyptus species (eucalypts)	Trees	62,949[c] 273,573[d]	9089	26,283	121[11]	195 (61%)	578	With the exception of *E. camaldulensis* and *E. saligna/grandis*, most species are not invasive (Forsyth et al. 2004). However, stands of eucalypts are often targeted for clearing in riparian zones to prevent impacts on water resources
Lantana camara (Lantana)	Shrub	69,268[c]	23,219	60,574	247	312 (26%)	1098	This shrub occurs in disturbed areas and along forest margins, in coastal and inland areas in the south and east of the country. Control efforts have been assisted by biological control
Parthenium hysterophorus (Parthenium Weed)	Annual herb	No estimate available	6598	10,607	15	89 (493%)	73	This species invades grasslands and savannas, and has recently spread rapidly. Control is problematic as this is an annual species
Pinus species (pine trees)	Trees	77,093[c] 480,331 uncondensed ha (Cape Floristic Region protected areas only, van Wilgen et al. 2016)	11,579	22,627	85[12] 70[13]	108 (27%) 95 (36%)	463	*Pinus* species are widespread invaders of the fynbos shrublands and grasslands in the Western and Eastern Cape. Control is problematic due to ongoing spread following repeated wildfires, invasion of rugged and inaccessible terrain, a lack of biological control, and resistance to the removal of forestry plantations that can act as seed sources for invasions (van Wilgen 2015)

Populus species (poplar trees)	Trees	15,253[c] 58,082[d]	1989	6765	185[14] 100[15]	197 (6%) 131 (31%)	151	Widely planted along drainage lines, spreads steadily by means of vegetative suckering
Prosopis species (Mesquite)	Trees	173,149[c] 1,473,951 uncondensed ha noted as invaded to some degree in the Northern Cape (van den Berg 2010)	17,767	61,428	40[16] 390[17]	112 (180%) 481 (23%)	765	Originally introduced and widely planted to provide an additional source of fodder for livestock, now a widespread invader such that the value of benefits is exceeded by the cost of impacts (Wise et al. 2012)

[a]Estimated area invaded (for the whole of South Africa unless otherwise noted) is "condensed ha", which is the equivalent area at a canopy cover of 100% (e.g. 50% cover over 100 ha = 50 condensed ha), based on data from [c]Versfeld et al. (1998) and [d]Kotze et al. (2010) except where otherwise specified. Note that the survey by Kotze et al. (2010) excluded most of the arid parts of South Africa (i.e. the Karoo, Grassland and arid Savanna Biomes in the west of the country).
[b]*QDGCs* quarter degree grid cells; QDGC data are from Henderson and Wilson (2017)
[1]*Acacia mearnsii*; [2]*Acacia saligna*; [3]*Acacia cyclops*;[4]*Acacia dealbata*; [5]*Acacia decurrens*; [6]*Cylindropuntia fulgida*; [7]*Cylindropuntia imbricata*; [8]*Opuntia engelmannii*; [9]*Opuntia ficus-indica*; [10]*Opuntia humifusa*; [11]*Eucalyptus camaldulensis*; [12]*Pinus pinaster*; [13]*Pinus radiata*; [14]*Populus alba/canescens*; [15]*Populus deltoids*; [16]*Prosopis glandulosa*; [17]*Prosopis* hybrids

Box 21.1 Co-ordinating Alien Plant Control Across Jurisdictions: Three examples from Biosphere Reserves in South Africa

Implementing effective alien plant control projects is difficult enough on individual protected areas or farms, but the complexity increases exponentially when an invasion occurs across multiple land parcels, owned or managed by different individuals or organisations, each with different purposes and levels of capacity. For the management of invasive alien plants to be effective, collaboration across different land parcels is needed. Biosphere reserves are a good model for how this can be achieved. Biosphere reserves are areas of terrestrial and coastal ecosystems that are internationally recognised within the framework of the United Nations Education, Scientific and Cultural Organisation's (UNESCO's) Man and Biosphere programme. Biosphere Reserves have core, buffer and transition zones that cater for strict conservation, limited sound ecological use, and ecologically-friendly development respectively. South Africa has eight Biosphere Reserves, and within some of these there have been attempts to co-ordinate alien plant control projects. An examination of three of these Biosphere Reserves (table below) reveals some common features:

- It is essential to have a dedicated and committed co-ordinator to provide direction and continuity;
- The disbursement of funds across multiple organisations increases the levels of bureaucracy, significantly slowing progress;
- There are no examples of comprehensive control plans that cover entire biosphere reserves, although there are attempts to foster collaboration;
- The funds required to address the problem over large areas are typically inadequate;
- The relative importance of different species differs according to land use, resulting in differences in priority across the area being managed; and
- Private landowners are obliged in terms of law to control invasive alien species, but the capacity to enforce the regulations is inadequate.
- The bulk of the funding comes from the government's Working for Water (WfW) programme.

Area and managing agencies	Features	Funding and planning	Achievements and challenges
Kogelberg Biosphere Reserve, Western Cape. Managed by: • CapeNature (provincial conservation agency) • Two municipalities (Cape Town and Overstrand) • Private landowners	The area covers 103,629 ha in the Fynbos Biome, Western Cape, including a core protected area (Kogelberg Nature Reserve), commercial forestry plantations with alien trees, and residential townships. The area is invaded by numerous alien plant species, mainly trees and shrubs in the genera *Pinus*, *Acacia* and *Hakea*.	The bulk of the funding comes from the government's Working for Water programme Municipalities contribute some funding additional to WfW. Several private landowners also contribute, but many do not There is a high-level alien plant control plan for the protected core area, and some groups of private landowners have developed separate plans for their own land.	Plant invasions in the core area have been brought down to a maintenance level; invasions on privately owned land remain a problem, and the cleared core area is at risk of re-invasion as a result Agreement has been reached between CapeNature and owners of pine forestry plantations for the systematic removal of plantations Uncontrolled wildfires frequently disrupt clearing operations, spreading the invasive plants Although some funding is available, it is insufficient to adequately address the problem.
Vhembe Biosphere Reserve, Limpopo. South African National Parks and provincial agencies are responsible for protected areas, but much of the 3 million ha of privately owned and communal land is not managed	The area covers ~3.7 million ha, with eight core protected areas totalling ~460,000 ha. It includes two national parks (northern region of Kruger National Park, and Mapungubwe National Park), and six provincial protected areas. There are over 120 invasive alien plant species in the area,	The bulk of the funding comes from the government's Working for Water programme Planning is largely carried out independently by landowners, with some coordinated planning between adjacent agencies	An Invasive Species Working and Network Group has been established for the Biosphere Reserve, but it has proved challenging to find an effective and dedicated co-ordinator Projects have been initiated to map the distribution of priority invasive alien species Species task teams have been established to co-ordinate alien plant control at a catchment or individual protected area level The complexity of

(continued)

	the most important of which are *Lantana camara* (Lantana), *Chromolaena odorata* (Triffid Weed) and *Senna* species in the lower catchment. Trees in the genera *Acacia* and *Eucalyptus* are prominent in the upper Soutpansberg		managing plant invasions over such a large area with varied land uses precludes effective coordination. However, coordinated communication at the biosphere scale, and cooperative planning at smaller scale, may overcome this
Kruger 2 Canyons Biosphere Reserve, Limpopo and Mpumalanga. South African National Parks, provincial conservation agencies, private conservation areas, municipalities, NGOs, and traditional leaders and communities. Alien plant control efforts across 21 organisations are monitored by a government environmental monitoring programme in an attempt to co-ordinate efforts	The area covers ~2.5 million ha, with a core area of ~898,300 ha of protected areas, including the central region of Kruger National Park, ten provincial protected areas, and large privately owned game farms (~400,000 ha). A wide variety of alien plants are present, the most important of which are *Parthenium hysterophorus* (Parthenium Weed) in lower-lying areas, and *Melia azedarach* (Syringa), and trees in the genera *Pinus, Eucalyptus* and *Acacia* in the upper catchments.	The bulk of the funding comes from the government's Working for Water programme SANParks and Kruger 2 Canyons collaborate to guide work in priority areas, but there are no formal management plans. Other stakeholders are kept informed through several committees.	Environmental monitors have been appointed to collect alien plant distribution data, for use by managers across different areas and projects A semi-formal supports the sharing of information Differing objectives (e.g. water conservation, biodiversity conservation, livestock production, and agriculture) result in different invasive species receiving priority in different areas. Buffer areas have been designated to allow close neighbours to align their objectives and efforts.

21.3 Alien Plant Eradication Projects

The Working for Water programme (WfW) explicitly took an area-based, as opposed to a species-based, approach to management (see Faulkner et al. 2020, Chap. 12, for a discussion on pathway-based approaches). Invaded areas were demarcated, estimates of the overall density of all alien plants in those areas were made, and contracts issued to clear all the alien plants in the specified areas. Classical biological control aside, individual species were not explicitly targeted. To address this gap, in 2008, the South African National Biodiversity Institute (SANBI) was contracted by WfW to develop species-specific control programmes focusing on alien plant species that were not yet widespread invaders (Wilson et al. 2013). Over time, this mandate has narrowed so that the focus of control efforts is on species where the possibility of national-scale eradication has not been ruled out (see Sect. 21.6 for a discussion of species-specific management programmes where the goal is not eradication).

There have been several documented plant eradication attempts in continental South Africa, all initiated by the government, but none of which succeeded (cf. work on the sub-Antarctic Islands, Greve et al. 2020, Chap. 8). Intensive programmes were initiated in the early 1960s to remove *Alhagi camelorum* (Camel Thorn) from irrigation schemes. However, the systemic herbicides available at the time were ineffective for dealing with the extensive underground root systems (Erasmus and Viljoen 1993; Jooste 1965). In 1968 an eradication programme was initiated against *Solanum elaeagnifolium* (Satansbos) (Wasermann et al. 1988). Despite some local successes, by 1972 the eradication campaign was cancelled; failure was ascribed to inadequate biological knowledge, ineffective herbicides and application techniques, and a lack of cooperation from many farmers. The most extensive eradication campaign in South Africa was against *Opuntia aurantiaca* (Jointed Cactus) (Moran and Annecke 1979), but despite significant governmental support, apparently not a single farm was fully cleared.

These efforts focussed on alien plants that were already widespread in the country, and basic requirements to achieve eradication were not always in place, e.g. no new immigration of propagules, all populations delimited, sufficient resources available to complete eradication, and adequate monitoring and evaluation in place (Wilson et al. 2017).

There are 42 alien plant species listed as Category 1a in South Africa's Alien & Invasive Species Regulations, published in 2014 under the National Environmental: Biodiversity Act (NEM:BA, Act 10 of 2004) (i.e. deemed as nation-wide eradication targets). On investigation, several of these species have been found to be present at many sites across the country (e.g. *Iris pseudacorus* (Yellow Flag) (Jaca and Mkhize 2015), and *Furcraea foetida* (Mauritian Hemp), although the formal process of documenting the evidence and transfer of these species to more appropriate management categories has not been completed yet (see Chap. 20; Kumschick et al. 2020). Several of the Category 1a species [*Cabomba caroliniana*, (Cabomba) and *Euphorbia esula* (Leafy Spurge)] have not been found again, and might simply not be in the country, while others are not known to have become invasive and were listed for precautionary purposes (Henderson and Wilson 2017). Several of the species are found in private gardens [e.g. *Triplaris americana* (Ant Tree)]. While this does not preclude eventual eradication, it complicates both control efforts and our ability to declare a species eradicated. A few Cactaceae species have (or will likely in future have) effective biocontrol agents, such that nation-wide eradication would probably not be required for adequate control to be effected. Consequently, only around a third of the species listed in Category 1a are still the focus of on-going eradication efforts. By contrast, many other taxa that are not yet listed in the regulations have been identified as eradication targets and are subject to control efforts [e.g. *Acacia viscidula* (Sticky Wattle), and *Melaleuca parvistaminea* (Rough-barked Honey Myrtle) (Magona et al. 2018; Jacobs et al. 2014)].

The mismatch between legal status and feasibility of eradication discussed above highlights the need to set eradication as the management goal only once a formal detailed assessment of eradication feasibility has been conducted. Such assessments require investment in delimitation and pilot control measures (Wilson et al. 2017). It is also clear that there is a substantial invasion debt in the country (Rouget et al. 2016)—many alien plants have only naturalised or invaded a few sites, and there are likely to be many that are still to be detected.

In the decade that the SANBI programme has been active, no alien plant species had been formally declared as eradicated. The project closest to achieving eradication is probably that against *Spartina alterniflora* (Smooth Cordgrass), a grass invading the Knysna estuary in the Western Cape. However, the conditions under which eradication can be declared have not been specified, nor is it clear why the plant was introduced in the first place, so the possibility of reintroduction has not been ruled out (Riddin et al. 2016). Detailed point patterns have been produced for a number of species (Wilson et al. 2013), and insights have been gained in terms of efforts to delimit populations (Jacobs et al. 2014), produce risk maps (Kaplan et al. 2014), estimate the costs of eradication (Moore et al. 2011), and the continuing need for morphological and molecular taxonomy (Magona et al. 2018; Jacobs et al. 2017).

The SANBI programme has funded postgraduate students to work on particular species or taxa, and produced an increasing number of published analyses of risk analyses, impact assessments, and estimates of eradication feasibility (Kumschick et al. 2020). However, the programme has suffered from similar issues to other projects funded under the WfW umbrella. The onus has been to report on input indicators (e.g. person days of employment), and few or no data are routinely collected on output indicators (e.g. the number of plants present). When assessed against the requirements set by the National Status Report on Biological Invasions (van Wilgen and Wilson 2018), the project planning was evaluated as being inadequate across the board. These are solvable issues, but will require a shift in approach to ensure that dedicated teams focus on specific targets year on year, that data are collected, and that monitoring data feed back into decision-making both at a project level and as input to the regulatory changes. If the global best practices regarding alien plant incursion response are applied (e.g. Wilson et al. 2017), then we can expect to see an increasing number of declared alien plant eradications in the next decade. For some taxa, particularly those with long-lived seed-banks (Zenni et al. 2009; Wilson et al. 2011), eradication might only be achieved far in the future, but it is feasible given persistence and effective monitoring.

21.4 Management Plans for Invasive Species

South Africa's National Environmental Management: Biodiversity Act (NEM:BA), Alien & Invasive Species (A&IS) Regulations, published in 2014, state that "*if an Invasive Species Management Programme has been developed in terms of section 75 (4) of the Act, a person must control the listed invasive species in accordance with such programme*". In many cases, the need for species-specific management programmes is clear, even for species where eradication is not feasible, but neither the Act nor the Regulations, provide guidance on which of the listed invasive species should be the subject of such a programme. The development of national-level, species-specific programmes for all listed species would be extremely onerous, but there has been little to no progress even on priority species. Species-specific strategies have been developed only for *Parthenium hysterophorus* (Parthenium Weed), and *Campuloclinium macrocephalum* (Pompom Weed) (Le Maitre et al. 2015; Terblanche et al. 2016; see also Table 21.1). These strategies recommended different management approaches for different administrative areas depending on the stage of invasion. In addition, two genus-level strategies [for Australian *Acacia* species and *Prosopis* (van Wilgen et al. 2011; Shackleton et al. 2017)], and one family-level strategy (for Cactaceae, Kaplan et al. 2017) have been developed.

None of these strategies has been formally adopted to date, and no entities have been established, as provided for in law, to co-ordinate and implement them (though the aim of the National Cactus Working Group is to facilitate the implementation of the Cactaceae strategy; Kaplan et al. 2017).

21.5 Management Plans for Invaded Areas

The successful implementation of invasive alien plant control projects relies on, among other things, careful planning that sets realistic goals, monitoring of progress towards those goals, and adapting management as new information comes to light. In South Africa, there are a number of statutory requirements to develop such plans. The management authorities of protected areas, and all other organs of state in all spheres of government are required in terms of the NEM:BA A&IS Regulations to prepare invasive alien species control plans; and in terms of the National Environmental Management: Protected Areas Act (Act 57 of 2003, NEM:PAA), the management authorities of all protected areas must submit a management plan for the protected area for ministerial approval. In turn, plans require accurate information on the extent and abundance of invasive species, so that the resources required to control them can be reliably estimated. These requirements have not been adhered to in practice, however. In terms of the NEM:BA requirement, submitted control plans covered only about 4% of the country, mainly in the Western Cape, and almost all of the plans failed to meet the required criteria (van Wilgen and Wilson 2018). Both the relatively small number of plans, and the inadequacy of many plans, was attributed to a lack of capacity or expertise within many organs of state (van Wilgen and Wilson 2018). Furthermore, while most protected areas have prepared management plans as required by NEM: PAA, the sections of these plans that deal with alien plant control are typically high-level, long-term statements of intent, and these have not been effectively carried forward into the more detailed medium to short-term plans that would be necessary for guiding control operations (van Wilgen et al. 2017).

Creating accurate maps of the distribution and abundance of alien plant invasions as a basis for realistic planning has also proved challenging up to now (Richardson et al. 2020, Chap. 3). At a national scale, there have been at least three attempts to map the extent of the problem. In 1993, the Council for Scientific and Industrial Research mapped invasive alien plants in South Africa, with the goal of estimating their impact at a national scale (Le Maitre et al. 2000). The mapping techniques used were coarse due to the paucity of reliable data, but a map at a 1:250,000 scale was produced, based primarily on the local knowledge of natural resource experts from across South Africa. The project estimated that invasive plants occupied 10.1 million ha (6.82% of South Africa and Lesotho). The longest-running project aimed at recording information on the national extent of alien plants is the Southern African Plant Invaders Atlas (SAPIA), which was initiated in 1994 (Henderson 2007). As of May 2016, SAPIA had over 87,000 geo-referenced records for 773 alien plant taxa that are present outside of cultivation in southern Africa, making it the most extensive source of information on the distribution of invasive plants in the region (see Richardson et al. 2020). In 2008, the Department of Environmental Affairs commissioned the Agricultural Research Council to develop and implement a repeatable sampling protocol to track trends in alien plant distribution and density across half of the country. This project has run for more than a decade, and has mapped the distribution of 27 alien plant taxa (species in the genera *Pinus*, *Eucalyptus* and some Australian *Acacia* species were mapped collectively). The project is ongoing, but no adequate description of the sampling methodology has been published to date, nor have any peer-reviewed papers that present the findings been published.

At finer scales, relatively detailed maps of the extent of invasion have been developed for some areas, mostly protected areas [e.g. Foxcroft et al. (2004, 2009) for Kruger National Park; Cheney et al. (2018) for Table Mountain National Park; and van Wilgen et al. (2016) for protected areas in the Cape Floristic region]. Cheney et al. (2018) compiled a map of invasive alien plants derived from fine-scale systematic sampling of the entire Table Mountain National Park (26,500 ha), and compared this to two other datasets in use for planning and management. They found that management datasets overestimated species cover by orders of magnitude, and that this resulted in questionable allocations of funding. They concluded that "*contrary to perception, fine-scale surveys are a cost-effective way to inform long-term monitoring programmes and improve programme effectiveness*". In addition, where plans are developed, they are not always followed (Kraaij et al. 2017). It appears, thus, that the level of planning for alien plant control in South Africa falls substantially short both of what is required by law and what is necessary for management to be effective.

21.6 National-Scale or Species-Specific Assessments of Management Effectiveness

The most comprehensive national-scale assessment of management efficacy at a species level to date (Henderson and Wilson 2017) was based on the Southern African Plant Invaders Atlas (SAPIA). The 773 alien plant taxa recorded in SAPIA was an increase of 172 taxa over the last assessment in 2006 (Henderson and Wilson 2017). Between 2000 and 2016 there was also an approximately 50% increase in the broad-scale documented range of alien plants in SAPIA. Several species (*Campuloclinium macrocephalum*, *Parthenium hysterophorus*, *Opuntia engelmannii*, *Cryptostegia grandiflora*, *Pennisetum setaceum*, *Tecoma stans*, *Sagittaria platyphylla*, *Gleditsia triacanthos*, and *Trichocereus spachianus*) were considered to be of particular concern as they had increased substantially in distribution over the past decade. Henderson and Wilson (2017) reported further that approximately 126 taxa were targeted for clearing by the Department of Environmental Affairs' Natural Resource Management (NRM) programmes (formerly "Working for Water") between 2000 and 2012, although most effort was focussed on a relatively small number of widespread taxa (Table 21.1). Examination of the data suggested that whether a species was targeted for control or not made little difference, as both targeted and neglected species continued to spread at comparable rates. Henderson and Wilson (2017) concluded that this outcome was perhaps not surprising, given the lack of evidence of a general strategic approach to NRM's activities, and the absence of dedicated strategic efforts to contain specific invasive plants, or to reduce the rate at which they invade particular areas. By contrast, they found a clear signal that biological control had reduced rates of spread of several important invasive alien plant species. Notably, however, SAPIA was not designed as a tool to monitor management effectiveness, but rather as a means of collating information on alien plant distributions and how that distribution has changed over time. For reliable assessments of management efficacy over time, SAPIA would need to be augmented by monitoring specifically designed for this purpose. There have been few examples of such monitoring to date.

A species-specific study on the integrated control of *Hakea sericea* (Sweet Hakea) was conducted in the Western Cape by Esler et al. (2010). The control included a combination of felling and burning, augmented by biological control (van Wilgen et al. 1992). Data from two surveys, 22 years apart, suggested that the distribution of the species was reduced by 64%, from ~530,000 to ~190,000 ha between 1979 and 2001. The species either decreased in density, or was eliminated from 492,113 ha, while it increased in density, or colonised 107,192 ha. It was concluded that the initial mechanical clearing, integrated with the judicious use of prescribed burning, in the 1970s and 1980s by the then Department of Forestry (van Wilgen et al. 1992) was responsible for reducing the density and extent of infestations, and that biological control was largely responsible for the failure of the species to re-colonise cleared sites, or to spread to new areas following unplanned wildfires

(Hill et al. 2020b, Chap. 19, Sect. 19.3). Between 2000 and 2015, *H. sericea* increased its occurrence in quarter degree grid cells from 77 to 85, an increase of 10% (Henderson and Wilson 2017). During the same period, the ecologically similar pine trees [*Pinus pinaster* (Cluster Pine) and *P. radiata* (Monterey Pine), for which no biological control is available] increased from 85 to 108, and from 70 to 95 QDGCs, or 27% and 21% respectively.

Marais et al. (2004) reported that good progress had been made with clearing certain species (at a cost of ~ZAR 2.3 billion between 1996 and 2004, costs unadjusted for inflation), but also that at current estimated rates of clearing, many of the targeted species would not be brought under control within the next half century. They stressed that their estimates were preliminary, given the incomplete data on the project management system. In 2012, van Wilgen et al. (2012) reported that control operations were in many cases only applied to a relatively small portion of the estimated invaded area (2–5% depending on the species), despite substantial spending (ZAR 3.2 billion in 2012 values).

21.7 The Effectiveness of Management in Selected Areas

21.7.1 Monitoring of Control Effectiveness

The effectiveness of control measures in a particular area (for example a protected area, a catchment area, a farm, or a stretch of river) needs to be assessed against the goal of the management, with such assessments based on regular monitoring of outcomes. However, while almost all alien plant control projects in South Africa have an implicit goal of reaching a "maintenance level", this goal is seldom stated explicitly in terms of the desired final extent or density of invasion (van Wilgen et al. 2016; Fill et al. 2017). The concept of a maintenance level recognises that, for most invasions, eradication is infeasible, but that invasions can be reduced to a level where the negative impacts are negligible and control costs are relatively low in perpetuity. In the vast majority of South Africa's government-funded alien plant control projects, the indicators used to monitor progress and set targets include the amounts of money to be spent, the number of people to be employed, and the areas to be treated. These are input or output indicators, rather than outcomes in terms of changes in the levels of plant invasions (Wilson et al. 2018). In the absence of a monitoring programme that is focussed on outcomes, it is difficult to assess effectiveness objectively. However, several studies have been conducted, particularly over the past decade, in which the effectiveness of management has been assessed, and these are summarised here. These studies provide a limited basis from which to derive broad conclusions about the effectiveness of control measures.

21.7.2 Alien Plant Control Projects in the Cape Floristic Region

Most studies addressing the effectiveness of alien plant control measures in South Africa have been carried out in the Cape Floristic Region (CFR) in the Western and Eastern Cape provinces. The natural vegetation of the CFR comprises fynbos shrublands, and the most prominent invasive species are trees and shrubs (Australian *Acacia*, *Hakea*, and *Pinus* species in particular). Historical costs for control in CFR protected areas between 1996 and 2015 amounted to ZAR 564 million (2015 values; van Wilgen et al. 2016), additional amounts spent outside of formally protected areas. When assessed at the scale of individual projects, there is clear evidence that progress has been made. Macdonald et al. (1989) recorded marked declines in cover of all alien species in the Cape of Good Hope Nature Reserve following the implementation of a systematic clearing plan in the 1970s. Similar declines were recorded in the Berg River catchment (Fill et al. 2017; Fig. 21.2), the Vergelegen Estate (van Rensburg et al. 2017; Fig. 21.3), and along the Rondegat River (Fill et al. 2018, see Sect. 21.9). In the Hawequas mountains, where control focussed on the removal of abandoned pine plantations, McConnachie et al. (2016) estimated that the cover of invasive trees would have been almost 50% higher had there been no control. They also concluded that control might have prevented a larger area from being invaded if it had focussed all of its effort on untransformed land, and not on abandoned plantations. However, the costs associated with many of these projects were much higher than originally estimated. McConnachie et al. (2012) concluded that the cost to clear the Krom and Kouga

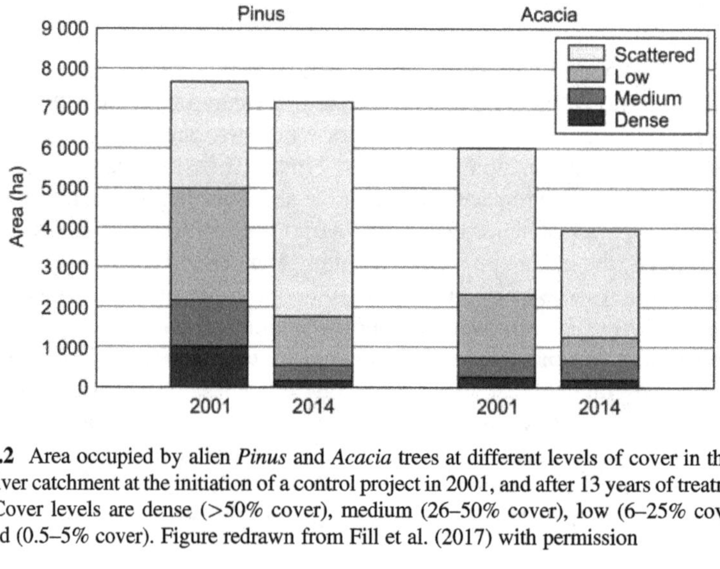

Fig. 21.2 Area occupied by alien *Pinus* and *Acacia* trees at different levels of cover in the upper Berg River catchment at the initiation of a control project in 2001, and after 13 years of treatments in 2014. Cover levels are dense (>50% cover), medium (26–50% cover), low (6–25% cover) and scattered (0.5–5% cover). Figure redrawn from Fill et al. (2017) with permission

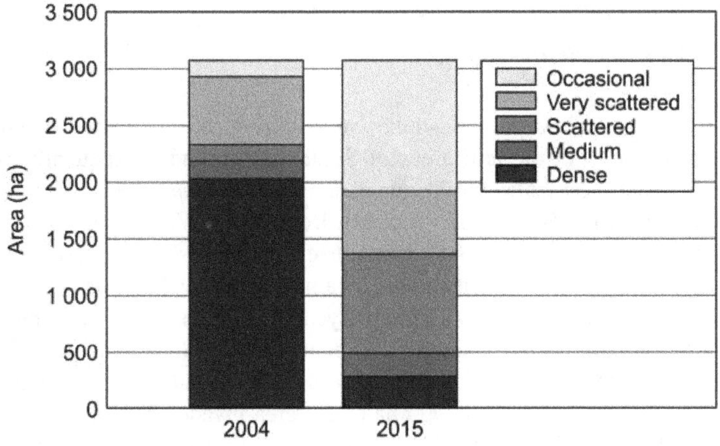

Fig. 21.3 Area occupied by invasive plants in six cover classes at Vergelegen Wine Estates in 2004 and 2015. The classes are occasional (<1% cover); very scattered (1–5% cover); scattered (5–25% cover); medium (25–50% cover); dense (50–75% cover); and closed (>75% cover). Figure redrawn from van Rensburg et al. (2017) with permission

catchments in the Eastern Cape was 2.4 times higher than the highest equivalent estimate made elsewhere in South Africa at the time. The cost to clear the Berg River catchment was estimated at ZAR 6 million in 1996 (2016 ZAR values; van Wilgen et al. 1997), but by 2016 the actual cost had reached ZAR 50 million (2016 ZAR values), 8.3 times the original estimate, without having reached a maintenance phase (Fill et al. 2017). Similarly, on privately-owned land at Vergelegen Estate, operations cost 3.6 times more than was originally estimated (ZAR 43.6 vs. 12.19 million respectively; van Rensburg et al. 2017). Much of this problem can be attributed to regular unplanned wildfires which necessitate large amounts of follow-up to clear seedlings that appear in dense stands after wildfires. However, in some cases, the additional costs may well be due to management inefficiencies. For example, McConnachie et al. (2012) found significant inefficiencies in the Krom and Kouga catchments, in the form of inaccurate records, where 25% of the areas recorded as having been cleared had in fact not been cleared; and Kraaij et al. (2017) found that the quality of many treatments in the Garden Route National Park was inadequate, with work done to standard in only 23% of the assessed area. The prognosis for gaining control of alien plant invasions in the CFR's protected area network was investigated by van Wilgen et al. (2016). The study concluded that, for scenarios in which control measures continued against all invasive plant species, the estimated required funding to achieve the goal of reducing invasions to a manageable level was up to 4.6 times greater than the amount spent over the past 20 years. Under many plausible future scenarios (for example 8% spread and current or reduced funding) the invaded area would continue to grow.

21.7.3 Management of Prosopis *Species in the Northern Cape*

Trees in the genus *Prosopis* (Mesquite) were introduced to provide a source of fodder for livestock in the arid areas of South Africa, and subsequently became invasive. Historical estimates for the rate of spread of *Prosopis* trees in South Africa ranged from 3.5 to 18% per year (van den Berg 2010; Wise et al. 2012), which implied that the invaded area could double every 5–8 years. In the Northern Cape, the estimated total invaded area increased by almost a million hectares between 2002 and 2007, which is equivalent to 27.5% per year, and this occurred at a time during which ZAR 390 million (2012 values) was spent on control (van Wilgen et al. 2012). A more recent update (RT Shackleton unpubl. data) found that the public works clearing projects had treated 203,000 ha of the area invaded by *Prosopis* between 2000 and 2015. Each site also received on average 2.7 follow up clearings. The cost of these measures amounted to ZAR 1.8 billion (unadjusted for inflation) over the same period. The project started in 1995, but cost estimates prior to the year 2000 are not available. Between 2000 and 2016, *Prosopis glandulosa* (Mesquite), and *Prosopis* hybrids increased their range from 40 to 112, and 390 to 481 quarter-degree grid cells, increases of 180% and 23% respectively (Henderson and Wilson 2017), suggesting that control is doing little to stop the spread of these trees.

21.7.4 Invasive Plant Control in the Kruger National Park (Mpumalanga and Limpopo Provinces)

Van Wilgen et al. (2017) provided a recent review of alien plant control in the Kruger National Park (KNP). There have been attempts at control in the KNP since the mid-1950s, but in the late 1990s these attempts were broadened, and between 1997 and 2016, over ZAR 300 million was spent on invasive alien plant control. Good progress was made with the control of several species, notably *Sesbania punicea* (Red Sesbania), *Opuntia stricta* (Australian Pest Pear), *Lantana camara* (Lantana) and two species of invasive alien aquatic plants. In all of these cases, progress with reducing populations of the invasive species was due to biological control. Nonetheless, much effort was also directed towards species that were subsequently recognised as being of lower priority. For example, 38% of available funds was spent on alien annuals between 1997 and 2016. Funds were sometimes directed towards these annuals to meet the goals of employment creation in areas where priority species were not present. The absence of documented assessments of the potential impact of various species also allowed managers to base their decisions on perceptions of the relative impact of candidate species. In addition, because management goals were focussed on inputs (funds disbursed, employment created) or outputs (area treated), there was a lack of monitoring the ecological outcomes of control operations.

21.7.5 Control of Chromolaena odorata *in the Hluhluwe-iMfolozi Park, KwaZulu-Natal*

In 1978, managers of the 90,000 ha Hluhluwe-iMfolozi Park (HiP) in KwaZulu-Natal first noticed the presence of the alien shrub *Chromolaena odorata* (Triffid Weed). By 2003, this species had increased in extent and covered almost half of the HiP (Dew et al. 2017). A concerted control programme was then implemented at a cost of ZAR 103 million, and by 2011 the invasions were reduced to acceptably low levels (Fig. 21.4). The success came about because the management team applied several aspects of best practice, including a dedicated "rapid response" team, the integration of fire and mechanical clearing, a focus on areas of low infestation, flexibility with regard to the deployment of teams, regular monitoring and generous funding. In addition, te Beest et al. (2017) reported that "*the team was only paid following completion of a contract and after a thorough inspection of the quality of the work by the Project Manager*". Many other control projects in South Africa unfortunately did not incorporate these features, and this may well account for the differences in success. This programme essentially focussed on a single species, and the control of other invasive taxa in the HiP was not documented.

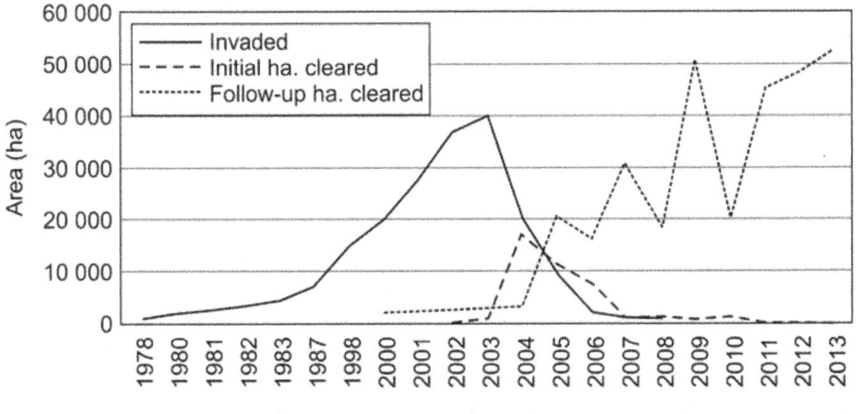

Fig. 21.4 Area invaded by *Chromolaena odorata* (Triffid Weed) in Hluhluwe-iMfolozi Park, and areas cleared and followed up between 2000 and 2013. Figure redrawn from te Beest et al. (2017) with permission

21.8 Managing Conflict Species

21.8.1 Conflict Species in South Africa

Conflict-generating invasive alien species are defined as species that have relatively high value for some people, while at the same time being capable of invading natural vegetation and generating high levels of negative impact (van Wilgen and Richardson 2014; Zengeya et al. 2017). There are several prominent examples of such species in South Africa (Table 21.2). The management of species that fall into this category is complicated because opposing value systems have to be accommodated. In South Africa, this issue has been addressed in a number of ways, discussed in the sections below.

21.8.2 Catering for Conflict Species in Regulations

Species listed as invasive in South Africa's A&IS Regulations have to be controlled, and may not be cultivated or traded. However, permits will be granted for some species (listed as Category 2) that have commercial value. These can be cultivated and traded under permit, but the permit-holder can be held liable for spread of the species. Some listed invasive species may be exempted from control requirements if there are many individual plants that have significant ornamental value (Category 3). These individual plants may be retained (e.g. in gardens), but may not be further cultivated, traded, or replanted (i.e. the species are phased out rather than attempting to actively remove them from private property). It is currently unclear whether or not these regulatory approaches are effective (van Wilgen and Wilson 2018).

21.8.3 Using Biological Control Agents to Reduce Seed Output

Proposals for the control of invasive Australian *Acacia* species were initially strongly resisted by the wattle industry (Stubbings 1977). Ecologists working in the field of biological control subsequently proposed the use of seed-feeding and gall-forming agents for these trees, and these were released following protracted negotiations with representatives of the wattle industry. These agents have been markedly successful in reducing seed output (Moran and Hoffmann 2012), and have substantially slowed the spread of these species in many areas (Henderson and Wilson 2017; Hill et al. 2020b, Chap. 19).

Table 21.2 Examples of conflict-generating invasive alien plant species or taxa in South Africa

Taxon	Growth form	Benefits	Impacts	Notes
Pinus species (pine trees)	Trees	Grown in plantations for timber	Reduces streamflow from catchment areas; negative impacts on biodiversity; increases in fire hazard	These trees make an important economic contribution in some parts of South Africa, and are regulated as Category 2 invasive alien species (may be cultivated under permit, otherwise must be controlled). They pose substantial threats to catchment areas in the Western and Eastern Cape.
Prosopis species (Mesquite)	Trees	Pods provide a source of fodder for livestock; firewood, charcoal and honey production	Reduces groundwater resources; negative impacts on biodiversity; forms impenetrable thickets and reduces rangeland carrying capacity	Widely promoted as fodder trees in arid parts of South Africa, where they are now serious invaders. Impacts probably outweigh benefits, and the impact: benefit ratio will grow as the trees spread (Wise et al. 2012). Regulated as Category 2 invasive alien species in the Northern Cape, and as Category 1 elsewhere (must be controlled).
Eucalyptus species (eucalypts)	Trees	Grown in plantations for timber; in woodlots for firewood; important source of pollen for bees	Reduces water resources from catchment areas and rivers; negative impacts on biodiversity	These trees make an important economic contribution in some parts of South Africa, and several are regulated as Category 2 invasive alien species. Some regulated species are not necessarily invasive (Forsyth et al. 2004), while others (e.g. *E. camaldulensis*) are widespread riparian invaders.
Australian *Acacia* species (wattles)	Trees	Grown in plantations for a wide range of purposes; prevention of erosion; firewood	Reduces water resources from catchment areas and rivers; negative impacts on biodiversity; forms impenetrable thickets and reduces rangeland carrying capacity	Proposals to introduce biological control for *Acacia* species in the 1970s met with stiff resistance from the wattle industry because of the commercial value of trees (Stubbings 1977). This has since been overcome through the deployment of non-lethal, seed-feeding and gall-forming insects.
Cactaceae (cacti)	Succulent shrubs	High ornamental value; planted as hedges; edible fruits	Negative impacts on biodiversity; harmful to livestock; reduces rangeland carrying capacity	Currently, 35 cactus species are listed as invasive aliens in South Africa; of these, 10 are targeted for eradication, and 12 are under substantial biological control. A strategic approach to the management of cacti has been developed (Kaplan et al. 2017).
Jacaranda mimosifolia (Jacaranda)	Tree	High ornamental value	Is invasive, but impacts not yet adequately documented	Iconic street tree of South Africa's capital city, Pretoria. Regulated as Category 3 invasive alien species (may be retained, but cannot be further traded or propagated).

21.8.4 Using Sterile Cultivars

The use of sterile cultivars (for example in the case of *Pinus* and Australian *Acacia* species used in commercial forestry) is sometimes proposed as a solution to the problem of invasions that originate from commercial plantations, but there are no documented cases of where this has been successful in South Africa. It has also been shown that large reductions in fecundity do not necessarily adequately reduce the population growth rates of long-lived species, which remain an invasion threat (Knight et al. 2011). In addition, while modern technologies such as genetic modification may be used to develop sterile varieties (Miao et al. 2012), forestry companies stand to lose environmental certification status as certifying bodies prohibit the use of genetically modified organisms (van Wilgen and Richardson 2012). Similarly, the use of sterile cultivars of horticultural species has been proposed as a means of reducing conflicts. There are, however, still several open research questions as to the nature and stability of sterility required to sufficiently reduce the risk, and whether sterility on its own would be sufficient to prevent invasions (Richardson and Petit 2005).

21.9 Returns on Investment from Control Measures

The economic costs of plant invasions, and the economic benefits of control, have also been the subject of a few studies in South Africa. One study (De Lange and van Wilgen 2010) suggested that the cost of some impacts (lost water, grazing and biodiversity) was currently about ZAR 6.5 billion per annum, but would become much higher as invasions grow. In the case of biological control of invasive plants, all studies have estimated very high returns on investment. By comparing the costs of biological control research and implementation to the benefits of restored ecosystem services, or avoided ecosystem degradation, and avoided ongoing control costs, biological control was shown to be extremely economically benefitial, with estimated benefit:cost ratios ranging from 8:1 up to 3726:1 (van Wilgen and De Lange 2011).

To estimate a return on investment from mechanical and chemical alien plant control measures, it would be necessary to know both the historic cost of control, and the value of impacts avoided due to control. There are no reliable estimates of the value of impacts avoided due to control. It is well known, though, that the cost per unit area to control an invasion rises exponentially as the density of the invasion increases (Marais et al. 2004). If invasions can be contained while they still occupy a smaller area, at relatively low densities, returns on investment from control operations should be positive. At some point, as yet unquantified, the cost of effective control would exceed the cost of the impact, and attempts to mechanically or chemically control invasions at this stage would deliver negative returns on investment (Fig. 21.5). More research is needed on this aspect to gain a better understanding.

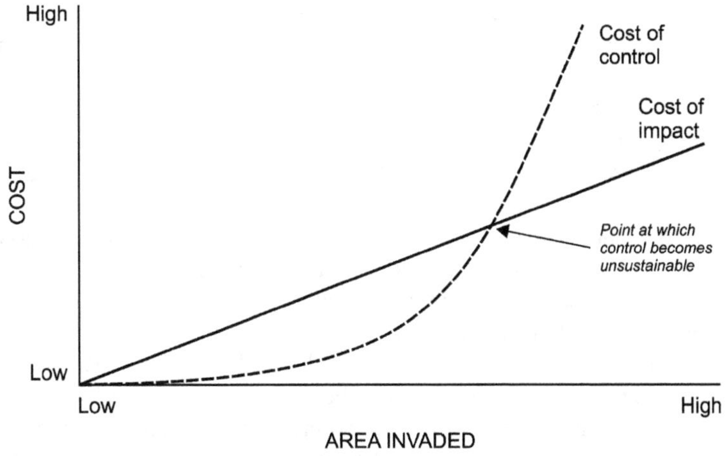

Fig. 21.5 Hypothetical representation of increases in the costs of impact, and the costs of control, associated with alien tree invasions. The cost of control increases exponentially as the invaded area and the density and size of trees increase. Control becomes economically unsustainable at the point at which the costs of control are exceeded by the costs suffered as a result of invasion (From van Wilgen and Richardson 2014). There are, of course, many other potential forms of these relationships—impact is often negligible at low densities, but rises exponentially once a threshold has been passed, while there are often fixed costs to controlling an area (related e.g. to issues of access and the minimum size of a control team). Of particular concern is that in general by the time an invasion has very obvious impacts, the cost of control is already several fold greater than the cost of pre-emptive management. If control costs vastly exceed impact costs (as per the far right of the graph), it has been proposed that control should not be attempted, and focus should rather be placed on deriving benefits from such "novel ecosystems". There are, however, various criticisms of this concept, e.g. that it might encourage managers to give up when it would be preferable to implement control

There have been mixed findings regarding returns on investment from alien plant control projects. In a cost-benefit analysis of six sites in the Eastern Cape, Hosking and Du Preez (2004) concluded that "*catchment management on all the sites carried out by the Working for Water Programme is inefficient*"; benefit:cost ratios ranged between 0.03 and 0.75, indicating negative returns on investment, although though the benefits of associated employment creation were not included. By modelling the spread of alien plants and their effects on water runoff, with and without attempts at control, in the Western Cape's Berg River catchment, van Wilgen et al. (1997) concluded that such control would be "*effective and efficient*". The estimated delivery cost of water, with and without the management of alien plants, was 57 and 59 c kl^{-1} respectively. The projected clearing costs used in arriving at this estimate were around ZAR 180,000 per year for initial clearing over 10 years, followed by about ZAR 25,000 per year for maintenance thereafter (1997 ZAR values). The actual costs eventually amounted to almost ZAR 50 million by 2015 (2015 ZAR; 8.3 times greater than the net present value of costs estimated in 1997, Fill et al. 2017). Despite considerable reductions in the cover of alien plants by 2015, the invasions were still present over much of the area, albeit at reduced densities.

The outcome that was projected in 1997 had therefore not been realised, because control methods were not effectively applied, and because the control costs were underestimated (Fill et al. 2017).

Finally, the potential returns on investment from invasive plant control operations have been the subject of several recent studies that compared the outcomes of various management scenarios (Vundla et al. 2016; Mudavanhu et al. 2016; Morokong et al. 2016; Nkambule et al. 2017). The scenarios included different rates of spread, included or excluded value-added products using biomass from invasive plants, and included or excluded private sector co-funding. The inclusion of co-funding and value-added products delivered more favourable returns on investment, and a failure to intervene at all would deliver negative outcomes. These operations could therefore be financially viable, but the accuracy of the predictions depends on whether or not the underlying assumptions will hold. These assumptions include that effective and professional clearing would continue into the future; that co-financing would be available; that due compensation for the services rendered and the value-added products produced would be realised; that the extent of the invasions was accurately known; and that the costs had been accurately estimated. Most, or even all, of the above assumptions will not hold, however, because alien plant invasions are rarely accurately mapped (see, for example, Cheney et al. 2018); the costs of control are routinely under-estimated by a factor of 3–7 times; there are low levels of efficiency associated with control work; and including value-added products could lead to unintended consequences. Consequently, there can only be a low level of confidence in these predictions of the return on investment from control projects.

21.10 Synthesis

There are a number of points that can be made with regard to the effectiveness of mechanical and chemical control measures. Firstly, as widespread invasions by alien plants can bring about substantial costs, it would obviously be beneficial to reduce invasions as far as possible. In South Africa, the largest proportion of funding for control operations comes from the Working for Water programme (WfW) within the Department of Environmental Affairs. Between 1995 and 2017, WfW spent ZAR 15 billion (unadjusted for inflation) on alien plant control, but this has only been enough to deal with between 2 and 5% of the estimated extent of invasions each year, and so the most important invasive species continue to spread (van Wilgen et al. 2012; Henderson and Wilson 2017).

Control interventions have nonetheless succeeded in reducing the extent of invasions in some areas. Early work demonstrated that the systematic implementation of a careful plan resulted in the reduction of populations of invasive alien trees and shrubs to maintenance levels (Macdonald et al. 1989). Where concerted efforts have been made to remove invasive trees from fynbos catchment areas, marked declines in the density have been achieved (Fill et al. 2017; van Rensburg et al. 2017). One estimate (McConnachie et al. 2016) suggested that the invaded area in the fynbos-clad Hawequas mountains would have been almost 50% higher if control operations had not been carried out. Ongoing control has also reduced the extent of invasions of several species in savanna ecosystems, including *Lantana camara*, *Opuntia stricta*, and *Chromolaena odorata* (van Wilgen et al. 2017; Dew et al. 2017; te Beest et al. 2017). At several localised sites, therefore, control measures have been effective. The picture changes when progress is assessed at a national scale, however, because plant invasions have generally continued to grow, some substantially (Henderson and Wilson 2017). Meaningful progress in reducing widespread invasions to a maintenance level, therefore, can arguably only be made if the available funding is focused on priority sites and species (Box 21.2). Essentially, the conscious practice of conservation triage (Bottrill et al. 2008) will need to be introduced, and this will require agreement on which species, and which areas, to target for control. Because alien plants spread more rapidly than they are being removed, current control efforts could fail if funds are spread too thinly, as suggested by modelling exercises (Higgins et al. 1997; van Wilgen et al. 2016). If adequate funding were re-directed to agreed priority areas, then the chances of achieving control in those areas would increase. Similarly, by focussing on priority species, scarce funds could be concentrated where they would be most effective. For example, funding for the control of *Pinus* and Australian *Acacia* species in fynbos is divided equally between these two taxa (van Wilgen et al. 2016). *Pinus* species, however, will eventually cover a much larger area than *Acacia* species if allowed to spread. If funds were diverted from *Acacia* species (which are under more effective biological control) to *Pinus*, then the eventual outcome would be far more favourable. Although prioritisation studies have already been initiated (e.g. Forsyth et al. 2012), it is going to be challenging to get managers to accept the need for triage, because terminating projects where funds have already been expended will understandably meet with resistance (see also Foxcroft et al. 2020, Chap. 28, Sect. 28.7). In conclusion, the implementation of focused, well-funded and well-managed control measures should bring invasions down to a maintenance level in many priority areas. Such interventions should bring very attractive returns on investment, but they will require some fundamental changes to the current *modus operandus*.

Box 21.2 Wisdom from the Past

Alien plant control has been considered, and practiced, in South Africa for many decades. The question arises as to whether we can learn from this experience. Two quotes, in particular, seem relevant.

In his report on the conservation of the vegetation of the Cape Floristic Region, Prof. C.L. Wicht noted that *"it seems, at present, that unless enormous sums of money are expended on their [invasive alien plant] eradication or control they will become dominant everywhere except in nature reserves and other selected areas where they will constantly be destroyed"*

(continued)

Box 21.2 (continued)
(Wicht 1945). There are two important aspects to this quote. First, there was an assumption that attempts to bring alien plant invasions down to maintenance levels would focus only on protected areas, and second, that a focus on protected areas would be the only way in which at least a representative portion of the unique Cape vegetation could be retained. In essence, this was an early call for the need to practice conservation triage. This concept (whereby some invaded land is abandoned to invasions, with control focussing on areas where progress can be made) has recently re-emerged (Bottrill et al. 2008; van Wilgen et al. 2016), but is controversial.

In the 1970s, the Department of Forestry embarked on an ambitious campaign aimed at eliminating invasive alien Hakea shrubs from vast areas in the southwestern mountains of what was then the Cape Province. John Fenn, regional director of forestry in the Western Cape, noted in 1980 that *"By using new methods and techniques, the brush-cutter became the most useful and effective machine for these operations. The costs of eradicating dense areas of Hakea dropped dramatically. Labour units used per hectare dropped from 22 to 5, and in certain areas, dropped as low as 1.5 units per hectare. All of a sudden the clearing of these vast areas of Hakea no longer looked impossible"* (Fenn 1980). This is a clear indication that mechanised techniques are essential if the goals of reaching a maintenance level are to be achieved. However, the current practice of relying on manual and labour-intensive clearing (to maximise employment opportunities) has reduced efficiency levels, resulting in a lack of progress towards the goals of reaching a maintenance level (Fill et al. 2017).

Acknowledgements BvW and JRW thank the DSI-NRF Centre for Invasion Biology and the South African National Research Foundation for support (grants 109467, 85412 and 86894). LCF acknowledges South African National Parks, the DSI-NRF Centre for Invasion Biology, Stellenbosch University, and the National Research Foundation of South Africa (Grant Numbers IFR2010041400019 and IFR160215158271). The South African Department of Environment, Forestry, and Fisheries (DEFF) are thanked for funding the South African National Biodiversity Institute noting that this publication does not necessarily represent the views or opinions of DEFF or its employees.

References

Attwell J (1985) Alien plant control by volunteer groups in the fynbos biome. In: Macdonald IAW, Jarman ML, Beeston P (eds) Management of invasive alien plants in the fynbos biome. South African national scientific programmes report no. 111, CSIR, Pretoria

Bottrill MC, Joseph LN, Carwardine J et al (2008) Is conservation triage just smart decision making? Sci Soc 23:649–654. https://doi.org/10.1016/j.tree.2008.07.007

Cheney C, Esler KJ, Foxcroft LC et al (2018) The impact of data precision on the effectiveness of alien plant control programmes: a case study from a protected area. Biol Invasions 20:3227–3241. https://doi.org/10.1007/s10530-018-1770-8

De Lange WJ, van Wilgen BW (2010) An economic assessment of the contribution of biological control to the management of invasive alien plants and to the protection of ecosystem services in South Africa. Biol Invasions 12:4113–4124. https://doi.org/10.1007/s10530-010-9811-y

Dew LA, Rozen-Rechels D, le Roux E et al (2017) Evaluating the efficacy of invasive plant control in response to ecological factors. S Afr J Bot 109:203–213. https://doi.org/10.1016/j.sajb.2016.12.007

Erasmus DJ, Viljoen BD (1993) Screening and sequential application of herbicides for the control of *Alhagi camelorum* Fisch. Agricultural Research Council, Pretoria

Esler KJ, van Wilgen BW, te Roller KS et al (2010) A landscape-scale assessment of the long-term integrated control of an invasive shrub in South Africa. Biol Invasions 12:211–218. https://doi.org/10.1007/s10530-009-9443-2

Faulkner KT, Burness A, Byrne M et al (2020) South Africa's pathways of introduction and dispersal and how they have changed over time. In: van Wilgen BW, Measey J, Richardson DM, Wilson JR, Zengeya TA (eds) Biological invasions in South Africa. Springer, Berlin, pp 311–352. https://doi.org/10.1007/978-3-030-32394-3_12

Fenn JA (1980) Control of Hakea in the western Cape. In: Neser S, Cairns ALP (eds) Proceedings of the third national weeds conference of South Africa. Balkema, Cape Town, pp 167–173

Fill JM, Forsyth GG, Kritzinger-Klopper S et al (2017) An assessment of the effectiveness of a long-term ecosystem restoration project in a fynbos shrubland catchment in South Africa. J Environ Manag 185:1–10. https://doi.org/10.1016/j.jenvman.2016.10.053

Fill JM, Kritzinger-Klopper S, van Wilgen BW (2018) Challenges to the restoration of ecosystems through multiple alien species control: lessons from the Rondegat River, South Africa. Restor Ecol 26:434–438. https://doi.org/10.1111/rec.12585

Forsyth GG, Richardson DM, Brown PJ et al (2004) A rapid assessment of the invasive status of Eucalyptus species in two South African provinces. S Afr J Sci 100:75–77

Forsyth GG, Le Maitre DC, van Wilgen BW et al (2012) The prioritisation of invasive alien plant control projects using a multi-criteria decision model informed by stakeholder input and spatial data. J Environ Manag 103:51–57. https://doi.org/10.1016/j.jenvman.2012.01.034

Foxcroft LC, Freitag-Ronaldson S (2007) Seven decades of institutional learning: managing alien plant invasions in the Kruger National Park, South Africa. Oryx 41:160–167. https://doi.org/10.1017/S0030605307001871

Foxcroft LC, Rouget M, Richardson DM et al (2004) Reconstructing 50 years of *Opuntia stricta* invasion in the Kruger National Park: environmental determinants and propagule pressure. Divers Distrib 10:427–437. https://doi.org/10.1111/j.1366-9516.2004.00117.x

Foxcroft LC, Richardson DM, Rouget M et al (2009) Patterns of alien plant distribution at multiple spatial scales in a large national park: implications for ecology, management and monitoring. Divers Distrib 15:367–378. https://doi.org/10.1111/j.1472-4642.2008.00544.x

Foxcroft LC, van Wilgen BW, Abrahams B, Esler KJ, Wannenburgh A (2020) Knowing-doing continuum or knowing-doing gap? Information flow between researchers and managers of biological invasions in South Africa. In: van Wilgen BW, Measey J, Richardson DM, Wilson JR, Zengeya TA (eds) Biological invasions in South Africa. Springer Berlin, pp 827–850. https://doi.org/10.1007/978-3-030-32394-3_28

Fugler SR (1983) The control of silky hakea in South Africa. Bothalia 14:977–980. https://doi.org/10.4102/abc.v14i3/4.1272

Greve M, von der Meden CEO, Janion-Scheepers C (2020) Biological invasions in South Africa's offshore sub-Antarctic territories. In: van Wilgen BW, Measey J, Richardson DM, Wilson JR, Zengeya TA (eds) Biological invasions in South Africa. Springer, Berlin, pp 205–226. https://doi.org/10.1007/978-3-030-32394-3_8

Henderson L (2007) Invasive, naturalized and casual alien plants in southern Africa: a summary based on the Southern African Plant Invaders Atlas (SAPIA). Bothalia 37:215–248. https://doi.org/10.4102/abc.v37i2.322

Henderson L, Wilson JRU (2017) Changes in the composition and distribution of alien plants in South Africa: an update from the Southern African Plant. Bothalia 47:a2172. https://doi.org/10.4102/abc.v47i2.2172

Higgins SI, Turpie JK, Costanza R et al (1997) An ecological economic simulation model of mountain fynbos ecosystems—dynamics, valuation and management. Ecol Econ 22:155–169. https://doi.org/10.1016/S0921-8009(97)00575-2

Hill MP, Coetzee JA, Martin GD et al (2020a) Invasive alien aquatic plants in South African freshwater ecosystems. In: van Wilgen BW, Measey J, Richardson DM, Wilson JR, Zengeya TA (eds) Biological invasions in South Africa. Springer, Berlin, pp 95–112. https://doi.org/10.1007/978-3-030-32394-3_4

Hill MP, Moran VC, Hoffmann JH et al (2020b) More than a century of biological control against invasive alien plants in South Africa: a synoptic view of what has been accomplished. In: van Wilgen BW, Measey J, Richardson DM, Wilson JR, Zengeya TA (eds) Biological invasions in South Africa. Springer, Berlin, pp 549–568. https://doi.org/10.1007/978-3-030-32394-3_19

Hosking SG, Du Preez M (2004) A cost-benefit analysis of the Working for Water Programme on selected sites in South Africa. Water SA 30:143–152. https://doi.org/10.4314/wsa.v30i2.5059

Jaca T, Mkhize V (2015) Distribution of *Iris pseudacorus* (Linnaeus, 1753) in South Africa. BioInvasions Rec 4:249–253. https://doi.org/10.3391/bir.2015.4.4.03

Jacobs LEO, Richardson DM, Wilson JRU (2014) *Melaleuca parvistaminea* Byrnes (Myrtaceae) in South Africa: invasion risk and feasibility of eradication. S Afr J Bot 94:24–32. https://doi.org/10.1016/j.sajb.2014.05.002

Jacobs LEO, Richardson DM, Lepschi BP et al (2017) Quantifying errors and omissions in the listing of alien species: *Melaleuca* in South Africa as a case study. NeoBiota 32:89–105. https://doi.org/10.3897/neobiota.32.9842

Jooste WJ (1965) Experiments on the chemical control of *Alhagi camelorum* Fisch. S Afr J Agric Sci 8:287–288

Kaplan H, van Niekerk A, Le Roux JJ, Richardson DM, Wilson JRU (2014) Incorporating risk mapping at multiple spatial scales into eradication management plans. Biol Invasions 16:691–703. https://doi.org/10.1007/s10530-013-0611-z

Kaplan H, Wilson JRU, Klein H et al (2017) A proposed national strategic framework for the management of Cactaceae in South Africa The history and status of Cactaceae in South Africa. Bothalia 47:a2149. https://doi.org/10.4102/abc.v47i2.2149

Knight TM, Havens K, Vitt P (2011) Will the use of less fecund cultivars reduce the invasiveness of perennial plants? Bioscience 61:816–822. https://doi.org/10.1525/bio.2011.61.10.11

Kotzé JDF, Beukes BH, Newby TS, van den Berg EC (2010) National invasive alien plant survey. Council, Institute for Soil, Climate and Water, Report no. GW/A/2010/21, 1–17. Agricultural Research Council—Institute for Soil, Climate and Water, Pretoria

Kraaij T, Baard JA, Rikhotso DR, Cole NS, van Wilgen BW (2017) Assessing the effectiveness of invasive alien plant management in a large fynbos protected area. Bothalia 47:a2105. https://doi.org/10.4102/abc.v47i2.2105

Kumschick S, Foxcroft LC, Wilson JR (2020) Analysing the risks posed by biological invasions to South Africa. In: van Wilgen BW, Measey J, Richardson DM, Wilson JR, Zengeya TA (eds) Biological invasions in South Africa. Springer, Berlin, pp 569–592. https://doi.org/10.1007/978-3-030-32394-3_20

Le Maitre DC, Versfeld DB, Chapman RA (2000) The impact of invading alien plants on surface water resources in South Africa: a preliminary assessment. Water SA 26:397–408

Le Maitre DC, Forsyth GG, Wilson JRU (2015) Guidelines for the development of national species-based invasive alien management programmes: setting geographically differentiated goals. Report no. CSIR/NRE/ECOS/ER/2015/0030/A. Council for Scientific and Industrial Research, Stellenbosch

Lukey P, Hall J (2020) Biological invasion policy and legislation development and implementation in South Africa. In: van Wilgen BW, Measey J, Richardson DM, Wilson JR, Zengeya TA (eds) Biological invasions in South Africa. Springer, Berlin, pp 513–548. https://doi.org/10.1007/978-3-030-32394-3_18

Macdonald IAW (2004) Recent research on alien plant invasions and their management in South Africa: a review of the inaugural research symposium of the Working for Water programme. S Afr J Sci 100:21–26
Macdonald IAW, Ashton ER et al (1985) The role of the volunteer group. In: Macdonald IAW, Jarman ML, Beeston PM (eds) Management of invasive alien plants in the fynbos biome. South African National Scientific Programmes report 111: 117–119. CSIR, Pretoria, pp 117–119
Macdonald IAW, Clark DL, Taylor HC (1989) The history and effects of alien plant control in the Cape of Good Hope Nature Reserve, 1941–1987. S Afr J Bot 55:56–75
Magona N, Richardson DM, Le Roux JJ et al (2018) Even well-studied groups of alien species might be poorly inventoried: Australian *Acacia* species in South Africa as a case study. NeoBiota 39:1–29
Marais C, van Wilgen BW, Stevens D (2004) The clearing of invasive alien plants in South Africa: a preliminary assessment of costs and progress. S Afr J Sci 100:97–103
McConnachie MM, Cowling RM, van Wilgen BW et al (2012) Evaluating the cost-effectiveness of invasive alien plant clearing: a case study from South Africa. Biol Conserv 155:128–135. https://doi.org/10.1016/j.biocon.2012.06.006
McConnachie M, van Wilgen BW, Gaertner M et al (2016) Using counterfactuals to evaluate the cost-effectiveness of controlling biological invasions. Ecol Appl 26:475–483. https://doi.org/10.1890/15-0351
Miao S, Li Y, Guo Q, Ding J et al (2012) Potential alternatives to classical biological control: using native agents in invaded habitats and genetically engineered sterile cultivars for invasive plant management. Tree For Sci Biotechnol 6:17–21
Moore JL, Runge MC, Webber BL et al (2011) Contain or eradicate? Optimizing the management goal for Australian acacia invasions in the face of uncertainty. Divers Distrib 17:1047–1059. https://doi.org/10.1111/j.1472-4642.2011.00809.x
Moran VC, Annecke DP (1979) Critical reviews of biological pest control in South Africa: 3. Jointed cactus, *Opuntia aurantiaca* Lindley. J Entomol Soc South Afr 42:299–329
Moran VC, Hoffmann JH (2012) Conservation of the fynbos biome in the Cape Floristic Region: the role of biological control in the management of invasive alien trees. BioControl 57:139–149. https://doi.org/10.1007/s10526-011-9403-5
Moran VC, Hoffmann JH, Zimmermann HG (2013) 100 years of biological control of invasive alien plants in South Africa: history, practice and achievements. S Afr J Sci 109(9/10):Art. #a0022, 6 p. https://doi.org/10.1590/sajs.2013/a0022
Morokong T et al (2016) Clearing invasive alien plants as a cost-effective strategy for water catchment management: the case of the Olifants River Catchment, South Africa. S Afr J Econ Manag Sci 19:774–787. https://doi.org/10.4102/sajems.v19i5.1594
Mudavanhu S, Blignaut J, Nkambule N (2016) A cost-benefit analysis of using Rooikrans as biomass feedstock for electricity generation: a case study of the De Hoop Nature Reserve, South Africa. S Afr J Econ Manag Sci 19:788–813. https://doi.org/10.4102/sajems.v19i5.1602
Nkambule NP, Blignaut JN, Vundla T et al (2017) The benefits and costs of clearing invasive alien plants in northern Zululand, South Africa. Ecosyst Serv 27:203–223. https://doi.org/10.1016/j.ecoser.2017.04.011
Richardson DM, Petit R (2005) Pines as invasive aliens: outlook on transgenic pine plantations in the Southern Hemisphere. In: Williams CG (ed) Landscapes, genomics and transgenic conifers. Springer, Dordrecht, pp 169–188. https://doi.org/10.1007/1-4020-3869-0_10
Richardson DM, Foxcroft LC, Latombe G et al (2020) The biogeography of South African terrestrial plant invasions. In: van Wilgen BW, Measey J, Richardson DM, Wilson JR, Zengeya TA (eds) Biological invasions in South Africa. Springer, Berlin, pp 65–94. https://doi.org/10.1007/978-3-030-32394-3_3
Riddin T, van Wyk E, Adams J (2016) The rise and fall of an invasive estuarine grass. S Afr J Bot 107:74–79. https://doi.org/10.1016/j.sajb.2016.07.008
Rouget M, Robertson MP, Wilson JRU et al (2016) Invasion debt—quantifying future biological invasions. Divers Distrib 22:445–456. https://doi.org/10.1111/ddi.12408

Shackleton RT, Le Maitre DC, van Wilgen BW et al (2017) Towards a national strategy to optimise the management of a widespread invasive tree (*Prosopis* species; mesquite) in South Africa. Ecosyst Serv 27:242–252. https://doi.org/10.1016/j.ecoser.2016.11.022
Stubbings JA (1977) A case against controlling introduced acacias (ACACIA). S Afr For J 102:8–14
te Beest M, Howison O, Howison RA et al (2017) Successful control of the invasive shrub Chromolaena odorata in Hluhluwe-iMfolozi Park. In: JPGM C, Archibald S, Owen-Smith N (eds) Conserving Africa's mega-diversity in the Anthropocene: the Hluhluwe-iMfolozi Park story. Cambridge University Press, Cambridge, pp 358–382. https://doi.org/10.1017/9781139382793.020
Terblanche C, Kaplan H, Strathie LW et al (2016) An approach to the development of a national strategy for controlling invasive alien plant species: the case of *Parthenium hysterophorus* in South Africa. Bothalia 46:1–11. https://doi.org/10.4102/abc.v46i1.2053
van den Berg EC (2010) Detection, quantification and monitoring Prosopis spp. in the Northern Cape Province of South Africa using remote sensing and GIS. Masters Dissertation, North-West University, Potchefstroom
van Rensburg J, Richardson DM, van Wilgen BW (2017) The challenges of managing invasive alien plants on private land in the Cape Floristic Region: insights from Vergelegen Wine Estate (2004–2015). Trans R Soc S Afr 72:207–216. https://doi.org/10.1080/0035919X.2017.1288175
van Wilgen BW (2015) Plantation forestry and invasive pines in the Cape Floristic Region: towards conflict resolution. S Afr J Sci 111(7/8), Art. 114, 2 p. https://doi.org/10.17159/sajs.2015/a0114
van Wilgen BW, De Lange WJ (2011) The costs and benefits of biological control of invasive alien plants in South Africa. Afr Entomol 19:504–514. https://doi.org/10.4001/003.019.0228
van Wilgen BW, Richardson DM (2012) Three centuries of managing introduced conifers in South Africa: benefits, impacts, changing perceptions and conflict resolution. J Environ Manag 106:56–68. https://doi.org/10.1016/j.jenvman.2012.03.052
van Wilgen BW, Richardson DM (2014) Challenges and trade-offs in the management of invasive alien trees. Biol Invasions 16:721–734. https://doi.org/10.1007/s10530-013-0615-8
van Wilgen BW, Wannenburgh A (2016) Co-facilitating invasive species control, water conservation and poverty relief: achievements and challenges in South Africa's Working for Water programme. Curr Opin Environ Sustain 19:7–17. https://doi.org/10.1016/j.cosust.2015.08.012
van Wilgen BW, Wilson JR (eds) (2018) The status of biological invasions and their management in South Africa in 2017. South African National Biodiversity Institute, Kirstenbosch and DST-NRF Centre of Excellence for Invasion Biology, Stellenbosch
van Wilgen BW, Bond WJ, Richardson DM (1992) Ecosystem management. In: Cowling RM (ed) The ecology of fynbos: nutrients, fire and diversity. Oxford University Press, Cape Town, pp 345–371
van Wilgen BW, Little PR, Chapman RA et al (1997) The sustainable development of water resources: history, financial costs and benefits of alien plant control programmes. S Afr J Sci 93:404–411
van Wilgen BW, Dyer C, Hoffmann JH et al (2011) National-scale strategic approaches for managing introduced plants: insights from Australian acacias in South Africa. Divers Distrib 17:1060–1075. https://doi.org/10.1111/j.1472-4642.2011.00785.x
van Wilgen BW, Forsyth GG, Le Maitre DC et al (2012) An assessment of the effectiveness of a large, national-scale invasive alien plant control strategy in South Africa. Biol Conserv 148:28–38. https://doi.org/10.1016/j.biocon.2011.12.035
van Wilgen BW, Fill JM, Baard J et al (2016) Historical costs and projected future scenarios for the management of invasive alien plants in protected areas in the Cape Floristic Region. Biol Conserv 200:168–177. https://doi.org/10.1016/j.biocon.2016.06.008
van Wilgen BW, Fill JM, Govender N et al (2017) An assessment of the evolution, costs and effectiveness of alien plant control operations in Kruger National Park, South Africa. NeoBiota 35:35–59. https://doi.org/10.3897/neobiota.35.12391

Versfeld DB, Le Maitre DC, Chapman RA (1998) Alien invading plants and water resources in South Africa: a preliminary assessment. WRC Report no. TT 99/98, Water Research Commission, Pretoria

Vundla T, Blignaut J, Nkambule N et al (2016) The opportunity cost of not utilising the woody invasive alien plant species in the Kouga, Krom and Baviaans catchments in South Africa. S Afr J Econ Manag Sci 19:774–787. https://doi.org/10.4102/sajems.v19i5.1603

Wasermann VD, Zimmermann HG, Neser S (1988) The weed silverleaf bitter apple ('Satansbos') (*Solanum elaeagnifolium* Cav.). Tech Comm 214. Department of Agriculture and Water Supply, Pretoria

Wicht CL (1945) Report of the committee on the preservation of the vegetation of the South Western Cape. Royal Society of South Africa, Cape Town

Wilson JRU, Gairifo C, Gibson MR et al (2011) Risk assessment, eradication, and biological control: global efforts to limit Australian acacia invasions. Divers Distrib 17:1030–1046. https://doi.org/10.1111/j.1472-4642.2011.00815.x

Wilson JRU, Ivey P, Manyama P et al (2013) A new national unit for invasive species detection, assessment and eradication planning. S Afr J Sci 109(5/6), Art. #0111, 13 p. https://doi.org/10.1590/sajs.2013/20120111

Wilson JR, Panetta FD, Lindgren C (2017) Detecting and responding to alien plant incursions. Cambridge University Press, Cambridge. https://doi.org/10.1017/CBO9781316155318

Wilson JRU, Faulkner KT, Rahlao SJ et al (2018) Indicators for monitoring biological invasions at a national level. J Appl Ecol 55:2612–2620. https://doi.org/10.1111/1365-2664.13251

Wise RM, van Wilgen BW, Le Maitre DC (2012) Costs, benefits and management options for an invasive alien tree species: the case of mesquite in the Northern Cape, South Africa. J Arid Environ 84:80–90. https://doi.org/10.1016/j.jaridenv.2012.03.001

Zengeya T, Ivey P, Woodford DJ et al (2017) Managing conflict-generating invasive species in South Africa: challenges and trade-offs. Bothalia 47:a2160. https://doi.org/10.4102/abc.v47i2.2160

Zenni RD, Wilson JRU, Le Roux JJ et al (2009) Evaluating the invasiveness of *Acacia paradoxa* in South Africa. S Afr J Bot 75:485–496. https://doi.org/10.1016/j.sajb.2009.04.001

22

Major Takeaways from Alien and Invasive Animal Control Projects

Sarah J. Davies, Martine S. Jordaan, Minette Karsten, John S. Terblanche, Andrew A. Turner, Nicola J. van Wilgen, Ruan Veldtman, Tsungai A. Zengeya and John Measey

S. J. Davies (✉) · J. Measey
Centre for Invasion Biology, Department of Botany and Zoology, Stellenbosch University, Stellenbosch, South Africa
e-mail: sdavies@sun.ac.za

M. S. Jordaan
CapeNature Biodiversity Capabilities Directorate, Cape Town, South Africa

Centre for Invasion Biology, South African Institute for Aquatic Biodiversity, Makhanda, South Africa

M. Karsten · J. S. Terblanche
Centre for Invasion Biology, Department of Conservation Ecology and Entomology, Stellenbosch University, Stellenbosch, South Africa

A. A. Turner
CapeNature Biodiversity Capabilities Directorate, Cape Town, South Africa

Department of Biodiversity and Conservation Biology, University of the Western Cape, Bellville, Cape Town, South Africa

N. J. van Wilgen
Centre for Invasion Biology, Department of Botany and Zoology, Stellenbosch University, Stellenbosch, South Africa

Cape Research Centre, South African National Parks, Cape Town, South Africa

R. Veldtman
Department of Conservation Ecology and Entomology, Stellenbosch University, Stellenbosch, South Africa

South African National Biodiversity Institute, Kirstenbosch Research Centre, Cape Town, South Africa

T. A. Zengeya
South African National Biodiversity Institute, Kirstenbosch Research Centre, Cape Town, South Africa

Centre for Invasion Biology, Department of Zoology and Entomology, University of Pretoria, Pretoria, South Africa

Abstract South Africa has a rich history of managing invasive alien animal populations. This chapter explores examples of animal control projects, their resourcing and degree of success or failure. Out of 1023 alien animal species present in South Africa, 80 are designated for compulsory control or eradication in national legislation, and 24 are currently being controlled with the aim of eradication or containment. Only two species have been successfully eradicated from mainland SA and its near-shore islands: *Otala punctata* (the Freckled Edible Snail) and *Trogoderma granarium* (the Khapra Beetle). These two projects took place in the late 1980s and early 1990s, and were rapid responses by small groups of role players to small infestations. In contrast, most current projects are larger, involving complex stakeholder management and considerable technical complexity. Three further invertebrate species are currently controlled through integrated pest management (*Bactrocera dorsalis*, the Oriental Fruit Fly) or nest removal (*Vespula germanica*, the German Wasp and *Polistes dominula*, the European Paper Wasp). No marine species are currently subject to control. Among vertebrates, 12 freshwater fish species have been controlled in localised areas, according to their specific listing in legislation and protected area management priorities; two amphibian, two bird and five mammal species are currently subject to control using a wide variety of techniques. Inter-institutional working groups have played a significant role in promoting the success of invasive alien species management in South Africa. Three working groups are actively addressing new and existing invasions, and promoting awareness and cooperation among a wide range of organisations, as well as recording the experience and learning of these groups.

22.1 Introduction

On average 50–100 alien animal taxa, including feral domestic animals, have established naturalised or invasive populations per quarter degree square in South Africa (Picker and Griffiths 2017). This tally belies a range of invasion levels of different taxa from very low in amphibians and marine fish, to very high in freshwater fish (van Rensburg et al. 2011). Vertebrates make up 30% of introduced animal species in South Africa (n = 309; Picker and Griffiths 2017), while insects, crustacea, annelids, molluscs and arachnids make up the remaining 70% (713 species; van Wilgen and Wilson 2018). Among vertebrates, freshwater fish species (17) and mammals (13) are most often introduced, while fewer birds (9) and reptiles (1) and no amphibian species have been introduced from outside the country. Consequently, there have been relatively few fully-fledged control projects undertaken, but a wide range of methods have been used, ranging from manual capture of Freckled Edible Snails (*Otala punctata*) and European Shore Crabs (*Carcinus maenas*) to the use of anaesthetic for Mallards (*Anas platyrhynchos*), toxic baits for House Crows (*Corvus splendens*) and piscicide such as rotenone for several fish species.

In South Africa, the management of alien and invasive species is guided by the Alien & Invasive Species Regulations (RSA 2014) of the National Environmental

Management: Biodiversity Act (Act 10 of 2004) (referred to as NEM:BA). The intent of the regulations is to reduce the risk of importing potentially invasive or harmful alien species, reduce the number of alien species becoming invasive, and limit the extent and impact of invasions (see van Wilgen et al. 2020, Chap. 1, Box 1.1).

The regulations prohibit the importation of alien species that are known to be invasive elsewhere in the world but are not yet in the country (prohibited species). Alien species that are already present but not widespread in the country can be targeted for eradication (Category 1a). In cases where invasive species are widespread the most practical option is often to contain further spread (Category 1b). The regulations have a provision for utilisation of listed alien species that provide benefits to human wellbeing, albeit with some strict conditions (Categories 2 and 3).

Current role-players in invasive and alien animal control programmes include national agencies such as South African National Parks (SANParks), provincial nature conservation agencies such as CapeNature (Western Cape Province), local authorities such as eThekwini Municipality (Durban) and a range of consultants and NGOs (e.g. Society for the Prevention of Cruelty to Animals). Research and higher education organisations such as universities, national research facilities and institutes provide information, best practice and expertise for monitoring alien and invasive species. A range of inter-institutional working groups on invasive taxa have been developed in South Africa, and these play a crucial role in advising government on management options and best practice and broadening participation and information flow. Invasive alien animal working groups in the Western Cape and KwaZulu-Natal (KZN) bring together role players to discuss priorities and species requiring control and to advise the responsible institutions on methods. Recently, a Marine Alien and Invasive Species Working Group has been constituted at a national level to advise on prioritisation of management of marine invasive species. This group will complement the work done by the terrestrial Western Cape and KZN working groups.

This chapter addresses the animal control operations of which we are aware, arranged by habitat and taxonomic groups. We describe the control projects that have been attempted in South Africa (except for offshore islands, see Greve et al. 2020, Chap. 8), including the methods used, when the operation started, how long the operation lasted, by whom it was undertaken, and what funding/resources were used. Where possible, we note the breadth of the stakeholder base, the relative success and current status of the programme. Biological control operations are included hypothetically, but none have yet been applied to animals on a large scale. The final section of the chapter addresses the learning points that have emerged and which could be used to maximise the effectiveness of alien and invasive animal management in South Africa.

22.2 Freshwater Invertebrates

22.2.1 Procambarus clarkii *(Red Swamp Crayfish)*

While 22 alien freshwater invertebrates have been documented in South Africa (Picker and Griffiths 2017), and several, particularly snails, are known to have impacts (e.g. see de Kock and Wolmarans 2008), there is no documentation on control of these species, aside from the Red Swamp Crayfish (see also Chap. 6).

Procambarus clarkii (the Red Swamp Crayfish) is native to North America but has been introduced in several African countries, including South Africa (see Weyl et al. 2020, Chap. 6). This species is important in aquaculture and has invaded parts of Africa via this pathway. By replacing native crab and snail populations, and consuming aquatic vegetation, red swamp crayfish exert trophic impacts and may also have structural impacts in rivers (Jackson et al. 2016). In 1988 this species was found in one or more dams on a farm in Mpumalanga, and in 1994 a control operation was undertaken by the provincial nature conservation authority to remove all *P. clarkii* from one dam, which was partially drained, and all crayfish were removed by hand or using dip-nets (Nunes et al. 2017). Unfortunately, this operation was poorly documented and monitoring data are not available. In a follow-up survey in 2015 and 2016, Nunes et al. (2017) used sweep nets, electrofishing and trapping, and established that *P. clarkii* was still present. No further follow ups have been conducted since 2016. *Procambarus clarkii* is listed as a prohibited species in NEM: BA (RSA 2014) but its presence in the country implies that its listing should be changed to Category 1a and that the species should be a target for national eradication (Table 22.1).

22.3 Marine Invertebrates

Information on introduced marine species has grown rapidly since the 1990s, when 15 introduced marine organisms were recognised (Griffiths et al. 1992); in 2011 a comprehensive review and survey identified 85 introduced and 39 cryptogenic marine organisms (Mead et al. 2011a, b). Currently, 95 introduced marine invertebrates are present in South Africa (Robinson et al. 2020, Chap. 9), although only one species (described below) has been subjected to control. Recognising the need for more research and management options for marine species, the Marine Alien and Invasive Species Working Group has recently been formed to advise on prioritisation of management of marine invasive species at a national level.

Table 22.1 The status of invasive alien animal control projects in South Africa. 'Status in NEM:BA' refers to the category in which the species is listed under the Alien & Invasive Species regulations. 'Assessment of control status in van Wilgen and Wilson (2018)' refer to assessment of control effectiveness reported in the National Status Report on Biological Invasions. The final column gives the updated assessment based on information included in this chapter. The regulatory category 'context specific' applies to species whose status in NEM:BA depends on the location of the population. See van Wilgen et al. (2020), Chap. 1 for definitions of the NEM:BA categories

Taxon	Species	Assessment of control status in van Wilgen and Wilson (2018)	Status in NEM:BA	Aim of control project	Status of control project
Freshwater invertebrate	*Procambarus clarkii* (Red Swamp Crayfish)	Failed	Prohibited	Eradication	Not subject to control; present at one or more sites in South Africa
Marine invertebrate	*Carcinus maenas* (European Shore Crab)	Not assessed	1b	Pilot project and data gathering	Not subject to control at present; control probably not worthwhile as not occupying undisturbed habitats and re-invasion in harbours is highly likely
Terrestrial invertebrate	*Otala punctata* (Freckled Edible Snail)	Successful	Unlisted	Eradication	No control necessary
Terrestrial invertebrate	*Trogoderma granarium* (Khapra Beetle)	Successful	1b	Eradication	No control necessary at present (prevention is in place)
Terrestrial invertebrate	*Bactrocera invadens* (Asian Fruit Fly)	Failed	1a	Containment	Not currently subject to control; potential for successful control (a related species *B. dorsalis* is controlled)
Terrestrial invertebrate	*Polistes dominula* (European Paper Wasp)	Under consideration	1b	Containment	No longer subject to control due to large extent of invasive population in urban areas of the Western Cape
Terrestrial invertebrate	*Vespula germanica* (German Wasp)	Under consideration	1b	Containment	Subject to control in local areas; eradication may be feasible in future

(continued)

Table 22.1 (continued)

Taxon	Species	Assessment of control status in van Wilgen and Wilson (2018)	Status in NEM:BA	Aim of control project	Status of control project
Freshwater fish	*Micropterus salmoides* (Largemouth Bass)	Not assessed	Context specific	Local extirpation	Has been controlled successfully in some areas
Freshwater fish	*Micropterus dolomieu* (Smallmouth Bass)	Not assessed	Context specific	Local extirpation	Has been controlled successfully in some areas
Freshwater fish	*Micropterus punctulatus* (Spotted Bass)	Not assessed	Context specific	Local extirpation	Has been controlled successfully in some areas
Freshwater fish	*Micropterus punctulatus* (Florida Bass)	Not assessed	Context specific	Local extirpation	Has been controlled successfully in some areas
Freshwater fish	*Oncorhynchus mykiss* (Rainbow Trout)	Not assessed	Not listed	Local extirpation	N/a; not regulated under NEM:BA
Freshwater fish	*Cyprinus carpio* (Common Carp)	Not assessed	Context specific	Local extirpation	Has been controlled successfully in some areas
Freshwater fish	*Clarias gariepinus* (Sharptooth Catfish)	Not assessed	Not listed (native species with some extra-limital populations)	Local extirpation	Has been controlled successfully in one project
Freshwater fish	*Gambusia affinis* (Mosquito Fish)	Not assessed	Context specific	Local extirpation	Has been controlled successfully in one project
Freshwater fish	*Tinca tinca* (Tench)	Not assessed	Context specific	Local extirpation	Has been controlled successfully in one project
Freshwater fish	*Lepomis macrochirus* (Bluegill Sunfish)	Not assessed	Context specific	Local extirpation	Has been controlled successfully in some areas
Freshwater fish	*Oreochromis mossambicus* (Mozambique Tilapia)	Not assessed	Not listed (native species with some extra-limital populations)	Local extirpation	Has been controlled successfully in one project
Freshwater fish	*Tilapia sparrmanii* (Banded Tilapia)	Not assessed	Not listed (Native species with some extra-limital populations)	Local extirpation	Has been controlled successfully in one project

Amphibian	*Sclerophrys gutturalis* (Guttural Toad)	Failed	Unlisted	Eradication	Control is ongoing in invaded areas of the City of Cape Town; aim of control may be re-evaluated in the near future
Amphibian	*Xenopus laevis* × *X. gilli* hybrids (African Clawed Frog × Cape Platanna)	Not assessed	1b	Local extirpation	Control is carried out by SANParks in Table Mountain National Park where *X. gilli* is sympatric
Bird	*Acridotheres tristis* (Common Myna)	Not assessed	3	Pilot project and data gathering	Not subject to formal control; occasional removal (Hart and Downs 2015; van Wilgen and Wilson 2018)
Bird	*Corvus splendens* (House Crow)	Ongoing	1a	Local extirpation	Control is ongoing with the aim of eradication in two cities
Bird	*Anas platyrhynchos* (Mallard Duck)	Not assessed	2	Containment	Subject to control in defined areas in the City of Cape Town
Mammal	*Hemitragus jemlahicus* (Himalayan Tahr)	Ongoing	1b	Eradication	Under ongoing control by SANParks in Table Mountain National Park as per NEM:BA status
Mammal	*Sus scrofa* (Feral Domestic Pig)	Kasteelberg population nearly eradicated	1b	Containment	Ongoing control in two populations
Mammal	*Felis catus* (Domestic Cat)	Successful	1a on islands	Eradication from island	Controlled on Robben Island; Marion Island population is covered in Chap. 8
Mammal	*Oryctolagus cuniculus* (European Rabbit)	Not assessed	1b on off-shore islands	Eradication from island	Control is ongoing; very few animals remain on Robben Island (not seen in several years)
Mammal	*Dama dama* (Fallow Deer)	Not assessed	2	Eradication from island	Control is ongoing; 20–30 animals remain on Robben Island

22.3.1 Carcinus maenas *(European Shore Crab)*

Carcinus maenas has a broad distribution on four continents, but is native to the north Atlantic and Baltic regions, and possibly to North Africa (see Robinson et al. 2020, Chap. 9; Hampton and Griffiths 2007). In South Africa, its distribution is restricted to two harbours on the Cape Peninsula—Table Bay and Hout Bay (Mabin et al. 2017). Between 2015 and 2016, a pilot management programme was conducted in Hout Bay harbour, as part of a PhD project on *C. maenas*. The results of the surveys suggested that although eradication may be feasible if sufficient management effort was to be applied, *C. maenas* is not spreading outside harbours, or into undisturbed habitats (Mabin et al. 2017). No further control is planned for this species.

22.3.2 Tetrapygus niger *(Chilean Black Urchin)*

Although not the result of a control operation, the presence and subsequent disappearance of *Tetrapygus niger* in South Africa deserves mention. The urchin was recorded at an onshore oyster farming facility in Alexander Bay on the north west coast of South Africa in 2007 (Haupt et al. 2010). Since the initial observation, oyster farming has ceased at Alexander Bay and the facility has been abandoned. In 2014 a re-survey of the two oyster farm dams in which the urchins has been recorded as well as nearby intertidal and subtidal habitats was performed by researchers (Mabin et al. 2015). No living *T. niger* were found in the dams or on the shore and the population is thought to be extinct (Mabin et al. 2015). The 2014 survey was funded by the Department of Environmental Affairs (DEA) through the South African National Biodiversity Institute (SANBI) Marine Programme. Haupt et al. (2010) judged the likely source of the urchins to be the importation of oyster spat from Chile. Since the farming operations have ceased, re-introduction through this pathway is unlikely.

22.4 Terrestrial Invertebrates

22.4.1 Otala punctata *(Freckled Edible Snail)*

Otala punctata is a helicid snail native to the Western Mediterranean and invasive in the United States, Argentina, Uruguay and Chile. Herbert and Sirgel (2001) recorded the eradication of *O. punctata* introduced to the Western Cape. The first population was found at Tygerberg Hospital near Cape Town in December 1986 and the second population at the Cape Town docks in January 1987 (see also Janion-Scheepers and Griffiths 2020, Chap. 7). An eradication programme was started promptly under the auspices of the Department of Agriculture and Water Supply, Winter Rainfall Region. The two populations were removed from 1987 to 1989, monitored through to August 1990 and no further presence of the species was detected thereafter (Herbert and Sirgel 2001). Control techniques included manual collection of snails and baiting with molluscicide (methaldehyde and methiocarb). Dense vegetation such as patches of rank grass were removed using herbicide and flame throwers so that snails could be detected more easily. The total cost of the programme was ZAR 215,000 over 3 years.

Herbert and Sirgel (2001) estimated that the Tygerberg colony initially covered about 4 ha, and that over 22,000 snails were removed from the area; the species has not been recorded subsequently. The eradication project was justified because no species of *Otala* had ever been reliably recorded in South Africa prior to 1986, and they were known to be invasive elsewhere. The species is not listed in NEM:BA as either an invasive or a prohibited species in South Africa. Since this is a relatively large (ca 35 mm diameter) polyphagous herbivore, there is a possibility it could be re-introduced in future, either for cultivation or by accident.

22.4.2 Trogoderma granarium *(Khapra Beetle)*

Trogoderma granarium is a dermestid beetle that feeds on grains and is a serious pest of stored grains such as wheat. Consumption of grain contaminated with this species has negative consequences for human health because of the numerous body parts and cast skins (Athanassiou et al. 2019). The first record of Khapra Beetle in South Africa was from imported malt in Pietermaritzburg in May 1954; sporadic records followed in disparate parts of the country until an outbreak in an intensive agricultural area in the Northern Cape Province in 1972 (Viljoen 1990). Various outbreaks continued during the 1990s in the Northern Cape and were dealt with by extensive pesticide application. It is not clear from published literature which agencies were involved in Khapra Beetle control, or what the operations cost.

According to South Africa's National Status Report (van Wilgen and Wilson 2018) this species has been eradicated. According to the European and Mediterranean Plant Protection Organisation's Global Database (EPPO 2019), this species is no longer present in SA, having been recorded at three different sites up to 1972 but failed to establish after control and quarantine measures were applied (Day and White 2016).

22.4.3 Fruit Flies (Tephritidae) as Exemplar Invasive Insect Species

True fruit flies (Tephritidae) have a long history of global invasions and South Africa is no exception. Among the most widely recognised globally invasive fruit fly species are the Mediterranean Fruit Fly (*Ceratitis capitata*) and the Oriental Fruit Fly (*Bactrocera dorsalis*). Fruit flies have a range of impacts from direct (e.g. fruit damage) to indirect, including restricting international market access and having knock-on socio-economic impacts, especially in Africa where small-scale farmers may be heavily reliant on fruit or commodity sales and trade. In South Africa, the Mediterranean Fruit Fly has long been the focus of pest management strategies, despite this region likely forming part of its historical native distribution. Although not classified as invasive within South Africa (Richardson et al. 2011; Karsten et al. 2015), the Mediterranean Fruit Fly is managed in South Africa as an agricultural pest and elsewhere (e.g. Europe) as an invasive and an agricultural pest (Karsten et al. 2018).

On the African continent, there is also a growing threat of pest and invasive fruit fly species that readily establish and can affect agricultural food security. South Africa is a signatory of the International Plant Protection Convention (IPPC), an agreement between 182 countries worldwide to protect cultivated and wild plants against invasive pest species. Trapping protocols for fruit fly detection at borders and in commercial fruit growing areas were established in South Africa in 2006 as part of a National Exotic Fruit Fly Action Plan (Barnes and Venter 2006). Several species in particular are being closely-monitored [including *B. dorsalis, B. zonata* and *Zeugodacus* (*Bactrocera*) *cucurbitae*] due to concerns surrounding their rapid spread across the African continent from their Asian origins. From its 2003 detection on the continent (Mwatawala et al. 2004; Drew et al. 2005), *B. dorsalis* was found at the northern South African border in 2007/2008, and only properly tackled in 2010 when detected again after no individuals were detected in 2009 (Manrakhan et al. 2009, 2011). However, in the previous year a steering committee involving industry partners and government was formed to address the growing threat of *B. dorsalis* to the South African food economy (Manrakhan et al. 2009). The detection of a confirmed *B. dorsalis* specimen in a trap precipitated the prescribed eradication plan. In short, a quarantine area (a circle about 80 m^2) would be constructed around this point and a delimiting survey initiated (Manrakhan et al. 2012). Eradication efforts were then undertaken and included orchard sanitation (all fruit collected is buried 50 cm below ground), Male Annihilation Technique (400 blocks per km^2) as well as protein bait sprays (Manrakhan et al. 2012). If, after 12 weeks, no further flies are detected, eradication is assumed to be successful, quarantine lifted and 'pest free' status assigned, which was the case here.

However, eventually *B. dorsalis* was detected in multiple new and previously eradicated areas and is currently established in several fruit-growing regions of South Africa (Mpumalanga, KwaZulu-Natal, Limpopo, North-West and Gauteng Provinces) (Manrakhan et al. 2015; Karsten et al. 2018), indicating failure of the eradication attempts. The current consensus aim is to manage the invasive species alongside the other pest fruit flies, and to employ a broader range of methods based on integrated pest management (IPM), for example various combinations of attract-and-kill, classical biological control, introduction of natural enemies, and pesticide applications, if necessary. Although a sterile insect technique (SIT) is currently employed for the Mediterranean Fruit Fly (*C. capitata*) in parts of the Western Cape Province (Barnes and Venter 2006), as yet there are no plans, to our knowledge, to adopt a SIT program for *B. dorsalis*. Suckling et al. (2014) assembled a specific database for fruit flies from the Global Eradication Database (http://b3.net.nz/gerda/), investigating eradication and response programmes and their outcomes in 17 tephritid species worldwide. They show that for the 108 programmes targeting 13 *Bactrocera* species there was a 12% failure rate, compared to programmes targeting *Anastrepha* or *Ceratitis* species included in the study with no reported failures. Furthermore, *B. dorsalis* (including *B. philippinensis*, *B. invadens* and *B. papayae* after synonomisation; see Schutze et al. 2015) specifically had more official responses than any of the other *Bactrocera* species (n = 63) with eradication declared in 39 cases. It seems therefore that *Bactrocera* species may be inherently more difficult to eradicate than *Anastrepha* or *Ceratitis* species, as failures are more frequently reported in *Bactrocera*, as was the case in South Africa's *Bactrocera* eradication programme recently.

Despite best intentions, heightened awareness among the local population, on-going surveillance and large-scale eradication efforts, *B. dorsalis'* spread into South Africa was not thwarted. The reasons for this are unclear, but there are several possible explanations. For example, it may have been a consequence of high propagule pressure in the form of multiple introductions from multiple smaller, cryptic locations, combined with a large informal across-border (or within-country) fruit trade that is typically poorly regulated. However, no studies, to our knowledge, have examined these or alternative hypotheses. With new invasive Tephritidae species already making an appearance at South African borders, it is time these knowledge gaps are addressed thoroughly and the management responses carefully monitored to learn from past mistakes. In our view, an across-border and within-country program to restrict informal fruit movement would be worth implementing to attempt to reduce the establishment of new invasive fruit fly species. Presently, two international networks (Addison et al. 2016), FRUITFLYNET (South Africa, Mozambique, Tanzania, Belgium) and the ERAfrica Fruit Fly project (South Africa, Ivory Coast, Belgium, La Réunion) are in place and both projects aim to improve trapping, surveillance and identification tools of Tephritidae pests in Africa.

22.4.4 Polistes dominula *(European Paper Wasp) and* Vespula germanica *(German Wasp)*

These two species are treated together here because they are both social wasp species that have global invasive distributions, and they have recently established invasive populations in South Africa. Both species are spreading and are recognised as invasive species in legislation (Category 1b invaders—RSA 2014).

Vespula germanica was first recorded in Newlands, Cape Town in 1974. For several years the range of *V. germanica* seems to have remained confined to the Cape Peninsula. However, in 2002 it was recorded in Somerset West more than 50 km from Cape Town, and by 2003 it was being detected in Stellenbosch, Elsenburg, southern

Paarl, Banhoek and Sir Lowry's Pass, as well as Franschhoek by 2004. These are all centres in agricultural landscapes that sequentially border the first Boland population detected in 2002 (Haupt 2015). It is thus clear that the species suddenly underwent a rapid range expansion after being isolated on the Peninsula for many years. The species now has a much larger distribution that can increase further over time.

Polistes dominula is a more recent invader but has already spread approximately 100 km from its point of introduction near Cape Town in 2008. It is now common throughout the Cape Town Metropolitan Region, Somerset West, Stellenbosch, Paarl, Wellington, Franschhoek and parts of Grabouw (Veldtman et al. 2012; Benadé et al. 2014; van Zyl et al. 2018).

From 2014 the City of Cape Town, in partnership with Stellenbosch Municipality, SANBI and Stellenbosch University researchers, initiated nest removal programmes for *V. germanica* and *P. dominula* (van Zyl et al. 2018). Due to the similar appearance of the two wasp species and the seriousness of their stings (particularly *V. germanica*), the public were invited to inform the City of nests on private property, and staff were dispatched to remove nests in response. The aim of the programme was not to eradicate, but to provide a service to residents who were experiencing problems with invasive wasps on their properties. In April 2015, the City of Cape Town Invasive Species Unit reported that 6142 nests had been removed and 691.5 person days worked. At that time, members of the Invasive Species Unit noted the high volume of requests and that the extent and density of the invasion was worse than expected, and concluded that a piecemeal approach of treating individual houses would never be effective. There are however far fewer cases of *V. germanica* reports and the destruction of nests as reported could help curb the impacts experienced by residents from this species. The *V. germanica* control programme is therefore still ongoing.

From 2016 November to February 2017, SANBI initiated a systematic trial to eradicate *P. dominula* at its range edge in the Franschhoek and Grabouw areas using contractors. After training by researchers these contractor teams (three teams of five people per area) set out to visit residential properties in a 3 month period in each area to systematically record and destroy all detected *P. dominula* nests. During this period 3708 homes were checked and 15,008 nests removed in Franschhoek, while in Grabouw 3370 homes were checked and 7029 nests were removed. This equates to 1890 persons days (30 people × 63 days) spent on inspecting 7078 homes and destroying 22,037 nests. The area-wide coverage and clearing effectiveness was better in Franschhoek than Grabouw, possibly because of communication difficulties in communities that speak several languages (Simakani 2017). There was no follow up work done in these areas to establish whether fewer nests were present the following season, so the effectiveness of the intervention is unknown.

For *V. germanica* 232 nests were destroyed between 2014 and 2018 across its distributional range (in 2014, 56; 2015, 12; 2016, 80; 2017, 23, and 2018, 61) by researchers based at Stellenbosch University, wine farms, pest control companies and the City of Cape Town. Spatial records for each of these nests will be used to model the fine scale distribution of this species (Veldtman et al. unpublished data) as a potential aid to future control strategies. The climate in the current range of

V. germanica is marginally suitable for the species, but if they are able to spread into the eastern coastal strip of South Africa they could potentially occupy a much larger climatically suitable range in the sub-tropical regions of the country (Tribe and Richardson 1994; Veldtman et al. 2012). De Villiers et al. (2017) have indicated that agricultural irrigation can extend the area of suitable climate for *V. germanica*, increasing the chance of this species spreading to the east of the country. Veldtman et al. (2012) assert that eradication of *V. germanica* is only possible with continued systematic control efforts.

The situation regarding *P. dominula* is more challenging due to the large populations already present. From the eradication trial it is clear that *P. dominula* is well established even at its range edge and containment would require substantial resources to be effective. However in New Zealand, where *V. germanica* has completely infiltrated and taken over native ecosystems, the use of technological methods such as gene drive (where RNA-guided gene drives—based on CRISPR/Cas9—are used to modify individuals that are then released to interfere with the reproduction of the invasive population) are being investigated as a means of eradicating their population (Lester and Beggs 2019). South Africa would benefit if New Zealand succeeds in applying and sharing this technology to develop a potential control method for *P. dominula*. Currently *P. dominula* occurs in most major towns of the Boland region. There are however high volumes of agricultural produce and equipment being moved between invaded and uninvaded areas which makes further spread via the Transport—Stowaway pathway (Convention on Biological Diversity 2014) likely.

22.5 Freshwater Fish

In South Africa, 27 alien freshwater fish taxa have been introduced into the wild and 16 of these are considered invasive (Ellender and Weyl 2014). Attempts to control alien fish have been restricted to a few catchments and species. While this indicates some progress in the management of invasive fish, the majority of invasive populations have not been subjected to control, and some control operations meet much resistance due to the recreational and subsistence value of certain species (Zengeya et al. 2017).

The formal management of alien fishes for biodiversity restoration was initiated in 2000 through the Cape Action for People and the Environment project (CAPE; Younge and Fowkes 2003). The CAPE project defined criteria for identifying rivers for fish control, subsequent identification of candidate rivers and identification of suitable eradication methods (Marr et al. 2012) within the Cape Fold Freshwater Ecoregion (Abell et al. 2008). The outcomes of the control projects are summarised below.

22.5.1 *Micropterus* Species (Black Basses)

Black Bass is the collective name for species of the genus *Micropterus*, and four species (*M. salmoides*, Largemouth Bass; *M. dolomieu*, Smallmouth Bass; *M. punctulatus*, Spotted Bass and *M. floridanus*, Florida Bass) have been introduced into South Africa (Ellender et al. 2014; Weyl et al. 2020, Chap. 6). These species were introduced for angling and today these fishes are ubiquitous throughout the country (Ellender et al. 2014). They prey on native fishes and can cause local extirpations of highly threatened endemic fishes (e.g. van der Walt et al. 2016). All four species are listed in NEM:BA (RSA 2016) in various categories depending on their location and are therefore regulated by area. For example, *M. dolomieu* is permitted in discrete systems (rivers, lakes, estuaries) where it is established but must be controlled and wherever possible removed from protected areas and mountain catchments. Control of Black Bass in tributary streams has been attempted using manual eradication methods as well as piscicides. Two mechanical extirpation projects have been implemented to date with different outcomes: *M. punctulatus* was successfully extirpated from the Thee River in the Cederberg region (van der Walt et al. 2018), but a similar project aimed at the extirpation of *M. dolomieu* failed in the Blindekloof Stream in the Swartkops system (Skelton 1993; Ellender et al. 2011).

Micropterus punctulatus was discovered in 2007 in the Thee River, and CapeNature's Invasive Alien Fauna Unit initiated a manual eradication project in 2010 (van der Walt et al. 2018). Removal efforts included netting, spearfishing and electrofishing as well as the construction of an invasion barrier. Annual surveys from 2015 to 2017 indicated that the population had been successfully extirpated and that native threatened fish species had recolonised the previously invaded reach of the river. A total of 442 person days (174 for actual eradication work) were recorded across all aspects of the project (van der Walt et al. 2018).

Micropterus dolomieu was successfully removed from the Blindekloof River in 1988 by means of angling and electrofishing (Skelton 1993). However, an invasion barrier was not constructed to prevent reinvasion and management was reliant on ongoing removal efforts. Management effort was not maintained and *M. salmoides* reinvaded the Blindekloof River (Ellender et al. 2011).

Stakeholder conflict and resistance to the two projects described above was largely absent, as the two bass populations were of low angling value. In contrast, significant opposition and negative media perception characterised the proposed use of the piscicide rotenone in the Rondegat River to remove *M. dolomieu* (Marr et al. 2012). This project was implemented by CapeNature in 2012–2013 and funded mainly through the DEA: NRM programme at a cost of approximately ZAR 1.4 million (Barrow 2014). The project was implemented under the guidance of staff from the California Department of Fish and Game using Standard Operating Procedures for the use of rotenone (Finlayson et al. 2010). *Micropterus dolomieu* were successfully extirpated from this river following two rotenone treatments 1 year apart and upgrading of an instream weir to prevent reinvasion from downstream sources (Weyl et al. 2014). Extensive post-treatment monitoring of non-target aquatic invertebrates illustrated that despite short-term negative effects, the invertebrate community recovered relatively quickly to pre-intervention levels (Woodford et al. 2013; Bellingan et al. 2015, 2019). Native fish from upstream of the treatment zone successfully recolonised the treatment zone, resulting in an increased area of occupancy for these species in the absence of *M. dolomieu* (Weyl et al. 2014; Marr et al. 2019).

Despite initial opposition from some stakeholders, perceptions changed considerably during the project from initial scepticism and resistance, to support. This was largely the result of an independent EIA process prior to treatment, rigorous and independent scientific monitoring of all aspects of the project such ecological impacts, project success and treatment regime, and a coordinated and ongoing media and stakeholder engagement effort.

22.5.2 **Oncorhynchus mykiss** ***(Rainbow Trout)***

Oncorhynchus mykiss were introduced to South Africa as an angling species in the late 1890s and are today considered a typical conflict species, i.e. a species with negative environmental impacts but which is also valuable from a socio-economic perspective (e.g. for angling and aquaculture) (Woodford et al. 2016; Zengeya et al. 2017; Weyl et al. 2020, Chap. 6). Due to its ability to invade headwaters, *O. mykiss* is especially problematic in the Cape Fold Ecoregion and in the high-lying areas of Mpumalanga and KwaZulu-Natal provinces, all of which are home to a number of endemic and highly threatened native fishes (e.g. Kleynhans 1987; Skelton 2001; Shelton et al. 2014; Weyl et al. 2015). Conflict with anglers about the removal of trout has resulted in a failure to list the species in the NEM:BA Alien and Invasive Species Regulations (RSA 2016) because of stakeholder opposition (Woodford et al. 2017). Until the impasse on the appropriate management options of *O. mykiss* is resolved, the species cannot be regulated under national legislation but is regulated under provincial conservation ordinances.

Only one control project has been implemented to date, namely a manual eradication attempt by CapeNature in the Krom River in the Cederberg (Shelton et al. 2017). The project aimed to remove *O. mykiss* from the headwaters of the Krom River, a tributary of the Olifants-Doring system, from 2013 to 2014; extirpation efforts focused on approximately 6 km of river, and used angling, fyke and gill nets. A significant reduction in trout numbers was achieved during the attempt, but the population returned to near its pre-removal abundance level within 2 years. The total cost of the project was approximately ZAR 150,000 which covered salaries, transport and equipment for the 40-day effort.

22.5.3 Cyprinus carpio *(Common Carp)*

Cyprinus carpio is a global invader that was introduced to South Africa in the 1850s. As with many other invasive fishes, it is now ubiquitous throughout the country (Ellender et al. 2014; Weyl et al. 2020, Chap. 6). It is generally associated with ecosystem-level impacts such as increased turbidity and disruption of substrates and aquatic plants through their bottom-feeding behaviour (de Moor and Bruton 1988). The management of *C. carpio* is therefore important for ecosystem service protection more than for biodiversity restoration. A number of extirpation projects aimed at improving ecosystem function have been implemented to date. The Century City wetland system in Cape Town was treated with rotenone in 2008 to remove *C. carpio* and improve ecosystem functioning and water quality. Treatment cost was estimated at ZAR 100,000 for a once-off treatment and was funded by the Century City Home Owners Association (D. Ferreira, pers. comm.). Post-treatment monitoring indicated that *C. carpio* were reduced to below detectable levels suggesting that the species had been successfully eradicated (B. Paxton, unpublished data). Subsequent monitoring detected no *C. carpio* for 5 years, but they were detected in very low numbers in 2016. It is not clear whether the individuals found in 2016 were survivors from the original project (i.e. the extirpation failed) or were re-introduced subsequently. The project was well-supported by the general public as well as government stakeholders (e.g. Department of Water and Sanitation), due to a well-coordinated and comprehensive communication plan and a strong management presence during the treatment (D. Ferreira, pers. comm.).

In 2017, rotenone treatment of a privately-owned farm dam near Nieuwoudtville, Northern Cape, removed *Cyprinus carpio* that posed an invasion risk to the Oorlogskloof River in a nearby protected area. The project was implemented jointly by the two provincial departments (CapeNature and the Northern Cape Department of Environment and Nature Conservation) using DEA: NRM funding. Post-treatment monitoring indicated that the *Cyprinus carpio* were successfully removed (Marr et al. 2019).

22.5.4 Clarias gariepinus *(Sharptooth Catfish)*

Clarias gariepinus is native to the northern parts of South Africa but several extra-limital and invasive populations exist in the Cape Fold Ecoregion (Cambray 2003; Weyl et al. 2020, Chap. 6). Only one extirpation project has been implemented to date, but it ultimately failed to achieve extirpation. Following an illegal introduction of catfish into Grey Dam near Makhanda, the Eastern Cape Nature Conservation agency, assisted by staff from Rhodes University and Albany museum, treated the dam with rotenone (Cambray 1995). The project was funded by Makhanda and while successful extirpation was reported post-treatment, subsequent reinvasion of this species was reported (Cambray 1995).

22.5.5 *Multi-species Extirpation Projects*

Many invasive and extra-limital freshwater fish species have been co-introduced and have been the subject of multi-species extirpation efforts, either for biodiversity restoration or improving ecosystem function. Paardevlei wetland outside Cape Town is an example where a large number of invasive species, including *C. carpio* (Common Carp), *Micropterus spp.* (Black Bass), *Tinca tinca* (Tench) and *Lepomis macrochirus* (Bluegill) had been stocked since the 1920s (Impson et al. 2005). The dominance of *C. carpio* over time resulted in declining water quality, algal blooms and altered ecosystem function. In 2005, the wetland was treated by aerial spraying of rotenone and over 35 tons of fish, mostly *C. carpio*, were removed after treatment (Impson et al. 2005). The project, including a coordinated stakeholder information campaign, was designed and implemented by an independent consulting firm and funded by the landowner. There was negligible stakeholder opposition, with the exception of concerns around animal welfare. The fish have not been detected since the termination of monitoring in 2014, so the operation appears to have succeeded.

Die Oog is a small urban wetland in Cape Town that forms part of a formal conservation area managed by the City of Cape Town. The wetland is home to alien fish species such as *C. carpio* and *Gambusia affinis* (Mosquito fish) and some extra limital species like *Tilapia sparrmanii* (Banded Tilapia) and *Oreochromis mossambicus* (Mozambique Tilapia). Ecological functioning of the wetland deteriorated, mostly as a result of the large *C. carpio* population. Attempts to remove the *C. carpio* by draining of the wetland were unsuccessful and rotenone treatment was initiated in 2005. This successfully removed all the alien fish species and restored ecosystem functioning of the wetland. Total costs are not known, but the treatment cost was reduced through partial drainage of the wetland, thereby reducing the volume of water to be treated. The fish have not been detected since and the wetland is in a good condition and supports a number of native bird and amphibian species, including the Endangered *Sclerophrys pantherina* (Western Leopard Toad; Measey et al. 2012).

An off-stream dam in the Lourens River catchment outside Cape Town was treated with rotenone in 2005 to remove *M. salmoides*, *T. sparrmanii* and *C. carpio*. The dam is part of the Helderberg Nature Reserve, managed by City of Cape Town. CapeNature implemented the project, and removal of the invasive species allowed for the establishment of a refuge population of native fish from the Lourens River. The extirpation was successful.

22.6 Amphibians

Two invasive amphibian species have been identified for control operations in the greater Cape Town area: *Xenopus laevis* (African Clawed Frog) and *Sclerophrys gutturalis* (Guttural Toad). Both species pose threats to endemic native species. The hybrids of *X. laevis* and *X. gilli* are listed as requiring control (Category 1b), and *S. gutturalis* is listed as requiring control (Category 1b in the Western Cape) under the Alien and Invasive Species Regulations (RSA 2016).

22.6.1 Sclerophrys gutturalis *(Guttural Toad)*

Sclerophrys gutturalis was first observed in the Western Cape in 2000 (De Villiers 2006), more than 1200 km from the native population to the north in Durban (Telford et al. 2019; Measey et al. 2020, Chap. 5). Environmental managers are concerned about this invasion because of the co-occurrence of endemic Endangered (IUCN 2017) *Sclerophrys pantherina* (Western Leopard Toad) in the same area. Guttural Toads have been subject to control operations in the City of Cape Town since 2010. The aim of the programme is to extirpate the species in the City and prevent its re-introduction. The City of Cape Town's Invasive Species Unit has contracted a private firm to locate and remove toads at a cost of approximately ZAR 200,000 per annum. The firm works closely with municipal staff to visit land owners, undertake nocturnal searches and remove toads from private property and green areas. All life stages were targeted for removal until demographic modelling revealed that removal of eggs and tadpoles had very little impact on population numbers and that effort was best expended on removing juvenile and adult toads (Vimercati et al. 2017a). During the 8 years of the control operation, the invaded area has increased from a few properties in one suburb to occupy about 5 km^2 in 2009 and over 8 km^2 by 2015 (Measey et al. 2017). Records of toads in outlying suburbs on the southern side of the Cape Peninsula mountain chain have raised concern that human-mediated jump dispersal could significantly increase the invaded area although such populations have not yet established.

Bureaucratic delays also impact negatively on the project's effectiveness. Each year, a new contract has to be issued and delays in the contract process and the availability of funds can cause delays of several months after the start of the breeding season in October/November. In most years, management only commences in December or January, providing ample opportunities for the toad population to increase and occupy new ponds in the first part of the breeding season. Gaining access to numerous private properties in this high-income, low density, housing area has been a major impediment to control, with models suggesting that missing 55% of properties undermines the total management effort with no net effect (Vimercati et al. 2017b; Potgieter et al. 2020, Chap. 11). The Guttural Toad management effort has been sustained until the present, but based on the recent data researchers and managers have discussed the possibility of changing the aim of the control operation from extirpation to containment, because without increased effort and dedicated long-term funding, it is unlikely that extirpation will be achieved.

22.6.2 Xenopus laevis *(African Clawed Frog)*

Xenopus laevis is widespread throughout southern Africa (Furman et al. 2015), and in the Western Cape it is sympatric with *X. gilli*, the endemic, range restricted Cape Platanna (Endangered—IUCN 2017). However, the original range of *X. laevis* remains unclear (Measey et al. 2017). Furthermore, research strongly suggests that the presence of *X. laevis* in the same water bodies reduces recruitment of *X. gilli* through competition and predation (De Villiers et al. 2016; Vogt et al. 2017; Thorp et al. 2019). Early research suggested that introgression was a threat to *X. gilli* (Picker et al. 1996), and this formed the basis of NEM:BA listing hybrids of *X. gilli* and *X. laevis* for removal. However, more recent genetic studies suggest that while F1 hybrids form between the species, there is no introgression (Furman et al. 2017).

From 1985 to 2000, several thousand *X. laevis* were removed from the Cape of Good Hope section of Table Mountain National Park (TMNP) with the aim of protecting the population of *X. gilli* from genetic introgression in the National Park (Picker and De Villiers 1989; De Villiers 2004). After a hiatus of 10 years in *X. laevis* population control and as part of a research project conducted in collaboration with SANParks protected area management staff between 2010 and 2014, a total of 2126 *X. laevis* were removed from the Cape of Good Hope section of TMNP (Measey et al. 2014; De Villiers et al. 2016; Measey et al. 2017). Annual operations are still underway to date (March 2019) in a collaboration between the Organisation for Tropical Studies, SANParks and the C·I·B. Extirpation is not considered possible as *X. laevis* regularly enter the park via streams from dams on adjacent private land.

22.7 Reptiles

No introduced reptile species are currently being controlled in South Africa.

22.8 Birds

Five invasive bird species have been considered for control or eradication in South Africa to date: *Acridotheres tristis* (Common Myna), *Columba livia* (Common Pigeon), *Corvus splendens* (House Crow), *Anas platyrhynchos* (Mallard) and *Psittacula krameri* (Rose-ringed Parakeet). Common Mynas are ubiquitous in many cities and towns around South Africa (see Measey et al. 2020, Chap. 5; Potgieter et al. 2020, Chap. 11), but no large-scale or systematic control has been attempted. However, in parts of KwaZulu-Natal province, *A. tristis* trapping trials have been conducted with the result that birds have been removed from local areas. Evidence for strong negative impacts on avian and other species (Hart and Downs 2015) by

A. tristis is lacking, but as *A. tristis* are fruit and nectar feeders and are known to damage fruit crops (Gumede and Downs 2019) it is likely that impacts do occur.

Columba livia is controversial because it exists as feral populations which interact with captive populations used as pets and for pigeon racing. Currently there are no systematic control programmes for this species (Measey et al. 2020, Chap. 5).

Corvus splendens and *A. platyrhynchos* are under ongoing control in Durban and Cape Town, in operations funded by DEA: NRM and supported by a range of institutions, principally local authorities. The two avian control projects are covered in more detail below.

Rose-ringed Parakeets *Psittacula krameri* are established as breeding colonies in metropolitan areas of Gauteng, Durban and Cape Town and smaller towns and cities Steytlerville and Pietermaritzburg (Hart and Downs 2014). The Rose-ringed Parakeet is listed in Category 2 of NEM:BA, but has only been controlled on a small scale in Somerset west, where birds were caught and removed from a single dwelling resulting in local extirpation (L. Stafford, pers. comm.).

22.8.1 Corvus splendens *(House Crow)*

The Cape Town population has been under continuous management since 2009 and has been reduced from a peak population of 10,000 in 2010 to about 300 individuals. The eThekwini (Durban) population was at very low levels (thought to be <10 individuals), but recovered following a suspension of management due to cessation of funding from the government in 2017. In early 2018 a new monitoring and control programme was funded by DEA: NRM and eThekwini Municipality, and is ongoing. Other populations of *C. splendens* occur in Richards Bay and East London. The Cape Town and eThekwini programmes aim to extirpate the species from these locations. Over ZAR 8 million has been spent on the programme over 8 years of active control using poison baits and trapping at House Crow roosting and feeding sites.

As is common in eradication projects, the last few animals are difficult to detect and remove, especially given the high intelligence of these birds which learn to avoid people and baiting areas (Suliman et al. 2011). It is thought that novel approaches will need to be employed to finally achieve extirpation in the four coastal cities that are currently invaded and funding for continued monitoring is likely to be required in perpetuity due to the high risk of reintroduction at these ports.

22.8.2 Anas platyrhynchos *(Mallard)*

Anas platyrhynchos hybridise with native *Anas undulata* (Yellow-Billed Ducks). It was thought that genetic introgression and *A. platyrhynchos* aggressive competitive behaviour may lead to suppression of native duck populations in urban and peri-urban areas and loss of genetic integrity and identity in these taxa. However, recent genetic studies have shown that introgression occurs from *A. undulata* into *A. platyrhynchos* (Stephens et al. 2020). The presence of *A. platyrhynchos* on farms and in rural settings also suggests that their populations should be reduced. Because of the widespread distribution of both *A. platyrhynchos* and *A. undulata* in South Africa, eradication is not an objective of this programme.

As *A. platyrhynchos* often swim in groups with *A. undulata*, the control programme opted to feed target mallards and putative hybrids bread soaked with anaesthetic (alpha-chloralose). All ducks in the urban area are habituated to people feeding them bread. Once ducks have succumbed to the anaesthetic, they are collected by kayakers waiting in the water, and transferred to a nearby facility where a duty vet euthanises *A. platyrhynchos* and hybrids. The carcasses are then sampled for DNA and removed by the City of Cape Town. The majority of birds removed (63%) were found to be hybrids (Stephens et al. 2020). The programme was efficiently identifying and removing hybrids, with 5% being pure yellow-billed ducks (Stephens et al. 2020). The goal of this project is containment—the removal of *A. platyrhynchos* and their hybrids in areas of high abundance of both alien and native ducks in order to conserve the genetic integrity of *A. undulata*.

22.9 Mammals

22.9.1 Hemitragus jemlahicus *(Himalayan Tahr)*

An isolated population of *Hemitragus jemlahicus* is established on Table Mountain in Cape Town (Measey et al. 2020, Chap. 5). Despite originating from only a few zoo escapees, 330 individuals were counted during a survey in the 1970s (Lloyd 1975; Skead et al. 2011). The broad dietary preferences of these animals, along with their tendency to aggregate, resulted in significant environmental degradation and erosion at high densities (Lloyd 1975). As such, the authorities in charge of Table Mountain (Department of Nature and Environmental Conservation, now CapeNature, followed by SANParks from 1998), have conducted Himalayan Tahr removal programmes since the early 1970s, with the view to restoring populations of native ungulates in their place (Lloyd 1975; Gaertner et al. 2016). Public concerns over animal welfare of the tahrs have received significant media attention over the years (Gosling 2002; Skead et al. 2011).

Management interventions reduced tahr densities to a point that sightings became a newsworthy occurrence (Skead et al. 2011). However, complete eradication has not been achieved. The rugged and inaccessible mountainous terrain, exacerbated by low detectability of individuals at low population densities, make it difficult to implement management interventions. Tahrs have also been introduced to New Zealand, where they are established and have been well-studied (e.g. Forsyth and Tustin 2001; Cruz et al. 2017). However, no research has been published on the ecology, space use or management of this species on Table Mountain. Research is required, particularly in light of the re-introduction of Klipspringers (*Oreotragus oreotragus*) to the mountain and the need to understand the interactions between the two species. *Hemitragus jemlahicus* is listed in NEM:BA as a Category 1a species, for which mandatory control is required ('must be combatted or eradicated'—RSA 2016).

22.9.2 Sus scrofa *(Domestic Pig)*

Invasive populations of *Sus scrofa* (feral Domestic Pigs and their hybrids) in South Africa are descendants of Eurasian Wild Boars released by the Department of Forestry in the 1920s and 1930s, and Domestic Pigs escaped from farms (Measey et al. 2020, Chap. 5). The original intention was to control Pine Emperor Moth (*Gonimbrasia cytherea*) populations in plantations of the Western Cape (Botha 1989). Today, three populations are present on farms and protected areas in the Western Cape (Skead et al. 2011). The feral pigs eat rare geophytic plants and the Critically Endangered Geometric Tortoise (*Psammobates geometricus*) adults, juveniles and eggs, making them a conservation concern. They also have a negative impact on agricultural crops.

In 2014, feral *S. scrofa* were listed as Category 1b invaders under NEM:BA. In 2011 CapeNature, the provincial conservation authority, produced a Feral Pig Management Strategy. The strategy covered effective control measures, monitoring of the effects of control, and prevention of re-introduction, and presented a comprehensive communication strategy to raise awareness of the presence and negative impacts of feral Domestic Pig populations. Starting in 2014, CapeNature conducted a pilot trial of feral Domestic Pig control by using baited traps and Judas pigs. This programme resulted in the removal of over 1200 pigs from the Kasteelberg and Porseleinberg populations, in addition to those removed by the land owners through hunting. The Kasteelberg population is close to extirpation (van Wilgen and Wilson 2018). The implementation of this project is carried out by a private contractor and funded by DEA: NRM.

22.9.3 Felis catus *(Domestic Cat)*

In South Africa, *Felis catus* is listed as Category 1a on islands, which means that they must be eradicated from islands. *Felis catus* has been successfully eradicated on sub-Antarctic Marion Island (Bester et al. 2002) and on Dassen Island, an inshore island off the west coast (Cooper and Dyer 2013). The successful eradication of *F. catus* from Marion island stands as an outstanding example of the eradication of an invasive vertebrate globally (see Greve et al. 2020, Chap. 8).

Felis catus has existed as feral populations on Robben Island since the late 1800s (Measey et al. 2020, Chap. 5). *Felis catus* is responsible for the deaths of native small vertebrates, invertebrates, terrestrial and sea birds, and likely poses a threat for the important colonies of threatened African Penguins (*Spheniscus demersus*) on Robben Island. Domestic Cats were also kept as pets, until this was banned by the Robben Island Museum management authority in 2009. As a result of the current control programme, which has removed 109 *F. catus* since its inception in 2009, the feral cat population is very low, estimated at five to ten individuals (C. Wilke, pers. comm). In terms of the Natural Environmental Policy (Robben Island Museum 2016) the introduction of *F. catus* to the Island is prohibited and all pet cats have been removed.

22.9.4 **Oryctolagus cuniculus** *(European Rabbit)*

Oryctolagus cuniculus were introduced to several near-shore islands along the east coast and west coast by the early Dutch settlers in the 1600s (Measey et al. 2020, Chap. 5). After several unsuccessful introductions, the populations became self-sustaining on five islands—Bird Island near Port Elizabeth and Robben, Dassen, Schapen and Malgas Islands along the west coast. *Oryctolagus cuniculus* have had serious negative impacts on the native vegetation of the islands, as has been documented for other oceanic islands (reviewed in Bergstrom et al. 2009). *Oryctolagus cuniculus* has been listed as Category 1b for islands (i.e. must be controlled) in NEM:BA (RSA 2016). However, the populations on Dassen Island off the west coast, and Schapen and Malgas Islands in Saldanha Bay are not under management at present and population sizes are not known. The population on Bird Island was eradicated in the 1980s (C. Wilke pers. comm.).

On Robben Island, the total population in 2003 was estimated to be 2137 ± 453 in 2003 and by November 2008 an estimated population of 24,229 (De Villiers et al. 2010). De Villiers et al. (2010) judged the population fluctuations to be induced by the availability of food resources and predation by *F. catus*. A control operation commenced in 2008, managed by Robben Island Museum. An estimated 13,600 *O. cuniculus* were removed during this time and, at the time of writing, *O. cuniculus* have not been seen on the island for several months, but are not yet extirpated (C. Wilke, pers. comm.). Initially, *O. cuniculus* were trapped and euthanised, but since 2009 night-shooting has been employed as this is more efficient, having accounted for 10,638 of the above mentioned animals, and can be conducted in conjunction with *F. catus* and *Dama dama* (Fallow Deer) control. The control operation will continue until the stated objective of eradication from the island is reached.

22.9.5 **Dama dama** *(Fallow Deer)*

Dama dama were introduced on Robben Island from Rhodes' Estate in 1963 (Chapman and Chapman 1980, see Measey et al. 2020, Chap. 5). Initially three animals were introduced, but by 1977 there were about 40 individuals on the Island. Hunting began in 1985, and the population was quickly reduced to 12 later that year. While the Island was managed as a prison, the Fallow Deer were regularly culled for sport and to provide meat, but since the facility has been managed as a museum and World Heritage Site (since 1996) the population has expanded significantly. A concerted culling and translocation operation in 2009 removed over 280 animals, and about 30–50 remained. In the ensuing years this population rebounded to over 300 animals by 2017 which prompted a renewed culling operation (C. Wilke pers. comm.) with the aim of eradicating Fallow Deer from the Island.

While there was initially some public opposition to control of Fallow Deer, monitoring by officers of the Society for Prevention of Cruelty to Animals (SPCA) and the condition of the deer on the island have added support for the project from an ethical perspective. *Dama dama* that have died on the island have been found to have ingested large amounts of waste plastic that has been deposited on the island from the ocean; this has resulted in the deaths of a number of deer, particularly adult females. Analysis of stomach contents of 255 deer culled since 2017 have shown that 37% of the population has ingested plastic (C. Wilke pers. comm.).

22.10 Synthesis

Invasive alien animal control projects are responses to complex problems that often defy simple, linear solutions (i.e. wicked problems—Woodford et al. 2016). In addition, the proposed solutions themselves, such as the projects described here, often become wickedly difficult to implement and are themselves subject to change as implementation takes place. This is a widely acknowledged problem in conservation practice, where large spatial scales, stakeholder diversity and multi-faceted contexts combine to confound solutions (Game et al. 2014). There is no right answer to many conservation problems, and one actor's optimal solution may be completely unacceptable or simply non-optimal to another actor, and the "multitude of conflicting perspectives, objectives, and management goals can make the problem almost impossible to characterise, let alone solve, to the satisfaction of all stakeholders" (Woodford et al. 2016). The acceptability of trade-offs such as animal welfare concerns is also sensitive to value systems, interests and cultural contexts, as shown in several of the projects described here.

22.10.1 Species Which Are Not Yet Under Adequate Control

Over 80 animal taxa (including some groups of hybrids) are listed as Category 1a or 1b in NEM:BA, thus placing an obligation on landowners to eradicate or control them, but only 28 have been or are currently controlled (Table 22.1). Much further action will be necessary to ensure that more of the Category 1 species are adequately controlled; to increase the impact of control operations, stakeholder management, authority engagement and collaboration and learning from existing experience will be highly beneficial.

22.10.2 *Stakeholder Management*

If invasive animal populations are going to be effectively controlled, strong collaborative relationships between institutions and between local government and civil society bodies and NGOs will be necessary. Three working groups (KwaZulu-Natal Invasive Alien Species Forum, CAPE Invasive Alien Animal Working Group in the Western Cape and the national Marine Alien and Invasive Species Working Group) are actively addressing issues related to the management of invasive alien animal populations. These groups are valuable for improving the flow of information between environmental managers in local and provincial government, researchers and NGOs, and contribute to networking and building and maintaining working relationships between individuals and institutions involved in invasive animal control. However, there is further work to be done in public engagement, media relations, social media, and the nature and timing of publicity in projects supported by the working groups.

The working groups also play an important role in supporting, and in some cases designating, champions for individual projects. For example, in the case of the Rondegat River *M. dolomieu* project, from the initiation of the project (environmental impact assessment) in 2008 to implementation in 2012–2013, the project was championed by the ichthyologist at CapeNature Scientific Services with the support of the CAPE IAA WG and other colleagues. The support of the working group was deemed essential to keeping the process on track through long delays and strong opposition from stakeholder groups (D. Impson pers. comm.).

22.10.3 *Conflict Management*

Public opposition to animal control operations can result in significant delays (Zengeya et al. 2017). On the other hand, where the public is supportive, this can greatly improve detections and improve relationships between stakeholder groups, as in the case of the invasive wasp species *V. germanica* and *P. dominula* in Cape Town. Among the projects described in this chapter, there have been a range of responses by public and stakeholder groups ranging from outright support (e.g. *V. germanica*, *P. dominula* and *C. splendens*) to strong conflict with multiple groups (e.g. *A. playrhynchos* and *H. jemlahicus*). Gaertner et al. (2016) constructed a framework for classifying invasive species into management classes based on their impacts and conflicts with particular reference to cities. Five possible management responses were then identified based on species impact, value and level of conflict experienced or expected. The management options are to tolerate, monitor, contain, control or eradicate. The authors point out that decision making in response to invasive species in cities has to be rapid, because of the spread rates and the rapidity with which invasive species populations become unmanageable (Gaertner et al. 2016). Similarly, conflicts also have to be managed rapidly, so that management can get off the ground as early as possible in the invasion process.

Participatory working groups, such as the three invasive species working groups established in South Africa to date, can assist invasive species managers to rapidly scope and initiate data gathering, stakeholder engagement and control operations. Since these groups contain a greater range of stakeholders than would typically be available for invasive species managers to consult, they broaden the debate and create awareness of the range of different perspectives that are likely to come from the public. Also, they allow other participants, for example, researchers, to understand the needs of managers and to target their research towards useful outcomes. A good example of this is the study that was initiated by a researcher involved with the CAPE IAAWG, who engaged a doctoral student to work on the Guttural Toad invasion. The study revealed through demographic modelling that collecting eggs and larvae of the toads has little impact on populations, and that management effort would be better expended on collecting adults. The range of experience held by the working group's network can be quickly brought to bear on new problems where there is initially very little definite information available (e.g. taxon-specific research studies) and urgent action is needed.

Gaertner et al. (2016) state that although stakeholders may have an incomplete understanding of the issues surrounding invasive species impacts and their interactions with native species and ecosystems, a wide range of views needs to be aired and considered during the decision-making process. From a study using cognitive hierarchy theory and risk perception frameworks, an expert group of assessors concluded that more conflicts arose over intrinsic values than utilitarian ones (Zengeya et al. 2017). This mirrors our findings, as the strongest conflicts experienced in South African cases (e.g. *A. platyrhynchos*, *D. dama* and *H. jemlahicus*) have been related to the ethics and techniques of removing and killing the animals. Alternative solutions such as leaving the animals where they are or translocating them to another habitat (potentially even back to the native range) have been endorsed by residents' groups and concerned individuals. In contrast, when the negative impacts of the species on humans were clearly recognised by stakeholders, no opposition was experienced by managers. The *V. germanica* and *P. dominula* projects have proceeded without strong public opposition.

In the context of the legislative requirement to manage invasive species (e.g. through the NEM:BA Alien and Invasive Species Lists), the effective management of conflicts to produce acceptable and practicable solutions is essential, as ultimately the control operation needs to proceed without excessive delays. Novoa et al. (2018) devised four categories of stakeholders based on impact: context setters, key players, crowd and subjects. These groups highlight the diversity that exists among invasive species conflicts and how different groups can be approached. For example, empowering stakeholders who currently have little influence or are 'hidden' from the public eye can build understanding, capacity and support for some initiatives (Novoa et al. 2018). This perspective may go some way to explaining why the projects involving wasps and House Crows did not result in significant opposition from stakeholder groups, as the societal costs of these species' presence in Cape Town were clear and did not require appreciation of impacts on ecosystems or the receiving environment (see Sect. 22.8.1).

22.10.4 Scaling Up

The acquisition of knowledge, better communication and awareness of the successes and failures of stakeholder engagement and options for improving practice are advancing but more work is urgently needed. Many of the control operations described above were undertaken at a local level—i.e. aiming to control a particular population or set of populations within a municipality or protected area—rather than on a national or subcontinental level. Conducting invasive and alien animal control operations at a local level is probably more effective—many of these operations were successful because they involved relatively small stakeholder groups and were flexible and efficient in response to changing circumstances and available techniques and tools. However, although local control operations are appropriate when the spatial scale of the invasion is small or contained, they can increase the chances of re-invasion from outside the control area if all populations are not addressed simultaneously. This is likely to be the case when the distribution of invasive populations is patchy and invasion pathways cannot be completely shut down.

House Crows are present in four widely spread in the port cities of Richards Bay, Durban, East London and Cape Town, which span the latitudinal extent of South Africa's coastline. Re-invasion through established pathways (Convention on Biological Diversity 2014) such as shipping (transport—stowaway) or the pet trade (release from confinement) is possible and likely in some cases. Therefore, South Africa requires an effective mechanism for scaling individual local operations up to work across provincial and perhaps national boundaries in cases where invasions are widespread. In particular, learning needs to be transferred between the levels of government, for example from local authority level to provincial and national agencies. Inter-institutional fora such as provincial and national working groups will be vital in establishing these broader channels of interaction.

22.10.5 Financial and Contract Management

Where external service providers are contracted to carry out operations, the administration and monitoring of these contracts becomes highly important. For example, several of the projects discussed here are carried out partly or entirely by private companies that specialise in invasive alien animal control. The projects on Guttural Toads and House Crows were delayed due to breaks in funding, resulting in lost opportunities and unchecked increases in the invasive populations. To avoid these breaks in funding, depending on the timing of financial cycles in the branch of government that is funding the operation, administrative processes may have to be initiated before the end of the previous financial year or bridging funds supplied to ensure that there is no break in resource availability.

22.10.6 Critical Assessment of Control Efforts to Date

This review has shown that the regional and eco-regional (e.g. marine) working groups have been highly effective in some cases where control or eradication, complementing the efforts of government and private partners. In the long-term, these groups must be supported to continue their work by the institutions that employ the members and support the secretariats. The question of whether the control projects that have been undertaken to date represent an adequate return on investment, is important, and needs to be addressed in future.

Ongoing re-assessment and monitoring of control efforts is important, and the goals of control projects should not be static but should be regularly re-evaluated. The value of advisory forums such as the CAPE IAAWG and the KZN Invasive Alien Species Forum is that there is a scheduled trigger for re-evaluation of project feasibility as regular meetings are held, bringing most or all of the required stakeholders together each time. At the same time, sustaining operations is important even when it seems that progress is sometimes not being made rapidly. The initiation costs of management and control projects are high, so current operations should be sustained despite temporary pressure to discontinue, such as funding delays.

The research reflected in this chapter has shown that the control of one species (*C. maenas*, European Shore Crab) is not necessary due to it not spreading outside harbours and the small chance of it becoming invasive on our highly energetic coastline. Species that can realistically be extirpated/eradicated by sustaining or increasing the current efforts are: Himalayan Tahrs (*H. jemlahicus*), Fallow Deer (*D. dama*), House Crows (*C. splendens*) and on islands Domestic Cats (*F. catus*) and European Rabbits (*O. cuniculus*). It is unlikely that the 12 freshwater fish species covered here will be eradicated, but ongoing control in priority catchments and protected areas is important for conservation purposes, given the high levels of endemism in the Western Cape. Eradication of the two species of wasps (*V. germanica* and *P. dominula*) may be feasible, depending on the project design—i.e. quick response to public reports of nest identification (Veldtman et al. 2012). In these two cases, large-scale efforts that span several local authorities and towns will have to be made, with guaranteed long-term funding from the national level.

Control of Mallards and Guttural Toads is in progress at a local authority level, but is not likely to lead to local extirpation of these species in the near term. It is likely that the goals of these projects may be re-evaluated in the near future. Species that are not under comprehensive control at present, such as *A. tristis* (Common Myna) require further research to determine the feasibility of control, some of which is being conducted at present. *Xenopus laevis* × *X. gilli* hybrids are being removed from one protected area (TMNP) but are not systematically controlled anywhere else in the range of *X. gilli*. Overall, the picture is one of varied, context-specific control projects being run at local scale. These efforts need to be scaled up to cover regional, national or larger areas if alien and invasive animals are to be controlled more effectively.

As shown in the examples and discussion above, managing stakeholders, resolving conflict, ensuring sustained funding and scaling efforts up to larger spatial and temporal scales all involve collaborative efforts that extend outside ordinary organisational and political boundaries. Involving role players as early as possible in the planning process is a key aspect of successful projects. Also, recognising the trade-offs among costs and benefits accruing to role players and "who loses, who pays, and who benefits" (Hirsch et al. 2010) can open the way to productive rather than conflictual relationships with stakeholders. Open and transparent communication among role players builds trust and facilitates learning from perceived successes and failures (Game et al. 2014).

Network governance (see Scarlett and McKinney 2016) has been used by several provincial and national invasive alien species working groups in South Africa to harness public, private and non-profit organisational expertise and carry out invasive alien species management projects. This way of actively collaborating achieves objectives that could not be achieved by one or two role players alone. Sustaining these approaches will advance future efforts to control invasive alien animal species in South Africa.

Acknowledgements We are grateful to the CAPE Invasive Alien Animals Working Group, Christopher Wilke, Sabelo Madlala, Deborah Winterton, Gavin Bell, Dean Impson, Louise Stafford, and Dean Ferreira for their constructive inputs. We would like to thank the DSI-NRF Centre of Excellence for Invasion Biology and the Natural Resources Management Programme of DEFF for their support. DEFF are also thanked for funding the South African National Biodiversity Institute noting that this publication does not necessarily represent the views or opinions of DEFF or its employees.

References

Abell R, Thieme ML, Revenga C et al (2008) Freshwater ecoregions of the world: a new map of biogeographic units for freshwater biodiversity conservation. Bioscience 58:403–414. https://doi.org/10.1641/B580507

Addison P, Cugula D, Daneel J-H et al (2016) Networking projects on fruit flies between Europe and Africa. In: Poster at 3rd international symposium of tephritid workers of Europe, Africa and the Middle East, Stellenbosch

Athanassiou CG, Phillips TW, Wakil W (2019) Biology and control of the Khapra Beetle, *Trogoderma granarium*, a major quarantine threat to global food security. Annu Rev Entomol 64:131–148. https://doi.org/10.1146/annurev-ento-011118-111804

Barnes BN, Venter J-H (2006) The South African fruit fly action plan—area-wide suppression and exotic species surveillance. In: Proceedings of the 7th international symposium of fruit flies of economic importance, pp 271–283

Barrow SB (2014) Contrasting impact of alien invasive sport fish in the Cape Floristic Region: a focus on *Micropterus dolomieu*. MSc Thesis, Stellenbosch University

Bellingan TA, Woodford DJ, Gouws J, Villet MH, Weyl OLF (2015) Rapid bioassessment of the effects of repeated rotenone treatments on invertebrate assemblages in the Rondegat River, South Africa. Afr J Aquat Sci 40:89–94. https://doi.org/10.2989/16085914.2014.984651

Bellingan TA, Hugo S, Woodford DJ et al (2019) Rapid recovery of macroinvertebrates in a South African stream treated with rotenone. *Hydrobiologia*. https://doi.org/10.1007/s10750-019-3885-z

Benadé PC, Veldtman R, Samways MJ et al (2014) Rapid range expansion of the invasive wasp *Polistes dominula* (Hymenoptera: Vespidae: Polistinae) and first record of parasitoids on this species and the native *Polistes marginalis* in the Western Cape Province of South Africa. Afr Entomol 22:220–225. https://doi.org/10.4001/003.022.0104

Bergstrom DM, Lucieer A, Kiefer K et al (2009) Indirect effects of invasive species removal devastate World Heritage island. J Appl Ecol 46:73–81. https://doi.org/10.1111/j.1365-2664.2008.01601.x

Bester MN, Bloomer JP, van Aarde RJ et al (2002) A review of the successful eradication of feral cats from sub-Antarctic Marion Island, Southern Indian Ocean. S Afr J Wildl Res 32:65–73

Botha SA (1989) Feral pigs in the Western Cape Province: failure of a potentially invasive species. South Afri For J 151:17–25. https://doi.org/10.1080/00382167.1989.9630500

Cambray JA (1995) *Sandelia bainsii* will it be next? Ichthos 45:2–3. https://doi.org/10.1007/BF00005228

Cambray JA (2003) The need for research and monitoring on the impacts of translocated sharptooth catfish, *Clarias gariepinus*, in South Africa. Afr J Aquat Sci 28:37–41. https://doi.org/10.2989/16085910309503786

Chapman NG, Chapman DI (1980) The distribution of fallow deer: a worldwide review. Mammal Rev 10:61–138. https://doi.org/10.1111/j.1365-2907.1980.tb00234.x

Convention on Biological Diversity (2014) Pathways of introduction of invasive species, their prioritization and management. UNEP/CBD/SBSTTA/18/9/ Add.1. Secretariat of the Convention on Biological Diversity, Montreal. https://www.cbd.int/doc/meetings/sbstta/sbstta-18/official/sbstta-18-09-add1-en.pdf. Accessed 19 Feb 2019

Cooper J, Dyer BM (2013) The eradication of feral cats from Dassen Island: a first for Africa? Aliens 33:35–37

Cruz J, Thomson C, Parkes JP et al (2017) Long-term impacts of an introduced ungulate in native grasslands: Himalayan tahr (*Hemitragus jemlahicus*) in New Zealand's Southern Alps. Biol Invasions 19:339–349. https://doi.org/10.1007/s10530-016-1283-2

Day C, White B (2016) Khapra beetle, *Trogoderma granarium* interceptions and eradications in Australia and around the world. SARE working paper 1609, School of Agricultural and Resource Economics, University of Western Australia, Crawley

de Kock KN, Wolmarans CT (2008) Invasive alien freshwater snail species in the Kruger National Park, South Africa. Koedoe 50:49–53. https://doi.org/10.4102/koedoe.v50i1.126

De Moor IJ, Bruton MN (1988) Atlas of alien and translocated indigenous aquatic animals of southern Africa. South African National Scientific Programmes Report 144

De Villiers A (2006) *Bufo gutturalis* Power, 1927. Guttural Toad. Introduced population. African Herp News 40:28–29

De Villiers AL (2004) *Xenopus gilli* Rose and Hewitt, 1927. In: Minter LR, Burger M, Harrison JA, Braack HH, Bishop PJ, Kloepfer D (eds) Atlas and red data book of the frogs of South Africa, Lesotho and Swaziland. SI/MAB Series #9. Smithsonian Institution, Washington, DC, pp 260–263

De Villiers MS, Mecenero S, Sherley RB et al (2010) Introduced European rabbits (*Oryctolagus cuniculus*) and domestic cats (*Felis catus*) on Robben Island: population trends and management recommendations. S Afr J Wildl Res 40:139–148. https://doi.org/10.3957/056.040.0205

De Villiers FA, de Kock M, Measey GJ (2016) Controlling the African clawed frog *Xenopus laevis* to conserve the Cape platanna *Xenopus gilli* in South Africa. Conserv Evid 13:17–17

De Villiers M, Kriticos DJ, Veldtman R (2017) Including irrigation in niche modelling of the invasive wasp *Vespula germanica* (Fabricius) improves model fit to predict potential for further spread. PLoS One 12:e0181397. https://doi.org/10.1371/journal.pone.0181397

Drew RAI, Tsuruta K, White IM (2005) A new species of pest fruit fly (Diptera: Tephritidae: Dacinae) from Sri Lanka and Africa. Afr Entomol 13:149–154

Ellender BR, Weyl OLF (2014) A review of current knowledge, risk and ecological impacts associated with non-native freshwater fish introductions in South Africa. Aquat Invasions 9:117–132. https://doi.org/10.3391/ai.2014.9.2.01

Ellender BR, Weyl OLF, Swartz ER (2011) Invasion of a headwater stream by non-native fishes in the Swartkops River system, South Africa. Afr Zool 46:39–46. https://doi.org/10.3377/004.046.0116

Ellender BR, Woodford DJ, Weyl OLF, Cowx IG (2014) Managing conflicts arising from fisheries enhancements based on non-native fishes in southern Africa. J Fish Biol https://doi.org/10.1111/jfb.12512 Available online at www.wileyonlinelibrary.com

EPPO (2019) EPPO global database (available online). Accessed 9 Jan 2019

Finlayson B, Schnick R, Skaar D et al (2010) Planning and standard operating procedures for the use of rotenone in fish management. American Fisheries Society, Bethesda, MD

Forsyth DM, Tustin KG (2001) Advances in New Zealand mammalogy 1990–2000: Himalayan tahr. J R Soc N Z 31:251–261. https://doi.org/10.1080/03014223.2001.9517653

Furman BL, Bewick AJ, Harrison TL et al (2015) Pan-African phylogeography of a model organism, the African clawed frog *Xenopus laevis*. Mol Ecol 24:909–925. https://doi.org/10.1111/mec.13076

Furman B, Cauret C, Colby G et al (2017) Limited genomic consequences of hybridization between two African clawed frogs, *Xenopus gilli* and X. *laevis* (Anura: Pipidae). Sci Rep 7:1091. https://doi.org/10.1038/s41598-017-01104-9

Gaertner M, Larson BMH, Irlich UM et al (2016) Managing invasive species in cities: a framework from Cape Town, South Africa. Landsc Urban Plan 151:1–9. https://doi.org/10.1016/j.landurbplan.2016.03.010

Game ET, Meijaard E, Sheil D et al (2014) Conservation in a wicked complex world; challenges and solutions. Conserv Lett 7:271–277. https://doi.org/10.1111/conl.12050

Gosling M (2002) Friends of the Tahr slam SA National Parks for failing to find a humane solution. Cape Times, 5 July 2002

Greve M, von der Meden CEO, Janion Scheepers C (2020) Biological invasions in South Africa's offshore sub-Antarctic territories. In: van Wilgen BW, Measey J, Richardson DM et al (eds) Biological invasions in South Africa. Springer, Berlin, pp 205–226. https://doi.org/10.1007/978-3-030-32394-3_8

Griffiths CL, Hockey PAR, van Erkom Schurink C et al (1992) Marine invasive aliens on South African shores—implications for community structure and trophic functioning. S Afr J Mar Sci 12, 713–722. https://doi.org/10.2989/02577619209504736

Gumede T, Downs CT (2019) Sugar preference of invasive Common Mynas (*Sturnus tristis*). J Ornithol 160:71–78. https://doi.org/10.1007/s10336-018-1600-x

Hampton SL, Griffiths CL (2007) Why *Carcinus maenas* cannot get a grip on South Africa's wave-exposed coastline. Afr J Mar Sci 29:123–126. https://doi.org/10.2989/AJMS.2007.29.1.11.76

Hart LA, Downs CT (2014) Public surveys of rose-ringed parakeets, *Psittacula krameri*, in the Durban Metropolitan area, South Africa. Afr Zool 49:283–289. https://doi.org/10.1080/15627020.2014.11407644

Hart L, Downs C (2015) Winged invaders: bird introductions. Quest 11:38–41

Haupt K (2015) Assessment of the invasive German wasp, *Vespula germanica*, in South Africa. M. Sc. thesis, Stellenbosch University, Stellenbosch

Haupt TM, Griffiths CL, Robinson TB et al (2010) Oysters as vectors of marine aliens, with notes on four introduced species associated with oyster farming in South Africa. Afr Zool 45:52–62. https://doi.org/10.3377/004.045.0101

Herbert DG, Sirgel WF (2001) The recent introduction of two potentially pestiferous alien snails into South Africa and the outcomes of different pest management practices: an eradication and a colonization. S Afr J Sci 97:301–304

Hirsch PD, Adams WM, Brosius JP et al (2010) Acknowledging conservation trade-offs and embracing complexity. Conserv Biol 25:259–264. https://doi.org/10.1111/j.1523-1739.2010.01608.x

Impson ND, Harding B, Steyn G (2005) The use of rotenone as a tool for restructuring fish populations in inland waters of the fynbos biome. Piscator 137:63–69

IUCN SSC Amphibian Specialist Group and South African Frog Re-assessment Group (SA-FRoG) 2017. *Xenopus gilli*. The IUCN Red List of Threatened Species 2017: e.T23124A77164368. Downloaded on 11 January 2019. https://doi.org/10.2305/IUCN.UK.2017-2.RLTS.T23124A77164368.en

Jackson MC, Gery J, Miller K et al (2016) Dietary niche constriction when invaders meet natives: evidence from freshwater decapods. J Anim Ecol 85:1098–1107. https://doi.org/10.1111/1365-2656.12533

Janion-Scheepers C, Griffiths CL (2020) Alien terrestrial invertebrates in South Africa. In: van Wilgen BW, Measey J, Richardson DM et al (eds) Biological invasions in South Africa. Springer, Berlin, pp 183–204. https://doi.org/10.1007/978-3-030-32394-3_7

Karsten M, Jansen van Vuuren B, Addison P et al (2015) Deconstructing intercontinental invasion pathway hypotheses of the Mediterranean fruit fly (*Ceratitis capitata*) using a Bayesian inference approach: are port interceptions and quarantine protocols successfully preventing new invasions? Divers Distrib 21:813–825. https://doi.org/10.1111/ddi.12333

Karsten M, Barton M, Addison P et al (2018) A synthesis for managing invasions and pest risks simultaneously for tephritid fruit flies in South Africa. Entomol Exp Appl 166:344–356. https://doi.org/10.1111/eea.12664

Kleynhans CJ (1987) A preliminary study of aspects of a rare minnow *Barbus treurensis* Groenewald, 1958 (Pisces: Cyprinidae) from the Eastern Transvaal, South Africa. J Limnol Soc S Afr 13:7–13. https://doi.org/10.1080/03779688.1987.9634537

Lester PJ, Beggs JR (2019) Invasion success and management strategies for social *Vespula* wasps. Ann Rev Entomol 64:51–71. https://doi.org/10.1146/annurev-ento-011118-111812

Lloyd PHT (1975) A study of the Homalaayan Tahr (*Hemitragus jemlahicus*) and its potential effects on the ecology of the Table Mountain range. Cape Department of Nature and Environmental Conservation, Cape Town

Mabin CA, Wilson JRU, Robinson TB (2015) The Chilean black urchin, *Tetrapygus niger* (Molina, 1782) in South Africa: gone but not forgotten. BioInvasions Rec 4:261–264. https://doi.org/10.3391/bir.2015.4.4.05

Mabin CA, Wilson JRU, Le Roux JJ, Robinson TB (2017) Reassessing the invasion of South African waters by the European shore-crab *Carcinus maenas*. Afr J Mar Sci 39:259–267. https://doi.org/10.2989/1814232X.2017.1363818

Manrakhan A, Grout T, Hattingh V (2009) Combating the African invader fly *Bactrocera invadens*. SA Fruit J 8:57–61

Manrakhan A, Hattingh V, Venter J-H et al (2011) Eradication of *Bactrocera invadens* (Diptera: Tephritidae) in Limpopo Province, South Africa. Afr Entomol 19:650–659. https://doi.org/10.4001/003.019.0307

Manrakhan A, Venter J-H, Hattingh V (2012) Action Plan for the Control of the African Invader Fruit Fly, Bactrocera invadens Drew, Tsuruta and White. Department of Agriculture, Forestry and Fisheries, Pretoria

Manrakhan A, Venter J-H, Hattingh V (2015) The progressive invasion of *Bactrocera dorsalis* (Diptera: Tephritidae) in South Africa. Biol Invasions 17:2803–2809. https://doi.org/10.1007/s10530-015-0923-2

Marr SM, Impson ND, Tweddle D (2012) An assessment of a proposal to eradicate non-native fish from priority rivers in the Cape Floristic Region, South Africa. Afr J Aquat Sci 37:131–142. https://doi.org/10.2989/16085914.2012.666654

Marr SM, Bellingan TA, Dalu T et al (2019) Rotenone policy support and capacity development: Part 1: impact and recovery of native biota in one river and two dams following alien fish removals using rotenone, WRC project no. K5/2538. Water Research Commission, Pretoria

Mead A, Carlton JT, Griffiths CL et al (2011a) Introduced and cryptogenic marine and estuarine species of South Africa. J Nat Hist 45:2463–2524. https://doi.org/10.1080/00222933.2011.595836

Mead A, Carlton JT, Griffiths CL et al (2011b) Revealing the scale of marine bioinvasions in developing regions: a South African re-assessment. Biol Invasions 13:1991–2008. https://doi.org/10.1007/s10530-011-0016-9

Measey J, Dorse S, Faraday A (2012) Conservation of the western leopard toad by a dedicated multi-stakeholder group in the City of Cape Town. FrogLog 103:14–16

Measey GJ, Annecke W, Davies SJ, Dorse C, Stafford L, Tolley KA, Turner AA (2014) Cape collaborations for amphibian solutions. FrogLog 109:46–47

Measey J, Davies SJ, Vimercati G et al (2017) Invasive amphibians in southern Africa: a review of invasion pathways. Bothalia 47:a2117. https://doi.org/10.4102/abc.v47i2.2117

Measey J, Hui C, Somers MJ (2020) Terrestrial vertebrate invasions in South Africa. In: van Wilgen BW, Measey J, Richardson DM et al (eds) Biological invasions in South Africa. Springer, Berlin, pp 113–150. https://doi.org/10.1007/978-3-030-32394-3_5

Mwatawala MW, White IM, Maerere AP et al (2004) A new invasive *Bactrocera* species (Diptera: Tephritidae) in Tanzania: short communication. Afr Entomol 12:154

Novoa A, Shackleton R, Canavan S et al (2018) A framework for engaging stakeholders on the management of alien species. J Environ Manag 205:286–297. https://doi.org/10.1016/j.jenvman.2017.09.059

Nunes AL, Hoffman AC, Zengeya TA et al (2017) Red swamp crayfish, *Procambarus clarkii*, found in South Africa 22 years after attempted eradication. Aquat Conserv Mar Freshwat Ecosyst 27:1334–1340. https://doi.org/10.1002/aqc.2741

Picker MD, Griffiths CL (2017) Alien animals in South Africa—composition, introduction history, origins and distribution patterns. Bothalia 47:2311–9284. https://doi.org/10.4102/abc.v47i2.2147

Picker MD, De Villiers AL (1989) The distribution and conservation status of *Xenopus gilli* (Anura: Pipidae). Biol Conserv 49:169–183. https://doi.org/10.1016/0006-3207(89)90034-7

Picker MD, Harrison JA, Wallace D (1996) 4. Natural hybridization between *Xenopus laevis laevis* and *X. gilli* in the south-western Cape Province, South Africa. In: Tinsley RC, Kobel HR (eds) The biology of Xenopus. Oxford University Press, London, pp 1960–1999

Potgieter L, Douwes E, Gaertner M et al (2020) Biological invasions in South Africa's urban ecosystems: patterns, processes, impacts and management. In: van Wilgen BW, Measey J, Richardson DM et al (eds) Biological invasions in South Africa. Springer, Berlin, pp 273–310. https://doi.org/10.1007/978-3-030-32394-3_11

Richardson DM, Pyšek P, Carlton JT (2011) A compendium of essential concepts and terminology in invasion ecology. In: Richardson DM (ed) Fifty years of invasion ecology. The legacy of Charles Elton. Wiley-Blackwell, Oxford, pp 409–420. https://doi.org/10.1002/9781444329988.ch30

Robben Island Museum (2016) Natural environmental policy. Department of Arts and Culture, Cape Town

Robinson TB, Peters K, Brooker B (2020) Coastal invasions: the South African context. In: van Wilgen BW, Measey J, Richardson DM et al (eds) Biological invasions in South Africa. Springer, Berlin, pp 227–246. https://doi.org/10.1007/978-3-030-32394-3_9

RSA (Republic of South Africa) (2014) National environment management: biodiversity act, 2004 (Act no. 10 of 2004). Alien and Invasive Species Regulations. Government Gazette No. 37886

RSA (Republic of South Africa) (2016) National environmental management: biodiversity act, 2004 (Act no. 10 of 2004); Alien and Invasive Species Lists. Government Notice R864, Government Gazette No. 40166

Scarlett L, McKinney M (2016) Connecting people and places: the emerging role of network governance in large landscape conservation. Front Ecol Environ 14:116–125. https://doi.org/10.1002/fee.1247

Schutze MK, Aketarawong N, Amornsak W et al (2015) Synonymization of key pest species within the *Bactrocera dorsalis* species complex (Diptera: Tephritidae): taxonomic changes based on a

review of 20 years of integrative morphological, molecular, cytogenetic, behavioural and chemoecological data. Syst Entomol 40:456–471. https://doi.org/10.1111/syen.12113
Shelton JM, Samways MJ, Day JA (2014) Predatory impact of non-native rainbow trout on endemic fish populations in headwater streams in the Cape Floristic Region of South Africa. Biol Invasions 17:365–379. https://doi.org/10.1007/s10530-014-0735-9
Shelton J, Weyl O, van der Walt J et al (2017) Effect of an intensive mechanical removal effort on a population of non-native rainbow trout *Oncorhynchus mykiss* in a South African headwater stream. Aquat Conserv Mar Freshwat Ecosyst 27:1–5. https://doi.org/10.1002/aqc.2752
Simakani S (2017) "BEEWARE of the wasp": the role of language in studying and managing the European Paper Wasp. M.A. thesis, Stellenbosch University, Stellenbosch
Skead CJ, Boshoff A, Kerley GIH et al (2011) Introduced (non-indigenous) species: a growing threat to biodiversity. In: Skead CJ, Boshoff A, Kerley GIH, Lloyd P (eds) Historical incidence of the larger land mammals in the broader Northern and Western Cape, vol 1. Centre for African Conservation Ecology, Nelson Mandela Metropolitan University, Port Elizabeth
Skelton PH (2001) A complete guide to the freshwater fishes of Southern Africa. Struik Publishers, Cape Town
Skelton PH (1993) On bass in Blindekloof—the impact of an alien predator on a wilderness stream. Naturalist 37:21–27
Stephens K, Measey J, Reynolds C et al (2020) Occurrence and extent of hybridization between the invasive Mallard Duck and native Yellow-billed Duck in South Africa. Biol Invasions 22:693–707. https://doi.org/10.1007/s10530-019-02122-6
Suckling DM, Kean JM, Stringer LD et al (2014) Eradication of tephritid fruit fly pest populations: outcomes and prospects. Pest Manag Sci 72:456–465. https://doi.org/10.1002/ps.3905
Suliman AS, Meier GG, Haverson PJ (2011) Eradication of the house crow from Socotra Island, Yemen. In: Veitch CR, Clout MN, Towns DR (eds) Island invasives: eradication and management. IUCN, Gland, pp 361–363
Telford N, Channing A, Measey J (2019) Origin of invasive populations of the Guttural toad Sclerophrys gutturalis. Herpetol Conserv Biol 14(2):380–392
Thorp CJ, Vonesh JR, Measey J (2019) Cannibalism or congeneric predation? The African clawed frog (*Xenopus laevis* Daudin) preferentially predate on larvae of Cape platannas, *X. gilli* Rose and Hewitt. Afr J Ecol 57:59–65. https://doi.org/10.1111/aje.12577
Tribe GD, Richardson DM (1994) The European wasp, *Vespula germanica* (Fabricius) (Hymeoptera, Vespidae), in southern Africa and its potential distribution as predicted by ecoclimatic matching. Afr Entomol 2:1–6
van der Walt JA, Weyl OLF, Woodford DJ et al (2016) Spatial extent and consequences of black bass (*Micropterus spp.*) invasion in a Cape Floristic Region river basin. Aquat Conser Mar Freshwat Ecosyst 26:736–748. https://doi.org/10.1002/aqc.2589
van der Walt JA, Marr SM et al (2018) Successful mechanical eradication of spotted bass (*Micropterus punctulatus* (Rafinesque, 1819)) from a South African river. Aquat Conser Mar Freshwat Ecosyst. https://doi.org/10.1002/aqc.3035
van Rensburg BJ, Weyl OLF, Davies SJ et al (2011) Invasive vertebrates of South Africa. In: Pimentel D (ed) Biological invasions: economic and environmental costs of alien plant, animal, and microbe species, 2nd edn. CRC Press, Boca Raton, pp 325–378. https://doi.org/10.1201/b10938-23
van Wilgen BW, Wilson JR (eds) (2018) The status of biological invasions and their management in South Africa in 2017. South African National Biodiversity Institute, Kirstenbosch and DST-NRF Centre of Excellence for Invasion Biology, Stellenbosch
van Wilgen BW, Measey J, Richardson DM et al (2020) Biological invasions in South Africa: an overview. In: van Wilgen BW, Measey J, Richardson DM et al (eds) Biological invasions in South Africa. Springer, Berlin, pp 3–30. https://doi.org/10.1007/978-3-030-32394-3_1
van Zyl C, Addison P, Veldtman R (2018) The invasive Vespidae in South Africa: potential management strategies and current status. Afr Entomol 26:267–285. https://doi.org/10.4001/003.026.0267

Veldtman R, Addison P, Tribe GD (2012) Current status and potential future impact of invasive vespid wasps (*Vespula germanica* and *Polistes dominulus*) in South Africa. IOBC/WPRS Bull 75:217–221

Viljoen JH (1990) The occurrence of *Trogoderma* (Coleoptera: Dermestidae) and related species in Southern Africa with special reference to *T. granarium* and its potential to become established. J Stored Prod Res 26:43–51. https://doi.org/10.1016/0022-474X(90)90036-R

Vimercati G, Hui C, Davies SJ et al (2017a) Integrating age structured and landscape resistance models to disentangle invasion dynamics of a pond-breeding anuran. Ecol Model 356:104–116. https://doi.org/10.1016/j.ecolmodel.2017.03.017

Vimercati G, Davies SJ, Hui C et al (2017b) Does restricted access limit management of invasive urban frogs? Biol Invasions 19:3659–3674. https://doi.org/10.1007/s10530-17-1599-6

Vogt S, De Villiers FA, Ihlow F et al (2017) Competition and feeding ecology in two sympatric *Xenopus* species (Anura: Pipidae). PeerJ 5:e3130. https://doi.org/10.7717/peerj.3130

Weyl OLF, Finlayson B, Impson ND et al (2014) Threatened endemic fishes in South Africa's Cape Floristic Region: a new beginning for the Rondegat River. Fisheries 39:270–279. https://doi.org/10.1080/03632415.2014.914924

Weyl OLF, Ellender BR, Wasserman RJ et al (2015) Unintended consequences of using alien fish for human benefit in protected areas Koedoe 57, Art. #1264, 5 p. https://doi.org/10.4102/koedoe.v57i1.1264

Weyl OLF, Ellender B, Wassermann RJ et al (2020) Alien freshwater fauna in South Africa. In: van Wilgen BW, Measey J, Richardson DM et al (eds) Biological invasions in South Africa. Springer, Berlin, pp 151–182. https://doi.org/10.1007/978-3-030-32394-3_6

Woodford DJ, Barber-James HM, Bellingan TA et al (2013) Immediate impact of piscicide operations on a Cape Floristic Region aquatic insect assemblage: a lesser of two evils? J Insect Conserv 17:959–973. https://doi.org/10.1007/s10841-013-9578-4

Woodford DJ, Richardson DM, MacIsaac HJ et al (2016) Confronting the wicked problem of managing invasive species. Neobiota 31:63–86. https://doi.org/10.3897/neobiota.31.10038

Woodford DJ, Ivey P, Jordaan MS et al (2017) Optimising invasive fish management in the context of invasive species legislation in South Africa. Bothalia 47:a2138. https://doi.org/10.4102/abc.v47i2.2138

Younge A, Fowkes S (2003) The Cape Action Plan for the Environment: overview of an ecoregional planning process. Biol Conserv 112:15–28

Zengeya T, Ivey P, Woodford DJ et al (2017) Managing conflict-generating invasive species in South Africa: challenges and trade-offs. Bothalia 47:a2160. https://doi.org/10.4102/abc.v47i2.2160

23

Biological Invasion and Restoration Interventions

Patricia M. Holmes, Karen J. Esler, Mirijam Gaertner, Sjirk Geerts, Stuart A. Hall, Mlungele M. Nsikani, David M. Richardson and Sheunesu Ruwanza

Abstract Invasive alien plant species can be a major cause of ecosystem degradation in South Africa, and ecosystem recovery may require restoration interventions beyond controlling the target alien species. Active restoration interventions are usually required if legacy effects result from the invasion. Legacy effects may induce regime shifts when thresholds to autogenic recovery are breached. In such cases, active restoration interventions will be required to manipulate the ecosystem along a trajectory to recovery. In some cases, alien control measures may be sufficient to restore a structurally and functionally representative ecosystem, provided that implementation occurs early in the invasion process and that the control methods do not hamper spontaneous regeneration. It is important that key stakeholders discuss and set realistic restoration goals at the project planning stage. Studies on the costs and benefits of ecological restoration indicate that when important services are improved, benefits outweigh the costs of alien clearing (assuming spontaneous regeneration of the native ecosystem). The costs of moderate, active restoration interventions are

---

P. M. Holmes (✉) · K. J. Esler · S. A. Hall
Centre for Invasion Biology, Department of Conservation Ecology and Entomology, Stellenbosch University, Stellenbosch, South Africa
e-mail: holmes.patricia.m@gmail.com

M. Gaertner
Nürtingen-Geislingen University of Applied Sciences, Nürtingen, Germany

Centre for Invasion Biology, Department of Botany and Zoology, Stellenbosch University, Stellenbosch, South Africa

S. Geerts
Centre for Invasion Biology, Department of Conservation and Marine Sciences, Cape Peninsula University of Technology, Cape Town, South Africa

M. M. Nsikani · D. M. Richardson
Centre for Invasion Biology, Department of Botany and Zoology, Stellenbosch University, Stellenbosch, South Africa

S. Ruwanza
Centre for Invasion Biology, Department of Environmental Science, Rhodes University, Makhanda, South Africa

economically viable, whereas the costs of fully restoring ecosystem structure, functioning and composition in highly degraded ecosystems are rarely deemed economically justifiable. Valuations of specific biodiversity components, such as threatened ecosystems and species, remain problematic to assess, and these components could be under-valued in such studies. South African researchers have made significant contributions to the theory and practice of restoration ecology globally and have produced local guidelines for ecological restoration. However, there has been limited uptake in implementing active restoration projects at larger scales. This apparent knowing-doing gap may have three causes: firstly, insufficient co-production by all stakeholders in planning restoration projects, including prioritisation and goal setting; secondly, shifting beyond clearing invasive alien species to restoring ecosystems; and thirdly, insufficient resources to implement active restoration projects at the necessary scale. To achieve Convention on Biological Diversity and the UN Sustainable Development Goals, interventions must shift from controlling invasive alien species alone to restoring native ecosystem structure and functioning.

23.1 Introduction

Many invasive alien plant species in South Africa cause substantial ecosystem changes, and are the focus of expensive management operations due to the perceived negative impacts on ecosystem services and biodiversity (Richardson and van Wilgen 2004; Gaertner et al. 2011; Le Maitre et al. 2011). Other main causes of ecosystem degradation include over-grazing by livestock (Carrick and Krüger 2007) and inappropriate fire regimes (Kraaij and van Wilgen 2014). These degrading forces sometimes act synergistically, since alien plants may exploit recruitment opportunities created by other drivers of vegetation change (Sher and Hyatt 1999). Invasive alien plants may also disrupt ecological processes, for example resulting in reduced palatable forage or altered fire regimes (Brooks et al. 2004). Such synergistic forces may further accelerate the loss of biodiversity and ecosystem services.

To reverse ecosystem degradation and control invasions, it is sometimes necessary to not only optimise control of the invasive alien species, but also to actively restore the altered ecosystem. According to the Society for Ecological Restoration, "*Ecological restoration is the process of assisting the recovery of an ecosystem that has been degraded, damaged or destroyed*" (Clewell et al. 2004). This is generally interpreted to mean implementing actions that will set an ecosystem on a trajectory towards recovery. Managers therefore require a deep understanding of the ecosystems they manage, including the extent and intensity of change that has occurred, the recovery potential of the native ecosystem and the interventions required to promote recovery. Key questions include: "Will the ecosystem self-repair following removal of the invasive alien species?"; "Which removal method best promotes natural ecosystem recovery?" and, if the ecosystem is degraded beyond a state where self-repair is likely, "Is active restoration feasible and affordable?"

Here we describe the contexts for ecological restoration following alien plant invasions, summarise theoretical and applied research, and outline South Africa's contributions to this field. Conceptual frameworks such as the "restorative continuum" (McDonald et al. 2016a, b) are useful in addressing ecological restoration requirements along a gradient of habitat change or degradation, and for visualising desired outcomes. Restorative interventions may be seen as a continuum from the bare minimum of reducing the causes of degradation in permanently modified habitats, to restoring to an appropriate natural reference ecosystem (McDonald et al. 2016a, b). When the potential for recovery (or progress to some desired state) is determined to have been compromised by invasive alien species, active restoration by manipulating the abiotic and biotic ecosystem components may be required to achieve goals. In contrast, removal of the invasion-mediated impact through appropriate control measures alone may be sufficient for ecosystems deemed to have good recovery potential. In practice, the latter generally refers to invaded ecosystems with intact native seed banks and/or surviving remnant native plant populations. Here we illustrate some outcomes of spontaneous regeneration and active restoration applications using South African case studies.

23.2 Global and National Contexts

23.2.1 The Need for Ecological Restoration

Restoration ecology is a relatively new science that gained prominence in the 1980s, and has since expanded in response to the human-induced, rapid rate of habitat transformation and degradation across the globe (Hobbs and Richardson 2011). Global land modification and degradation estimates range from 0.99–6 billion ha (up to 66% of global land surface) and the annual, economic costs arising from these impacts are estimated to be ~ 10–17% of current global Gross Domestic Product (GDP) (Crossman et al. 2017). Various international instruments, to which South Africa is a signatory, seek to reverse this degradation, including the Bonn Challenge and Sustainable Development Goal 15 (Mohieldin and Caballero 2015) and the Convention on Biological Diversity Aichi Targets (2010). Within South Africa, the National Biodiversity Framework (Government Gazette No. 32474, 2009) summarises the actions required to conserve and restore the country's natural ecosystems.

23.2.2 Restoration Ecology

Empirical research in restoration ecology has promoted a deeper understanding of ecosystem structure and functioning (Pretorius et al. 2008; Gaertner et al. 2011; Zaloumis and Bond 2011). This has led to the development of conceptual models that link agents of change and ecosystem responses to potential restoration actions and outcomes (Carrick and Krüger 2007; Standish et al. 2007; Holl et al. 2011; Le Maitre et al. 2011; Davies et al. 2018). Threshold models have played a significant conceptual role. Initially they were applied in semi-arid rangelands to describe sudden shifts in ecosystem structure or functioning, where the degraded alternative ecosystem state was resistant to natural recovery (Milton and Siegfried 1994; Whisenant 1999). Subsequent work showed that sudden regime shifts (sometimes termed tipping points) also operate in many other types of ecosystems previously assumed to follow linear degradation or successional trajectories (Gaertner et al. 2012b; Richardson et al. 2007). Traditional restoration approaches in degraded lands may fail due to constraints such as local loss of native species, seed dispersal limitations, shifts in dominance, altered biogeochemical processes and altered feedbacks that entrench an undesirable stable state (Beisner et al. 2003; Norton 2009; Suding et al. 2004). Threshold models that incorporate alternative ecosystem states and feedbacks can be applied to assist in habitat management decision-making (Standish et al. 2007; Gaertner et al. 2012b; Suding and Hobbs 2009).

It is reasoned that restoring ecosystem structure (including functional group diversity) will simultaneously restore ecosystem functioning (Holmes and Richardson 1999). Mori et al. (2013) explored the links between biodiversity, ecosystem stability and functionality, and how degradation may affect these. They suggested that traditional measures such as species diversity do not adequately capture aspects that are pivotal for ecosystem resilience, which requires the recovery of guilds representing the range of functional responses required to drive key processes. Sudden shifts to alternative ecosystem states coincide with a rapid loss of functional diversity that results from an aggregated loss of response diversity. Mori et al. (2013) emphasise that perspectives incorporating functional effects and responses of biodiversity are essential for developing restoration management strategies. However, the recovery of functions may lag behind structure in some ecosystems such as riparian forests (Matzek et al. 2016). Furthermore, where abiotic thresholds have been crossed, such as under climate change, the situation is likely to be more challenging. Climate change might reduce the effect of some stressors whilst enhancing others, further complicating restoration efforts (Rohr et al. 2018, 2013).

Implementing ecological restoration is complex: managers need to assess the impacts of degradation on ecosystem functions and recovery potential as well as the likely impacts of different management interventions. Lack of consideration for all aspects involved in restoration has likely resulted in limited success of implementation (Kettenring and Adams 2011). In addition, global change impacts such as habitat loss, fragmentation, biological invasions, and climate change may influence restoration outcomes. In South Korea, an approach using flexible restoration targets was applied: this considered ecosystem functions and functional trait diversity, rather than historic precedents, to create model restoration projects and implement adaptive management (Temperton et al. 2014). Because there is a need to improve restoration outcomes, share information from different restoration projects, and acknowledge stakeholder contexts, James and Carrick (2016) advocated the use of quantitative systems models. These models link ecological processes, and potentially social processes, that influence the desired outcome. Data gathered from empirical studies can be tested against a quantitative systems model to iteratively improve restoration outcomes for a particular ecosystem.

A major complication for restoration is the incorporation of larger-scale ecological processes, such as hydrological regimes and climatic factors that may limit the feasibility or trajectory of a desired outcome (Richardson et al. 2007). This may entail relaxing restoration goals based on historically informed reference systems (Prins et al. 2004; Aronson et al. 2017; Balaguer et al. 2014) to more flexible ones based on the new combination of environmental factors (Hobbs et al. 2009). This requires a detailed understanding of how local and regional processes influence population and community dynamics in the target ecosystem (Holmes and Richardson 1999; Suding and Leger 2012).

23.2.3 Biological Invasions and Restoration Ecology

Studies have shown that invasions of alien plants cause substantial changes to ecosystems by breaching biotic or abiotic thresholds to recovery (Suding and Hobbs 2009; Gaertner et al. 2012b). Biological invasions can also be the driver of ecosystem change during the restorative process (Norton 2009), resulting in altered ecosystem composition and structure. This implies that removal of the invaders alone may not lead to recovery, and indeed that such actions may divert successional processes in unwanted directions that may require additional interventions (active restoration measures) to fix. Despite the well-documented problem of impacts caused by invasive alien plants (Vilà et al. 2011; Downey and Richardson 2016; Le Maitre et al. 2011), a review of ecosystem restoration studies found that only 8% had control of invasive alien species as their main objective (Gaertner et al. 2012a). In those 8% of studies, the prevalent cause for degradation was invasive alien species outcompeting and replacing native species, indicating that a biotic threshold to recovery had been crossed. Measures other than invasive alien control were implemented in 65% of those cases (Gaertner et al. 2012a).

Invasive aliens with potential for causing regime shifts should be prioritised for control, as these species can modify ecosystems to their own benefit and suppress native species through reinforcing feedback processes that present barriers to recovery (Gaertner et al. 2014). Feedbacks likely to result in regime shifts were related to processes associated with seed banks, fire and nutrient cycling (Gaertner et al. 2014). A recent review of soil legacy effects resulting from invasion by alien N_2-fixing woody species identified several potential barriers to restoration following alien control (Nsikani et al. 2018). Biotic barriers included altered soil microbial communities, depleted native seed banks, secondary invasions of alien species, and weedy native species dominance. Altered soil properties, especially those involving N, C and moisture, potentially thwart restoration in some ecosystems.

These barriers to restoration point to the need for active restoration to facilitate recovery. There are numerous potential interventions that can be applied on their own or in combination. However, for most ecosystems, more experimental work is required to test optimal treatment combinations. Examples of where this has been explored for the Fynbos Biome in South Africa include: (1) highly degraded, alien grass-dominated renosterveld shrubland, where a combination of fire or tillage and herbicide treatments followed by active restoration sowing met the desired outcomes (Waller et al. 2016); (2) mountain fynbos densely invaded by woody alien pines and wattles, where alien clearance resulted in good spontaneous regeneration from the soil-stored seed bank, but ecosystem structure was improved by sowing the seeds of additional species that had been displaced (Holmes 2001b); and (3) lowland fynbos that was densely invaded by alien wattles, where native seed banks were too depleted to support spontaneous regeneration, necessitating active sowing to restore vegetation structure after alien clearance (Hall 2018). These examples illustrate that even within the same biome, different approaches may be required, depending on the ecosystem affected, the invasive alien species and the type of degradation that has occurred at a site. However, resource-intensive, active restoration interventions, such as comprehensive species sowing, may only be justifiable for priority threatened ecosystems.

Richardson et al. (2007) reviewed studies of riparian vegetation restoration following alien plant invasions. They recommended that a framework for restoration should consider biogeographical processes at different spatial scales, and specific relationships between invasive alien plants, resilience and ecosystem functioning. For example, large-scale, human-mediated changes, such as impoundments, can alter downstream river geomorphology in favour of the invasive alien species and limit the outcomes of reach-scale restoration (Rouwntree 1991).

23.2.4 Biological Invasions and Restoration Ecology in South Africa

South Africa has been a prominent contributor to global invasion science since the 1980s (Macdonald et al. 1986; Richardson et al. 1997; Wilson et al. 2014), and the management of invasions has been a strong component of this research (van Wilgen 2018). Given the importance of catchment areas for providing water and conserving the country's rich biodiversity, coupled with the escalating threat posed by alien plant invasions in these areas, it is not surprising that ecological restoration has developed here as a complementary research stream to invasion biology (van Wilgen et al. 2016). The past two decades have seen a closer integration of the two disciplines, in both theoretical and empirical research aspects (Gaertner et al. 2012a). Restoration ecology studies have enabled us to test our understanding of ecosystem structure and functioning (Gaertner et al. 2011; Holmes and Richardson 1999), and to explore ways of improving restoration when managing alien-invaded ecosystems (Holmes 2001b; Waller et al. 2016).

Disproportionate research attention has focussed on the Fynbos Biome, which covers only 4% of South Africa. Early analyses showed that this Mediterranean-climate biome was more severely invaded than other biomes, though not necessarily more susceptible to alien plant invasions (Macdonald 1984). An unusual feature of plant invasions in the fynbos is the prominence of alien trees and shrubs originating from other Mediterranean-climate regions of the globe (Richardson et al. 1997). Owing to the negative impacts of these alien species on ecosystem services, particularly water supply, and locally endemic and highly threatened biodiversity (the Fynbos Biome has 38% of the country's critically endangered ecosystems and 67% of threatened plant taxa; Raimondo et al. 2009), research to document these impacts was intensified (Richardson and van Wilgen 2004). A motivation to the national government by stakeholders in the Fynbos Biome for a large-scale intervention for alien control resulted in the "Working for Water" programme being initiated in 1995 (Marais and Wannenburgh 2008). This programme stimulated continued research into the impacts of invasion and invasive alien control methods, including restoration.

A significant discovery that revolutionised restoration in fynbos and other fire-prone ecosystems is that chemicals in smoke stimulate germination in many taxa (Brown 1993). This cue promotes seedling recruitment in the immediate post-fire environment and can be used as a pre-treatment for seeds, in combination with heat shock for some taxa, to optimise native species establishment in restoration projects (Hall et al. 2017).

23.3 Restoration After Biological Invasion

23.3.1 The Restorative Continuum

An overarching goal of managing biological invasions is to halt or slow the spread of the invaders, which entails intercepting invasion pathways and managing invaded landscapes in an integrated way. For ecological restoration, however, the overarching goal is to optimise ecosystem recovery, i.e. structure and functioning, using spontaneous regeneration or active interventions as the situation requires (McDonald et al. 2016a, b; Aronson et al. 2017). These two goals can be synergistic, and optimal restoration outcomes are most likely to result from good planning and implementation that integrates the research findings from both invasion biology and restoration ecology.

If one considers the restorative continuum in relation to biological invasions, the costs of ecosystem repair increase as a function of the extent, duration and intensity of environmental damage caused by the invasive alien (Milton et al. 2003; Holmes et al. 2007). It is therefore important to act early in the invasion process, before biotic and abiotic thresholds have been crossed, as costs increase according to the number of interventions required to restore a resilient, functional ecosystem (Aronson et al. 2007). In the case of highly modified habitats, such as road embankments or quarries, removal of invasive alien species may be the only action needed to attain the optimal goal, i.e. to halt spread of the invader. Active restoration interventions such as native species re-introduction may not be warranted in such situations, owing to the high costs of procuring native propagules and/or the high likelihood of failed introductions owing to the highly modified biophysical conditions. However, if re-invasion is likely, there may be a case for revegetating modified areas with alternative, non-invasive species to pre-empt re-invasion. In highly modified sites, there may be examples where replacing the invasive alien with resilient native species could improve landscape functions, such as reconnecting fragmented areas of natural habitat. In less modified habitats where there is potential for autogenic recovery through spontaneous regeneration, alien control may be sufficient to re-instate a structurally and functionally representative ecosystem (Gaertner et al. 2011; Mostert et al. 2017). However, where key structural or functional guilds have been severely depleted, active restoration through native propagule re-introduction, and possibly other actions, may be required. Where the conservation status of an ecosystem is a high priority, active restoration may be justified to further improve community composition and restore viable populations of threatened species (Morgan 1999; Hitchcock et al. 2012). Ensuring that key ecosystem processes (e.g. fire and herbivory) are initiated and maintained are important for optimising recovery.

23.3.2 Ecological Restoration Following Invasion: A Conceptual Framework

Gaertner et al. (2012b) modified models for dryland ecosystem degradation and repair (Milton and Siegfried 1994; Whisenant 1999, 2002) to develop a conceptual framework for restoring ecosystems degraded by invasive alien plants. They outlined a three-threshold conceptual model (Fig. 23.1) and linked the concept of ecosystem resilience to degradation thresholds. Resilience in this context is the ability of an ecosystem to spontaneously regenerate following invasive alien plant controls. A threshold is the point beyond which the ecosystem cannot self-repair following alien control alone, has lost resilience, and has formed an alternative state. The first threshold along the invasion-degradation gradient is usually biotic, and is indicative of a shift in structural biotic interactions within the invaded community for which recovery requires manipulation of vegetation components. In ecosystems where the biotic threshold has been reached, restoration interventions should include the re-introduction of missing plant guilds in addition to invasive alien vegetation control.

The second transitional threshold is reached after a longer duration of invasion and is harder to reverse. It is controlled by abiotic limitations that result from amplified biotic interactions and restoration requires manipulation of the physical environment in addition to the biotic components. Where alien species incur legacy effects such as N enrichment (Nsikani et al. 2018), this may be a barrier to restoration and may require the physical removal of nutrient-rich biomass (Marchante et al. 2009), the manipulation of soil nutrients (Zink and Allen 1998; Török et al. 2000), or the use of fire to volatilise excess nutrients. In many cases, biotic and abiotic

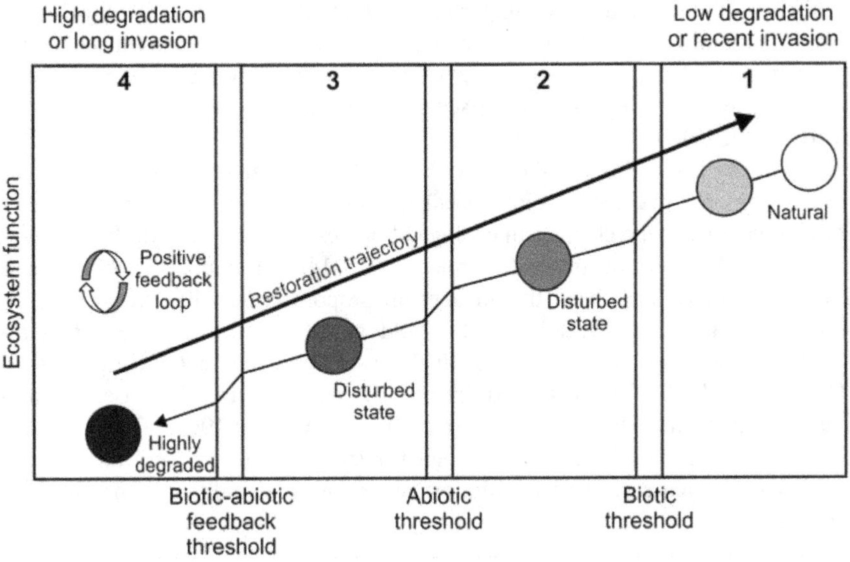

Fig. 23.1 Conceptual stepwise degradation model indicating three thresholds along an invasion intensity gradient (after Gaertner et al. 2012a; Stanturf et al. 2014). Thresholds indicate break points between alternative ecosystem states that require specific restoration interventions to ensure ecosystem recovery: (1) natural ecosystem state where no threshold is reached (e.g. altered species composition but above-ground vegetation and/or seed banks intact); (2) alternative ecosystem state where biotic threshold is reached (e.g. altered species composition and structure, depleted seed banks); (3) alternative ecosystem state where abiotic and biotic thresholds are reached (e.g. altered water and nutrient availability); and (4) alternative ecosystem state where positive feedback loops entrench a highly degraded state (e.g. changed fire regime favours persistence of the invader and prevents re-establishment of native species)

threshold effects are closely linked and may occur simultaneously. At this point management recommendations for ecological restoration are complex and experiments in the field may be required to determine the optimal combination of restoration treatments required, the interaction effects among treatments (e.g. use of fire, herbicides, seeding and planting) and the optimal timing of these interventions (Waller et al. 2016). However, if the duration of invasion can be reduced by timely alien control, changes to ecological functions may be avoided and an abiotic threshold may not have been reached. This research is useful to guide local ecological restoration projects while advancing our ecological knowledge of the system being restored, including seed ecology, community dynamics, ecosystem-level changes and mutualisms, to name a few.

Control of plant invaders can create bare ground that can be re-invaded by the same alien species or by secondary plant invader species (Pearson et al. 2016). This effect can be amplified if both biotic and abiotic thresholds have been crossed, as the altered conditions may favour competitive, weedy species. An example of secondary invasion is the colonisation by weedy herbaceous species in areas cleared of N-fixing alien acacia trees (Richardson et al. 2000; Yelenik et al. 2004; Nsikani et al. 2017).

Minimising re-invasion is an important restoration objective and may require manipulation of both the biotic and abiotic components of the degraded ecosystem. Actively re-establishing canopy cover of native species may also be required to suppress the weedy, light-requiring secondary invader species (Falk et al. 2013; Herron et al. 2013).

The third degradation threshold may be reached at a later stage of invasion and results from positive biotic-abiotic feedback loops, whereby the persisting alien invader entrenches the changes in ecological processes that in turn further promote the invader above native species (Gaertner et al. 2014; Vilà et al. 2011; Suding et al. 2004). Examples of where the invader disproportionately benefits, ultimately resulting in the third degradation threshold being reached and preventing the re-establishment of native species, includes altered nutrient-cycling patterns (Gaertner et al. 2012b) and altered fire regimes resulting from increased biomass and/or changed fuel distribution (D'Antonio and Vitousek 1992). Once the third degradation threshold is reached, ecological restoration may be very difficult and expensive; in such cases less ambitious goals, such as rehabilitation, may be more realistic.

23.4 Best Practice: Restoration Planning

Whatever the context for an ecological restoration project, good planning and stakeholder involvement are essential if restoration goals are to be achieved. Following the four principles outlined by Suding et al. (2015), planned restoration projects strive to increase ecological integrity, i.e. either through alien control and spontaneous regeneration, or through additional interventions in tandem with alien control. Secondly, restoration is best planned for long-term sustainability. For example, practitioners can ensure that sufficient resources are available for follow-up alien control, monitoring and native propagule re-introduction if required, before embarking on initial control of the alien. The altered state of the ecosystem must be thoroughly assessed to ascertain whether restoration is likely to result in a sustainable, positive change. This would also assist in assessing priorities and setting realistic goals (D'Antonio and Meyerson 2002). Thirdly, most projects have goals based on a historically-informed reference ecosystem (Aronson et al. 2017). Nevertheless, it is also recommended to consider the future, and whether landscape-scale or global changes and external factors such as public perception could thwart desired restoration goals. Fourthly, restoration interventions should benefit and engage society.

Restoration planning should involve invasion and restoration ecologists, practitioners and other stakeholders who jointly develop restoration goals that are ecologically and economically feasible (Gaertner et al. 2012a). A participatory scenario approach for planning, implementing and monitoring restoration is one such inclusive approach (Metzger et al. 2017). At this stage, research questions may be addressed to fill knowledge gaps and data collected to better inform restoration best practice. Once the restoration project has been implemented, monitoring of outcomes should iteratively feed back into improving restoration actions, possibly using a quantitative systems modelling approach as advocated by James and Carrick (2016).

Stakeholders must be engaged early in planning to identify issues of concern that may compromise the ecological restoration goals (Metzger et al. 2017; Reyers et al. 2009; Urgenson et al. 2013). Ignoring such concerns could result in challenges to, or failure of, the project. For example, an alien species may provide an important resource for certain stakeholders and its removal could undermine their livelihoods (Kull et al. 2011). Engaging with such stakeholders may result in the project proceeding, provided that some alternative, non-invasive species resource is provided as part of the restoration project.

The Robust Offsetting restoration planning tool (RobOff) was applied to a large community reforestation project at a landfill site in Durban, to examine different restoration goals for the buffer zone (former sugar cane fields supporting invasive alien plants, with patches of native forest and grassland; Mugwedi et al. 2017). These goals included carbon storage, biodiversity and employment across a mosaic of habitats with varying levels of degradation and a limited budget (Mugwedi et al. 2018). The current restoration action was compared to three restoration intervention alternatives—spontaneous regeneration, carbon action and biodiversity action. RobOff indicated "biodiversity action" as the most beneficial in maximising the three goals, and that investing in biodiversity action would be preferable to the *status quo* (Mugwedi et al. 2018). Challenges included an increase in invasive alien plants, and alien control was included as a necessary restoration intervention (Mugwedi et al. 2017). Results from tools such as RobOff can help to inform stakeholder planning workshops of potential costs and benefits of different restoration interventions. In landscape-scale restoration projects, social and political processes may be more important to long-term success than ecological factors, therefore it is key to include multi-party stakeholders and specialists from the beginning (Aronson et al. 2017).

23.4.1 Ecological Restoration Goals

It has been argued that to become effective, adaptive and able to compete with other projects for resources, ecological restoration must become evidence-based (Ntshotsho et al. 2011). The three criteria required as evidence are baseline information, clearly defined goals, and monitoring. In a review of ten South African restoration programmes, Ntshotsho et al. (2011) found that of these three criteria only one, baseline information, was adequately addressed. Although both ecological and socioeconomic goals were set, these were found to be insufficiently clear. There was little monitoring of programme outcomes and monitoring of ecological indicators was inconsistent. To learn from early restoration attempts, it is important to define clear goals based on measurable parameters at the planning stage. These goals may be based on a historically-informed reference ecosystem (Prins et al. 2004), but must also be realistic—i.e. they must consider the intensity of past degradation and the ecological interventions necessary to promote recovery. Social and budgetary constraints should also be considered to ensure that realistic goals are set.

There are many different contexts and goals for ecological restoration of alien-invaded sites. In degraded ecosystems that retain intact soils, the first restoration action is usually to remove the invasive species, which may differ in structure and functioning from the dominant components of the native vegetation, and to re-instate ecosystem structure, either by stimulating the residual native seed bank or by re-introducing propagules of representative guilds. In so doing, it is anticipated that ecosystem functions and ecosystem services linked to these structural components will self-restore. However, it is important to collect baseline data on key functions and services and to monitor changes after the interventions, to test these assumptions and to modify the interventions as required. This approach is particularly important for invaded sites in areas identified as ecological infrastructure (e.g. water catchments) or high importance for ecosystem-based adaptation to climate change. In areas of high biodiversity importance, such as critically endangered ecosystems in global biodiversity hotspots, longer-term goals following the control of invasive alien plants may include the restoration of community composition and viable species populations.

23.4.2 Prioritisation

Resources are limited, so prioritisation is key to efficient restoration. An objective protocol that incorporates best-practice knowledge should be used, such as the Analytical Hierarchical Process (AHP), to select those restoration projects of the highest priority and/or likelihood of success (Mostert et al. 2018). AHP uses inputs from key stakeholders, who may participate in defining and scoring criteria for prioritisation, thus promoting a co-production approach during the planning process. An important principle in prioritising restoration projects is to first target the least degraded areas that have the highest potential for autogenic recovery (Holmes and Cowling 1997; Strydom et al. 2017).

23.4.3 Costs and Benefits of Restoration Projects

Intact ecosystems comprise the natural capital that in economic terms represents the stocks (i.e. component species) and the flows of ecosystem services upon which society depends. Degradation by invasive alien species disrupts these stocks and flows, and ecological restoration is required to reverse the impacts. However, what are the costs and benefits of doing so, and are the costs justified? Examples are not restricted to the improved ecosystem services delivered by the restored ecosystems, but include value-added industries related to alien biomass removal, such as firewood, charcoal and biomass to electricity industries (Stafford and Blignaut 2017). Managing and valuing ecosystem services remains a challenge in South Africa and globally, with native ecosystems typically undervalued, and is not limited to the field of restoration ecology. The accuracy of measuring restoration success is complicated because policy is still dictated by narrow, mainstream economic ideas (Costanza et al. 2017). Furthermore, information, databases and models related to ecosystem services in South Africa and elsewhere could be developed further, integrated and made available to stakeholders to ensure ecosystem services are optimised (Turner et al. 2016). One well-studied example is the negative impact of invasive alien trees on water supply and the benefits of restoring water catchments (Marais and Wannenburgh 2008).

Mountain fynbos ecosystems are important for the delivery of clean water to downstream agricultural and urban areas and for wild flower harvesting and ecotourism, among other services. Invasion by alien trees has severely disrupted these services (Holmes et al. 2007). Recovery potential is initially high in these ecosystems, owing to persistent soil-stored native seed banks, but this declines with the duration and density of invasion (Galloway et al. 2017; Holmes and Cowling 1997). For example, the benefit-cost ratio for total alien clearance (initial plus follow-ups) was 8:1 for medium-dense alien stands compared to 3:1 (at 4% discount rate) for long-term, closed alien stands (Holmes et al. 2007). Therefore, in an attempt to guide decisions on when restoration activities may be feasible, Crookes et al. (2013) coupled ecological restoration with system dynamics and portfolio mapping. They showed how restoration costs and derived benefits varied across sites, but that those projects in South Africa with the highest expectations of success and high payoffs were those associated with protection of water resources—justifying the focus on alien-clearing related restoration efforts in catchment areas.

Turpie et al. (2008) suggested that these valuable ecosystem services, such as water, should act as 'umbrella services' and that doing so would enhance broader conservation goals. In South Africa, payments for ecosystem services only started with the establishment of the Working for Water programme in 1995.

From an ecosystem services viewpoint, an increased investment in restoration activities often is warranted (Anderson et al. 2017). However, the effectiveness of these investments will differ depending on their costs and aims. For example, Currie et al. (2009) determined the cost-effectiveness of restoration by using the costs of invasive alien plant clearing, erosion control and revegetation as the input costs and water and tourism as the benefits. Importantly, they compared three different restoration options (comprehensive, moderate, basic) and three economic scenarios (optimistic, realistic, pessimistic). They found that comprehensive restoration was not worth the input costs, basic restoration was always economically viable, whilst for moderate restoration it depended on the economic scenario selected. Naturally, these results are highly dependent on the invasive alien species, the ecosystem type and potential benefits, highlighting the importance of conducting such studies in a variety of ecosystems. Gaertner et al. (2012c) showed that for flower harvesting, active fynbos restoration is financially feasible for flower harvesters compared to spontaneous regeneration, but mainly over the long term and in areas with low-density invasions.

Most restoration studies in South Africa are short term and are vegetation-orientated (e.g. Galatowitsch and Richardson 2005; Blanchard and Holmes 2008; Ruwanza et al. 2013a; Kerr and Ruwanza 2016; Ndou and Ruwanza 2016). Recovery of other taxa is less well quantified, largely because of the untested assumption that the recovery of native vegetation will facilitate the recovery of other taxa and subsequently ecosystem services and functions (Kaiser-Bunbury et al. 2017). Recent South African studies do indicate spontaneous recovery of other taxa and a link to ecosystem services (Mgobozi et al. 2008; Colvin et al. 2009; Magoba and Samways 2010; Samways et al. 2011; Maoela et al. 2016; Modiba et al. 2017). Invasive alien plant species were found to negatively impact most on seed dispersers and nectar-feeding bird pollinators (Mangachena and Geerts 2017). Although a few bird species were still absent 10 years after vegetation recovery, bird species richness and abundance recovered, and all feeding guilds—including pollinators and seed dispersers—were represented (JR Mangachena, unpublished data). However, this is likely to be context-dependant as invasive alien species may contribute additional food sources to the benefit of some pollinators (Geerts and Pauw 2009). Ecosystem processes such as pollination might not automatically be reinstated by restoring target plant species. But assessing these processes will provide an indication of the sustainability of restoration projects (Forup and Memmott 2005). A keystone species approach to restore ecosystem services might be a cost-effective way to achieve this (Traveset and Richardson 2006).

More studies are required to measure the extent to which the removal of invasive alien plant species will result in a restored ecosystem. In particular, clearing of invasive alien plant species can hamper native vegetation recovery through overuse of herbicides or the lack of follow-up clearing. This also adds additional cost to restoration, enhancing the uneconomical nature of some restoration interventions. Therefore, we argue that high quality alien control should be prioritised. Furthermore, invasive alien plant control is making very little progress, or none at all in many areas (e.g. McConnachie et al. 2012). We caution against claiming large potential ecosystem service benefits from the clearance of invasive alien plants (Stafford et al. 2017), since the desired goal of restoring ecosystem structure and function in support of these services might not materialise.

23.5 Best Practice: Restoration Implementation

23.5.1 Legacy Effects, Ecosystem Functions and Drivers

It is important to assess how different invasive alien species, and different levels of density and duration of invasion, affect restoration potential. For example, N-fixing alien *Acacia* species have higher negative impacts on the recovery potential of fynbos ecosystems than alien pines (Mostert et al. 2017). Acacias shade out native species quicker than do pines. Their legacy effects include large soil-stored seed banks, increased leaf litter and biomass accumulation, increased pressure from folivorous insects and phylopathogenic fungi and changes to soil chemistry (Maoela et al. 2016; Nsikani et al. 2017; Strydom et al. 2012; Yelenik et al. 2004), thus altering ecosystem functions such as nutrient-cycling. Their rapid growth to canopy closure halts native seed set, resulting in depletion of native seed banks (Holmes 2002; Mostert et al. 2017). By contrast, the legacy effects of pines are less pronounced, relating more to altered local hydrology through increased water use and increased biomass. However, long-duration, dense pine invasions cause the depletion of fynbos seed banks and reduced recovery potential (Galloway et al. 2017), and in grasslands the loss of the forb resprouter guild (Zaloumis and Bond 2011).

Where an alien species is adapted to fire in the invaded ecosystem, management interventions need to be carefully integrated to promote ecosystem recovery and avoid exacerbating the invasion through fire-stimulated germination and spread of the alien species. In the case of serotinous invaders such as *Hakea* and *Pinus* species, fire may be used after felling adult populations to kill their released seed, seedlings and saplings (van Wilgen et al. 1994; Table 23.1, Langeberg case study). The case study of Cape Flats Sand Fynbos restoration at Blaauwberg Nature Reserve highlights the legacy effects to be overcome following *Acacia* invasion for a critically endangered ecosystem in which restoring and conserving biodiversity is the ultimate goal (Table 23.1; Figs. 23.2 and 23.3).

In riparian ecosystems, invasions of alien *Acacia* species and River Red Gum, *Eucalyptus camaldulensis*, also induce legacy effects. Whereas nutrient enrichment was documented after *Acacia* invasions, concentrations of soil available nutrients did not change following *Eucalyptus* invasion along the Berg River, although soil pH and moisture decreased significantly, and soil water repellency increased (Tererai et al. 2015a; Ruwanza et al. 2013a, b). A decrease in soil nutrients was documented following clearance of *Acacia* species along the Palmiet River, Eastern Cape (Ndou and Ruwanza 2016). In the lowland Berg River case study (Table 23.1; Fig. 23.4), *E. camaldulensis* produced allelopathic chemicals that suppressed germination and growth of four native species (Ruwanza et al. 2015). The soil seed banks retained some potential to initiate recovery after alien control but contained a high proportion of alien herbaceous weedy species that caused secondary invasions post-control (Tererai et al. 2015b). By contrast, in mountain stream and foothill reaches, fynbos riparian ecosystems are relatively resilient to invasion and good spontaneous regeneration followed the removal of large alien trees (Blanchard and Holmes 2008). Although not all riparian species have soil-stored seeds, seed banks had potential to initiate recovery after alien control (Fourie 2008; Vosse et al. 2008), as was also found along the Sabie River in the Kruger National Park (Morris et al. 2008; Table 23.1; Fig. 23.5).

Most lowland rivers traverse highly modified landscapes, and suffer profound alterations to hydrological regimes and geomorphology, due to upstream impoundments and conversion of riparian zones to agriculture or dense stands of alien trees (Holmes et al. 2005). Such large-scale alterations to ecological processes limit the

Table 23.1 Case studies of ecosystem restoration at sites impacted by invasive alien plants in South Africa

Case study name; locality	Vegetation type; status; key drivers	Invasion scenario; alien species, duration, treatments	Key lessons learnt	References
Fynbos biome				
Flower Valley; Agulhas Plain, Western Cape (see Fig. 23.5)	Mountain fynbos, fire-driven	Stands of *Eucalyptus conferruminata and E. cladocalyx*, old field invaded by *Cenchrus clandestinus* (East African Kikuyu grass) and dense stands of alien legume trees dominated by *Acacia cyclops, A. longifolia, A. mearnsii, A. saligna* and *Paraserianthes lophantha.* Trees clear-felled and grass treated with herbicide. Some areas actively restored using native seed mixes.	Active restoration can be effective and financially feasible when compared to passive restoration, depending on the density of invasion. Because of the high costs involved, active restoration of densely invaded sites may only be justifiable if the target area is in a region of high conservation priority.	Gaertner et al. (2012b) and Gaertner et al. (2011)
Langeberg and Boland mountains, Western Cape	Mountain fynbos, fire-driven	Dense stands dominated by *Pinus pinaster* or *Hakea sericea* (both serotinous aliens); "fell and burn" treatments; summer versus winter slash burns	Fell tree stands at <10 yr to minimise cut biomass and expose seeds to granivory. Summer burn ≥ 1 yr later to kill seed, seedlings and saplings, and stimulate fynbos recruitment. Avoid felling older tree stands (>10 yr) as it creates high fuel loads and damaging summer fires. For higher fuel loads burning when soils are wet reduces soil and seed bank damage.	Holmes and Marais (2000), Holmes et al. (2000), and Holmes (2001a)
Blaauwberg Nature Reserve, Cape Town West Coast, Western Cape (see Figs. 23.2 and 23.3)	Lowland, Cape Flats Sand Fynbos; critically endangered ecosystem; fire-driven	Dense stands of *Acacia saligna*; invaded within the last 80 years. Clearing treatments: "fell and stack" versus "fell and burn" (passive restoration). Seed sowing in plots within fell and burn treatment (active restoration). Manual or chemical follow-up. Fungal biological control agents present.	Best treatment depends on level of degradation/surviving native cover and diversity at time of clearing. Fell and stack treatment sufficient if $>10\%$ cover intact fynbos is present. Fell and burn alone not effective due to depleted native seed bank: sowing necessary. Seed pre-treatment improves diversity and density of recruitment from seed.	Hall et al. (2017), Hall (2018), Krupek et al. (2016), Mostert et al. (2017), and Nsikani et al. (2017)

			Follow-up manual *Acacia* clearing is most effective but too expensive to be viable large-scale. Herbicide foliar spray kills some native species, but is less costly for dense *Acacia* recruitment. Herbicide must be applied to cut stems to prevent coppicing in *A. saligna* but must be done carefully to minimise collateral damage. Fynbos recovers poorly after >1 cycle of *Acacia* invasion compared to *Pinus* invasion. Legacy effects of *Acacia* results in medium-term altered soil chemistry and potential for secondary invasions.	
Riparian Vegetation; Berg River, Boland, Western Cape (see Fig. 23.4)	Historically Fynbos Riparian Scrub and Afrotemperate Forest, with adjacent Swartland Shale Renosterveld, critically endangered, now mainly under cultivation; floods and droughts are key drivers, with fire a historical driver	Dense stands of alien trees (mainly *Eucalyptus camaldulensis*, some *Acacia mearnsii*). Invaded at least 60 years ago. Fell and stack clearing treatment. Active restoration of some sites using native seed mix.	Passively restored sites had higher total plant species richness than invaded and near-natural sites, but significantly lower canopy cover. Secondary invasion of alien forbs and graminoids dominated cleared areas targeted for restoration. Passive restoration was limited by native seed banks which had reduced potential for recruitment. However soils were assessed to be close to the pre-invasion condition. Active restoration was affected by low native species germination and poor seedling establishment over the dry summer season. Bird assemblages in invaded sites are a subset of those in near-pristine areas. A decade after removal of invasive	Tererai et al. (2013), Tererai et al. (2015a, b), Ruwanza et al. (2013a, b), Ruwanza et al. (2015), Tererai et al. (2015a, b), Mangachena and Geerts (2017), and Ruwanza et al. (2018)

(continued)

Table 23.1 (continued)

Case study name; locality	Vegetation type; status; key drivers	Invasion scenario; alien species, duration, treatments	Key lessons learnt	References
			E. camaldulensis, bird species richness and abundance had largely recovered, but not all species returned. This is attributed to lower abundance and diversity of native trees in restored sites. Importantly, frugivores and nectarivores largely absent in invaded sites, mostly return. Re-introduction of keystone tree species is suggested to fully restore bird species complement.	
Afrotemperate Forest Biome				
Buffeljags River, Swellendam, Western Cape	Afrotemperate Forest; recruitment in small gaps created by e.g. wind-throw and lightning strike	Dense stand of *Acacia mearnsii*. Alien trees used as nurse plants, with periodic thinning of aliens to promote growth of colonising forest species. Eventual removal of all invasive alien trees.	Autogenic recovery expedited by selectively manipulating the invaded forest environment rather than clear felling all invasive species. Mature forest trees within the dense *A. mearnsii* stands promoted natural succession, reduced the need for planting, and significantly reduced costs of restoration. 28 forest species mostly with vertebrate-dispersed fleshy fruits (of potentially 40 species in the surrounding forest patches) colonised the thinned *A. mearnsii* stands.	Geldenhuys (2013) and Geldenhuys et al. (2017)
Grassland Biome				
Eastern shore of Lake St Lucia, Kwa-zulu-Natal	Maputaland coastal Grassland; critically endangered; fire-driven ecosystem	*Pinus elliottii* plantations of 40+ yrs duration removed over 17 yr period; grassland allowed to passively restore	Compared to uninvaded reference sites, restored grassland had altered structure and reduced diversity, particularly in the forb resprouter growth form. There was no trajectory towards full recovery along	Zaloumis and Bond (2011)

			the chronosequence despite similar edaphic conditions. The forb resprouter growth form is difficult to restore from seed and current methods of active restoration for this vegetation type are inadequate.	
Grassland and Savanna Biomes				
Sabie River: Graskop to Kruger National Park; Mpumulanga	Riparian scrub and Forest; river traverses various land-uses, including plantation forestry, before entering the national park; floods and droughts are key drivers.	Dense, tall *Eucalyptus grandis* in upper grassland reaches. Dense shrub and herb invasion in lower savanna reaches dominated by *Acanthospermum hispidium*, *Lantana camara*, *Senna obtusifolia*, *Senna occidentalis*, *Tagetes minuta* and *Xanthium strumarium*. Alien trees and shrubs felled.	Tall trees cleared, but a decade later invasion intensity remained high. Clearing operations reduced herbaceous cover, increased bare soil and litter, and reduced surface stability. The 2000 flood may have exacerbated secondary invasion by alien shrub and herbaceous species. Clearance reduced alien shrub and herb density by 80% and facilitated colonisation by native species, mostly herbaceous growth forms. Regular follow-up clearance, at least annually, is key, and must be timeous following a large flood event.	Beater et al. (2008) and Morris et al. (2008)

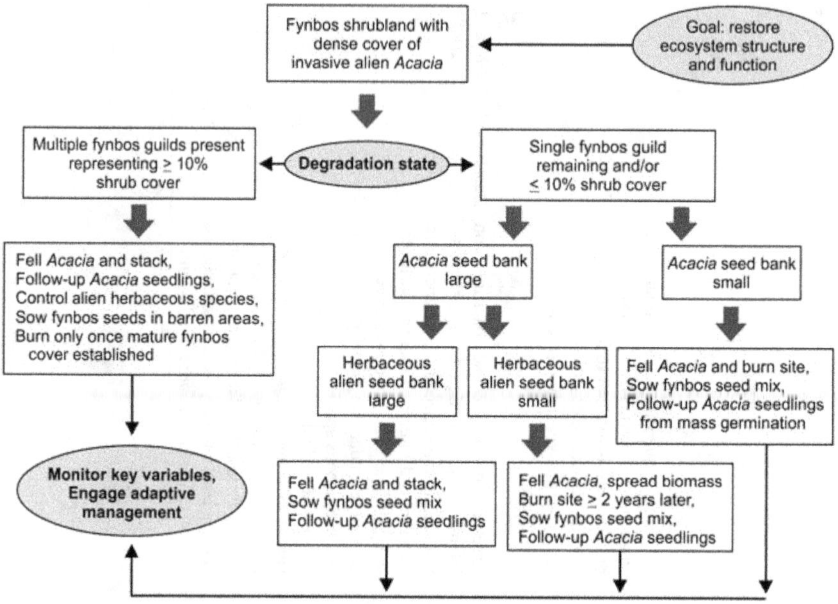

Fig. 23.2 A decision tree to guide ecological restoration in *Acacia*-invaded lowland fynbos shrubland (adapted from Hall 2018)

extent to which these ecosystems may be restored. For example, fires that historically swept across the lowland shrubland and riparian scrub communities no longer operate in an agricultural or urban matrix. Instead, remnants of riparian scrub have become colonised by vertebrate-dispersed species that are more typically present in fire-resistant sites (Meek et al. 2013). Such a community may be a more realistic target for long-term ecological restoration in lowland riparian ecosystems in a highly modified landscape. An additional challenge to restoration in lowland river systems is the altered flood regimes, mainly controlled by releases from upstream impoundments, that cause havoc with restoration plantings (J. van Biljon, pers. comm. August 2018). For many sites, restoration to anything resembling historical community structure and composition is unfeasible (Meek et al. 2013).

23.5.2 *Implementation: Spontaneous Regeneration*

Selecting the most appropriate alien control method to expedite spontaneous regeneration is important. A key consideration is the need to optimise recruitment by native soil-stored propagules and colonisation by wind or vertebrate-dispersed propagules that represent components of the historic ecosystem most suited to the local conditions. Although the most appropriate control method may be more expensive than standard methods, it may still be more cost-effective by eliminating

Fig. 23.3 Blaauwberg Nature Reserve Case Study. The photographs show: (**a**) A reference site for sand fynbos; (**b**) A worker clearing a dense stand of *Acacia* species; (**c**) Secondary invasion of a cleared site by herbaceous vegetation; (**d**) Recovery following clearing, with no fire applied (passive restoration); (**e**) Recovery by passive restoration following fire; and (**f**) Recovery following fire and sowing of seeds. Photographs courtesy of P.M. Holmes and S.A. Hall

the need for active restoration (Kimball et al. 2015). An important step is to assess which sites have moderate to good restoration potential. At such sites, control methods should attempt to minimise damage to native propagules, seedlings and adults, or attract dispersal agents, depending on the ecosystem (Table 23.1).

In grassland ecosystems invaded by Trifid Weed, *Chromolaena odorata*, integrated control using frequent fire after mechanical or chemical treatment was effective in controlling this species, as it is not fire-adapted. Furthermore, spontaneous regeneration by grassland species followed this treatment combination (Dew et al. 2017).

Fig. 23.4 Berg River Case Study. The photographs show: (**a**) A reference site for near-natural riparian vegetation; (**b**) Dense invasion of the riparian zone by *Eucalyptus camaldulensis*; (**c**) A site where the alien trees have been clear-felled following clearing; (**d**) Some recovery of native vegetation six years after clearing; and (**e**) Secondary invasion of the cleared site by grasses. Photographs courtesy of A. Rebelo and S. Ruwanza

In forest ecosystems densely invaded by alien trees, a phased approach to clearing the aliens better promoted conditions for *in situ* recruitment by vertebrate-dispersed forest species that germinate under a canopy and are shade-tolerant (Geldenhuys et al. 2017). The standard method of clear felling the aliens may leave large areas of bare ground without suitable frugivore perches or microclimatic conditions suited to dispersal, germination and establishment of native forest species. The Buffeljags River Forest case study (Table 23.1) illustrates the phased clearance approach, which

Fig. 23.5 Flower Valley Case Study. The photographs show: (**a**) A reference site for mountain fynbos; (**b**) A worker clearing a dense stand of *Acacia* species; (**c**) A dense invasion of the alien grass *Cenchrus clandestinus*; (**d**) A prescribed burn on the site following clearing; and (**e**) Seedling recruitment following sowing of seeds of native species. Photographs courtesy of M. Gaertner

was adopted as a four-stage alien thinning process to promote recovery in lowland riparian zones of the Berg River, where the restoration goal is thicket or forest (Ruwanza et al. 2013a, b).

For invasive alien species under effective biological control (Moran and Hoffmann 2012), integrated control is needed to best optimise the impacts of the biological control agents, as this will reduce future costs of restoration. For example in the case of the gall-forming rust fungus *Uromycladium tepperianum* on Port Jackson Willow, *Acacia saligna*, biological control substantially reduces the growth rate and canopy density of infected trees, allowing some fynbos species to persist in dense alien stands (Wood 2017). This slows the invasion process and maintains spontaneous regeneration potential in the invaded stands.

23.5.3 Implementation: Active Restoration

Where abiotic thresholds have been breached, interventions may be required to manipulate the abiotic environment before native species can be re-established. An example for nutrient-poor ecosystems includes reducing excess nutrients through removing or burning the litter layer and applying C-rich organic matter to immobilise nutrients (Zink and Allen 1998; Török et al. 2000; Gaertner et al. 2011). In riparian zones, steep banks and sediment accumulation resulting from stands of alien trees may first need to be re-contoured and flushed out, respectively, to create a more natural geomorphology before native wet and dry bank communities can be successfully re-established (Richardson et al. 2007). Exposure of bare soil and reduction in herbaceous cover following felling operations in riparian zones can result in secondary invasions (Beater et al. 2008; Table 23.1) which may be countered by planting native riparian species. If mutualistic microbial communities (e.g. specialist mycorrhizae) required for plant growth have been lost following long-term invasion or severe fires, soils or plants may first need to be inoculated to ensure successful re-introductions (Nsikani et al. 2018).

Native species may be restored from seed and propagated material. Seed is the preferred method for a number of reasons: it allows for more species and genetic diversity to be returned to the site, avoids the potential introduction of foreign soil, pests and parasites from nursery-grown stock, and potentially is more cost-effective, allowing larger areas to be restored. In grasslands, where the goal is to promote diversity, it is recommended to sow a propagule mix of grass species with low invasion potential (species with short stature, slow growth, low leaf mass and few tillers, Fynn et al. 2009). However, not all species are easily restored by seed. For example, fynbos obligate resprouter species produce few viable seeds (Marais et al. 2014) and in other species seeds are recalcitrant (Walters and Berjak 2013) thus cannot be collected and stored ahead of sowing. Some large-seeded species are nutritious and are highly parasitised or produce few viable seeds owing to granivory (PMH pers. obs.), while others are dependent on ants for dispersal and germination (Bond and Slingsby 1983). For threatened ecosystems there may be few intact remnants remaining as a source of suitable seeds for restoration, further limiting the available options. For some of the above examples, propagating new plants from cuttings and splits, or germinating from seed and first growing on before planting, may be more successful. In some cases, a combination of sowing and planting might be recommended to optimise the restoration outcomes.

In semi-arid ecosystems, restoration success can be limited by poor establishment from actively seeded species (Madsen et al. 2016). Seed enhancement technologies, such as polymer seed coatings containing germination stimulants (e.g. smoke extract), have been shown to improve restoration success (Turner et al. 2006). Embedding the broadcast seed into the soil by raking and sowing at the optimal time of year greatly improved recruitment response (Turner et al. 2006). Hydroseeding is one mechanised method commonly used in roadside revegetation projects, but can be useful, albeit expensive, in areas accessible by vehicles (Martin et al. 2002). If the application slurry includes additives such as germination stimulants, mulch and organic soil binders, this can improve restoration efficiency compared to dry broadcasting methods.

For both seed and cuttings, it is important to collect material from the nearest remaining natural remnant, and also to match the habitat type for edaphic, hydrological and climatic variables, to improve the chances of establishment. Propagated material should be grown to develop strong rooting systems relative to shoot growth and preferably be grown in local soil and hardened off before planting to minimise planting shock and optimise establishment. Many plant communities develop in clumps rather than as regularly-spaced individuals, as positive interactions among individuals such as simple sheltering effects, can promote establishment. Where clumps or windbreaks may facilitate establishment, artificial wind breaks or alien slash should be used to create shelters. Furthermore, careful species selections that maximise trait diversity, or match niche requirements of the invasive alien species likely to re-invade, can enhance the likelihood of restoration success (Funk et al. 2008; Laughlin 2014).

Increasing surface roughness after fire, for example by using felled alien branches, can help to reduce surface erosion by raising the boundary layer and promoting the trapping of wind-dispersed native seeds from neighbouring intact remnants. Transferring seed-bearing branches of native pioneer shrubs to a burnt riparian zone resulted in good recruitment and augmented a sowing treatment (Pretorius et al. 2008). Where soil erosion may be a potential problem on steep slopes after alien control, both physical and biotic interventions should be considered. Alien logs or biomass may be pegged or stacked across the slopes to trap sediment and short-lived, commercially-available, non-invasive species sown to provide short-term soil surface stability. An example of the latter is the use of commercial wheat in nutrient-poor fynbos ecosystems, as it does not grow too vigorously or produce a viable seed bank under such conditions.

Granivores such as mice and gerbils may be prominent and voracious following alien clearance, therefore faunal control could be considered. For small mammals, the encouragement of raptors through erecting suitable perches on the restoration site may have a positive impact. Herbivores such as antelope preferentially browse or graze young vegetation, targeting the establishing plants, and the use of exclosures may be justified to ensure establishment.

Restoration of highly-degraded ecosystems to sustainable, fully functional natural ecosystems could take a long time, representing several generations or recruitment events (e.g. 45+ years for 3 generations in fire-driven fynbos). The reasons for this include that community re-assembly may not be even for all guilds and some may require repeated re-introductions, or new methods of propagation, should the initial approaches fail, as was found for restoring the diverse forb resprouter guild in coastal grasslands (Zaloumis and Bond 2011; Table 23.1). Variations in annual rainfall result in good and poor years for establishment of re-introduced species, and repeated re-introductions should be planned. This reality poses challenges for monitoring, and for the restoration practitioner who may be tasked with ambitious short-term restoration goals. Nevertheless, statistical and modelling tools allow for an assessment of early progress and whether the degraded ecosystems are developing along the desired trajectory (Hall 2018). Ecological goals must be as specific as possible and outcomes carefully monitored if effectiveness of the restoration interventions is to be fully assessed (Ntshotsho et al. 2011). Results of the monitoring should feed back into decision tree frameworks (Fig. 23.2) for the different ecosystems and indicate to managers whether treatments should be modified or remedial actions required.

23.6 Conclusions

Restoration ecology is a relatively young science, and numerous South African studies have improved the understanding of local ecosystem dynamics and ecological restoration principles in general. However, there is still a dearth of research on how non-plant taxa recover, as well as active restoration examples for biomes other than fynbos. The stimulus for local ecological restoration research has been the large negative impacts of invasive alien species on ecosystem structure and functioning, including the impacts on economically-important ecosystem services. Restoration is economically viable for specific ecosystem services such as water, which can serve as an ‘umbrella service’ to enhance other conservation goals. Other ecosystem services delivered via restoration, such as pollination, deserve more attention.

Ecological restoration is a long-term process that greatly exceeds the time-spans of most post-graduate research studies (see Hill et al. 2020, Chap. 19, for similar arguments). Early field results (2–3 years post-intervention) may not accurately reflect the restoration trajectory as measured in subsequent years (Ruwanza et al. 2018). More long-term studies are required that build on, and monitor, earlier short-term field studies to further analyse the impacts of restoration interventions and to improve restoration guidelines. In addition, owing to the large diversity of ecosystems in South Africa, further applied research would be beneficial to test optimal combinations of treatments, including alien control, spontaneous regeneration and active restoration methods for different ecosystems. In assessing the outcomes, the potential of recent technologies should be considered for use, especially for scaling up implementation. These include specialised seed coatings, mechanised sowing, and aerial imagery (remote sensing, high resolution aerial imagery and drones; Dufour et al. 2013; Zahawi et al. 2015; Rebelo et al. 2017; Harris et al. 2018).

In South Africa, there has been limited implementation of ecological restoration results from published scientific papers, popular articles and guidelines, despite the need to scale up restoration interventions in the field. If we are to apply nature-based solutions to the urgent global challenges of invasive alien species, ecosystem degradation and climate change, ecological restoration efforts should be intensified. The limited action so far may relate to resource limitations, especially for active restoration interventions as these may be particularly resource-intensive. Funding streams for invasive plant management in South Africa currently focus on the removal of invasive alien plant stands, and a shift in mind-set is needed to incorporate restoration goals. For instance, spontaneous regeneration approaches that are less resource-intensive than active restoration can easily be applied at scale. An example is to plan and apply the optimal clearing methods and timing of interventions correctly to ensure invasive alien removal is successful without damaging native species or seed banks, thus promoting the probability of spontaneous regeneration and reducing long-term restoration costs.

It is important to promote stakeholder involvement in restoration projects as this is more likely to result in securing resources for larger-scale implementation. One example of this is the Blaauwberg Sand Fynbos project where researchers, conservation managers, volunteers and other organisations are involved. Ecological restoration is being implemented as and when resources are secured. Another example is the Berg River restoration project which involves government agencies, researchers, land-owners, restoration practitioners and other stakeholders who together plan and implement restoration interventions in the field.

Despite the above examples, South Africa has a long way to go to meet the required international targets, such as the Convention on Biological Diversity 2020 Aichi Target 14 (restore and protect ecosystems providing essential services) and Target 15 (restore 15% of degraded ecosystems to contribute to climate change mitigation and adaptation).

Acknowledgements Much of the work reviewed in this chapter was supported by the DSI-NRF Centre of Excellence for Invasion Biology and the Natural Resources Management programme of the Department Of Environmental Affairs as part of the collaborative research project on "Integrated management of invasive alien species in South Africa". Additional support was provided by the National Research Foundation (grant 85417).

References

Anderson SJ, Ankor BL, Sutton PC (2017) Ecosystem service valuations of South Africa using a variety of land cover data sources and resolutions. Ecosys Serv 27:173–178. https://doi.org/10.1016/j.ecoser.2017.06.001

Aronson J, Milton SJ, Blignaut JN (2007) Restoring natural capital: definitions and rationale. In: Aronson J, Milton SJ, Blignaut JN (eds) Restoring natural capital; science, business and practice. Island Press, Washington DC, pp 3–8

Aronson J, Blignaut JN, Aronson TB (2017) Conceptual frameworks and references for landscape-scale restoration: reflecting back and looking forward. Ann Miss Bot Gard 102:188–200. https://doi.org/10.3417/2017003

Balaguer L, Escudero A, Martín-Duque JF et al (2014) The historical reference in restoration ecology: re-defining a cornerstone concept. Biol Conserv 176:12–20. https://doi.org/10.1016/j.biocon.2014.05.007

Beater MMT, Garner RD, Witkowski ETF (2008) Impacts of clearing invasive alien plants from 1995 to 2005 on vegetation structure, invasion intensity and ground cover in a temperate to subtropical riparian ecosystem. S Afr J Bot 74:495–507. https://doi.org/10.1016/j.sajb.2008.01.174

Beisner BE, Haydon DT, Cuddington K (2003) Alternative stable states in ecology. Front Ecol Environ 1:376–382. https://doi.org/10.1890/1540-9295(2003)001[0376:ASSIE]2.0.CO;2

Blanchard R, Holmes PM (2008) Riparian vegetation recovery after invasive alien tree clearance in the Fynbos Biome. S Afr J Bot 74:421–431. https://doi.org/10.1016/j.sajb.2008.01.178

Bond WJ, Slingsby P (1983) Seed dispersal by ants in shrublands of the Cape Province and its evolutionary implications. S Afr J Sci 79:231–233

Brooks ML, D'Antonio CM, Richardson DM et al (2004) Effects of invasive alien plants on fire regimes. Bioscience 54:677–688. https://doi.org/10.1641/0006-3568(2004)054[0677:EOIAPO]2.0.CO;2

Brown NAC (1993) Promotion of fynbos seeds by plant-derived smoke. New Phytol 123:575–583. https://doi.org/10.1111/j.1469-8137.1993.tb03770.x

Carrick PJ, Krüger R (2007) Restoring degraded landscapes in lowland Namaqualand: lessons from the mining experience and from regional ecological dynamics. J Arid Environm 70:767–781. https://doi.org/10.1016/j.jaridenv.2006.08.006

Clewell A, Aronson J, Winterhalder K (2004) The SER international primer on ecological restoration. Ecol Restor 2:206–207

Colvin C, Riemann K, Brown C et al (2009) Ecological and environmental impacts of large-scale groundwater development in the Table Mountain Group (TMG) aquifer system. Report to Water Research Commission of South Africa 1327, p 170

Convention on Biological Diversity Aichi Targets (2010) Aichi biodiversity targets. https://www.cbd.int/sp/targets/. Accessed 10 Apr 2019

Costanza R, de Groot R, Braat L et al (2017) Twenty years of ecosystem services: how far have we come and how far do we still need to go? Ecosyst Serv 28:1–16. https://doi.org/10.1016/j.ecoser.2017.09.008

Crookes DJ, Blignaut JN, de Wit MP et al (2013) System dynamic modelling to assess economic viability and risk trade-offs for ecological restoration in South Africa. J Environ Manage 120:138–147. https://doi.org/10.1016/j.jenvman.2013.02.001

Crossman ND, Bernard F, Egoh B et al (2017) The role of ecological restoration and rehabilitation in production landscapes: an enhanced approach to sustainable development (Working paper for the UNCCD Global Land Outlook)

Currie B, Milton SJ, Steenkamp JC (2009) Cost-benefit analysis of alien vegetation clearing for water yield and tourism in a mountain catchment in the Western Cape of South Africa. Ecol Econ 68:2574–2579. https://doi.org/10.1016/j.ecolecon.2009.04.007

D'Antonio CM, Meyerson LA (2002) Exotic plant species as problems and solutions in ecological restoration: a synthesis. Restor Ecol 10:703–713. https://doi.org/10.1046/j.1526-100X.2002.01051.x

D'Antonio CM, Vitousek PM (1992) Biological invasions by exotic grasses, the grass/fire cycle, and global change. Annu Rev Ecol Syst 23:63–87. https://doi.org/10.1146/annurev.es.23.110192.000431

Davies KW, Boyd CS, Madsen MD et al (2018) Evaluating a seed technology for sagebrush restoration across an elevation gradient: support for bet hedging. Rangel Ecol Manag 71:19–24. https://doi.org/10.1016/j.rama.2017.07.006

Dew LA, Rozen-Rechels D, Roux E et al (2017) Evaluating the efficacy of invasive plant control in response to ecological factors. S Afr J Bot 109:203–213. https://doi.org/10.1016/j.sajb.2016.12.007

Downey PO, Richardson DM (2016) Alien plant invasions and native plant extinctions: a six-threshold framework. AoB Plants 8:plw047. https://doi.org/10.1093/aobpla/plw047

Dufour S, Bernez I, Betbeder J et al (2013) Monitoring restored riparian vegetation: how can recent developments in remote sensing sciences help? Knowl Manage Aquat Ecosyst 410:10. https://doi.org/10.1051/kmae/2013068

Falk AD, Fulbright TE, Smith FS et al (2013) Does seeding a locally adapted native mixture inhibit ingress by exotic plants? Restor Ecol 21:474–480. https://doi.org/10.1111/j.1526-100X.2012.00918.x

Forup ML, Memmott J (2005) The restoration of plant-pollinator interactions in Hay meadows. Restor Ecol 13:265–274. https://doi.org/10.1111/j.1526-100X.2005.00034.x

Fourie S (2008) Composition of the soil seed bank in alien-invaded grassy fynbos: potential for recovery after clearing. S Afr J Bot 74:445–453. https://doi.org/10.1016/j.sajb.2008.01.172

Funk JL, Cleland EE, Suding KN et al (2008) Restoration through reassembly: plant traits and invasion resistance. Trends Ecol Evol 23:695–703. https://doi.org/10.1016/j.tree.2008.07.013

Fynn RW, Wragg PD, Morris CD et al (2009) Vegetative traits predict grass species' invasiveness and the invasibility of restored grassland. Afr J Range For Sci 26:59–68. https://doi.org/10.2989/AJRFS.2009.26.2.2.845

Gaertner M, Richardson DM, Privett SDJ (2011) Effects of alien plants on ecosystem structure and functioning and implications for restoration: insights from three degraded sites in South African fynbos. Environ Manage 48:57–69. https://doi.org/10.1007/s00267-011-9675-7

Gaertner M, Fisher J, Sharma G et al (2012a) Insights into invasion and restoration ecology: time to collaborate towards a holistic approach to tackle biological invasions. NeoBiota 12:57–76. https://doi.org/10.3897/neobiota.12.2123

Gaertner M, Holmes PM, Richardson DM (2012b) Biological invasions, resilience and restoration. In: van Andel J, Aronson J (eds) Restoration ecology: the new frontier, 2nd edn. Wiley Blackwell, Oxford, pp 265–280. https://doi.org/10.1002/9781118223130.ch20

Gaertner M, Nottebrock H, Fourie H et al (2012c) Plant invasions, restoration, and economics: perspectives from South African fynbos. Perspect Plant Ecol Evol Syst 14:341–353. https://doi.org/10.1016/j.ppees.2012.05.001

Gaertner M, Biggs R, Te Beest M et al (2014) Invasive plants as drivers of regime shifts: identifying high-priority invaders that alter feedback relationships. Divers Distrib 20:733–744. https://doi.org/10.1111/ddi.12182

Galatowitsch S, Richardson DM (2005) Riparian scrub recovery after clearing of invasive alien trees in headwater streams of the Western Cape, South Africa. Biol Conserv 122:509–521. https://doi.org/10.1016/j.biocon.2004.09.008

Galloway AD, Holmes PM, Gaertner M et al (2017) The impact of pine plantations on fynbos above-ground vegetation and soil seed bank composition. S Afr J Bot 113:300–307. https://doi.org/10.1016/j.sajb.2017.09.009

Geerts S, Pauw A (2009) African sunbirds hover to pollinate an invasive hummingbird-pollinated plant. Oikos 118:573–579. https://doi.org/10.1111/j.1600-0706.2008.17167.x

Geldenhuys CJ (2013) Converting invasive alien plant stands to natural forest nature's way. Overview, theory, and practice. In: Jose S, Singh HP, Batish DR et al (eds) Invasive plant ecology. CRC, Boca Raton, pp 217–237. https://doi.org/10.1201/b13865-17

Geldenhuys CJ, Atsame-Edda A, Mugure MW (2017) Facilitating the recovery of natural evergreen forests in South Africa via invader plant stands. Forest Ecosys 4:21. https://doi.org/10.1186/s40663-017-0108-9

Hall SAW (2018) Restoration potential of alien-invaded Lowland Fynbos. PhD thesis. Stellenbosch University, Stellenbosch

Hall SA, Newton RJ, Holmes PM et al (2017) Heat and smoke pre-treatment of seeds to improve restoration of an endangered Mediterranean climate vegetation type. Austral Ecol 42:354–366. https://doi.org/10.1111/aec.12449

Harris D, Vlok J, van Niekerk A (2018) Regional mapping of spekboom canopy cover using very high resolution aerial imagery. J Appl Remote Sens 12:046022. https://doi.org/10.1117/1.JRS.12.046022

Herron CM, Jonas JL, Meiman PJ et al (2013) Using native annual plants to restore post-fire habitats in western North America. Int J Wildl Fire 22:815–821. https://doi.org/10.1071/WF11179

Hill MP, Moran VC, Hoffmann JH et al (2020) More than a century of biological control against invasive alien plants in South Africa: a synoptic view of what has been accomplished. In: van Wilgen BW, Measey J, Richardson DM et al (eds) Biological invasions in South Africa. Springer, Berlin, pp 549–568. https://doi.org/10.1007/978-3-030-32394-3_19

Hitchcock A, Cowell C, Rebelo T (2012) The lost fynbos of Tokai Park. Veld Flora 98:30–33

Hobbs RJ, Richardson DM (2011) Invasion ecology and restoration ecology: parallel evolution in two fields of Endeavour. In: Richardson DM (ed) Fifty years of invasion ecology. The legacy of Charles Elton. Wiley-Blackwell, Oxford, pp 61–69. https://doi.org/10.1002/9781444329988.ch6

Hobbs RJ, Higgs E, Harris JA (2009) Novel ecosystems: implications for conservation and restoration. Trends Ecol Evol 24:599–605. https://doi.org/10.1016/j.tree.2009.05.012

Holl KD, Zahawi RA, Cole RJ et al (2011) Planting seedlings in tree islands versus plantations as a large-scale tropical forest restoration strategy. Restor Ecol 19:470–479. https://doi.org/10.1111/j.1526-100X.2010.00674.x

Holmes PM (2001a) A comparison of the impacts of winter versus summer burning of slash fuel in alien-invaded fynbos areas in the Western Cape. South Afr For J 192:41–50. https://doi.org/10.1080/20702620.2001.10434132

Holmes PM (2001b) Shrubland restoration following woody alien invasion and mining: effects of topsoil depth, seed source, and fertilizer addition. Restor Ecol 9:71–84. https://doi.org/10.1046/j.1526-100x.2001.009001071.x

Holmes PM (2002) Depth distribution and composition of seed banks alien-invaded and uninvaded fynbos vegetation. Austral Ecol 27:110–120

Holmes PM, Cowling RM (1997) The effects of invasion by *Acacia saligna* on the guild structure and regeneration capabilities of South African fynbos shrublands. J Appl Ecol 34:317–332. https://doi.org/10.2307/2404879

Holmes PM, Marais C (2000) Impacts of alien plant clearance on vegetation in the mountain catchments of the Western Cape. South Afr For J 189:113–117. https://doi.org/10.1080/10295925.2000.9631286

Holmes PM, Richardson DM (1999) Protocols for restoration based on recruitment dynamics, community structure, and ecosystem function: perspectives from South African fynbos. Restor Ecol 7:215–230. https://doi.org/10.1046/j.1526-100X.1999.72015.x

Holmes PM, Richardson DM, van Wilgen BW et al (2000) Recovery of South African fynbos vegetation following alien woody plant clearing and fire: implications for restoration. Austral Ecol 25:631–639. https://doi.org/10.1111/j.1442-9993.2000.tb00069.x

Holmes PM, Richardson DM, Esler KJ et al (2005) A decision-making framework for restoring riparian zones degraded by invasive alien plants in South Africa. S Afr J Sci 101:553–564

Holmes PM, Richardson DM, Marais C (2007) Costs and benefits of restoring natural capital following alien plant invasions in the fynbos ecosystems in South Africa. In: Aronson J, Milton SJ, Blignaut JN (eds) Restoring natural capital; science, business and practice. Island Press, Washington DC, pp 188–197

James JJ, Carrick PJ (2016) Toward quantitative dryland restoration models. Restor Ecol 24:85–90. https://doi.org/10.1111/rec.12393

Kaiser-Bunbury CN, Mougal J, Whittington AE et al (2017) Ecosystem restoration strengthens pollination network resilience and function. Nature 542:223–227. https://doi.org/10.1038/nature21071

Kerr TF, Ruwanza S (2016) Does *Eucalyptus grandis* invasion and removal affect soils and vegetation in the Eastern Cape Province, South Africa? Austral Ecol 41:328–338. https://doi.org/10.1111/aec.12315

Kettenring KM, Adams CR (2011) Lessons learned from invasive plant control experiments : a systematic review and meta-analysis. J Appl Ecol 48:970–979. https://doi.org/10.1111/j.1365-2664.2011.01979.x

Kimball S, Lulow M, Sorenson Q et al (2015) Cost-effective ecological restoration. Restor Ecol 23:800–810. https://doi.org/10.1111/rec.12261

Kraaij T, van Wilgen BW (2014) Drivers, ecology, and management of fire in fynbos. In: Allsopp N, Colville JF, Verboom AG (eds) Fynbos: ecology, evolution, and conservation of a megadiverse region. Oxford University Press, Oxford, pp 47–72. https://doi.org/10.1093/acprof:oso/9780199679584.003.0003

Krupek A, Gaertner M, Holmes PM et al (2016) Assessment of post-burn removal methods for *Acacia saligna* in Cape Flats Sand Fynbos, with consideration of indigenous plant recovery. S Afr J Bot 105:211–217. https://doi.org/10.1016/j.sajb.2016.04.004

Kull CA, Shackleton CM, Cunningham PJ et al (2011) Adoption, use, and perception of Australian acacias around the world. Divers Distrib 17:822–836. https://doi.org/10.1111/j.1472-4642.2011.00783.x

Laughlin DC (2014) Applying trait-based models to achieve functional targets for theory-driven ecological restoration. Ecol Lett 17:771–784. https://doi.org/10.1111/ele.12288

Le Maitre DC, Gaertner M, Marchante E et al (2011) Impacts of invasive Australian acacias: implications for management and restoration. Divers Distrib 17:1015–1029. https://doi.org/10.1111/j.1472-4642.2011.00816.x

Macdonald IAW (1984) Is the fynbos biome especially susceptible to invasion by alien plants? A re-analysis of available data. S Afr J Sci 80:369–377

Macdonald IAW, Kruger FJ, Ferrar AA (eds) (1986) The ecology and management of biological invasions in southern Africa. Oxford University Press, Cape Town

Madsen MD, Davies KW, Boyd CS et al (2016) Emerging seed enhancement technologies for overcoming barriers to restoration. Restor Ecol 24:S77–S84. https://doi.org/10.1111/rec.12332

Magoba RN, Samways MJ (2010) Recovery of benthic macroinvertebrate and adult dragonfly assemblages in response to large scale removal of riparian invasive alien trees. J Insect Conserv 14:627–636. https://doi.org/10.1007/s10841-010-9291-5

Mangachena JR, Geerts S (2017) Invasive alien trees reduce bird species richness and abundance of mutualistic frugivores and nectarivores; a bird's eye view on a conflict of interest species in riparian habitats. Ecol Res 32:1–10. https://doi.org/10.1007/s11284-017-1481-0

Maoela MA, Jacobs SM, Roets F (2016) Invasion, alien control and restoration: legacy effects linked to folivorous insects and phylopathogenic fungi. Austral Ecol 41:906–917. https://doi.org/10.1111/aec.12383

Marais C, Wannenburgh AM (2008) Restoration of water resources (natural capital) through the clearing of invasive alien plants from riparian areas in South Africa – costs and water benefits. S Afr J Bot 74:526–537. https://doi.org/10.1016/j.sajb.2008.01.175

Marais KE, Pratt RB, Jacobs SM et al (2014) Postfire regeneration of resprouting mountain fynbos shrubs: differentiating obligate resprouters and facultative seeders. Plant Ecol 215:195–208. https://doi.org/10.1007/s11258-013-0289-4

Marchante E, Kjøller A, Struwe S et al (2009) Soil recovery after removal of the N2-fixing invasive *Acacia longifolia*: consequences for ecosystem restoration. Biol Invasions 11:813–823. https://doi.org/10.1007/s10530-008-9295-1

Martin A, Khater C, Mineau H et al (2002) Rehabilitation ecology by revegetation. Approach and results from two Mediterranean countries. Korean. J Ecol 25:9–17. https://doi.org/10.5141/JEFB.2002.25.1.009

Matzek V, Warren S, Fisher C (2016) Incomplete recovery of ecosystem processes after two decades of riparian forest restoration. Restor Ecol 24:637–645. https://doi.org/10.1111/rec.12361

McConnachie MM, Cowling RM, van Wilgen BW et al (2012) Evaluating the cost-effectiveness of invasive alien plant clearing: a case study from South Africa. Biol Conserv 155:128–135. https://doi.org/10.1016/j.biocon.2012.06.006

McDonald T, Gann GD, Jonson J et al (2016a) International standards for the practice of ecological restoration – including principles and key concepts. Retrieved from http://c.ymcdn.com/sites/www.ser.org/resource/resmgr/docs/SER_International_Standards.pdf

Mcdonald T, Jonson J, Dixon KW (2016b) National standards for the practice of ecological restoration in Australia. Restor Ecol 24:s4–s32. https://doi.org/10.1111/rec.12359

Meek CS, Richardson DM, Mucina L (2013) Plant communities along the Eerste River, Western Cape, South Africa: community descriptions and implications for restoration. Koedoe 55:1–14. https://doi.org/10.4102/koedoe.v55i1.1099

Metzger JP, Esler K, Krug C et al (2017) Best practice for the use of scenarios for restoration planning. Curr Opin Environ Sustain 29:14–25. https://doi.org/10.1016/j.cosust.2017.10.004

Mgobozi MP, Somers MJ, Dippenaar-Schoeman AS (2008) Spider responses to alien plant invasion: the effect of short- and long-term *Chromolaena odorata* invasion and management. J Appl Ecol 45:1189–1197. https://doi.org/10.1111/j.1365-2664.2008.01486.x

Milton SJ, Siegfried WR (1994) A conceptual model of arid rangeland degradation. Bioscience 44:70–76. https://doi.org/10.2307/1312204

Milton SJ, Dean WRJ, Richardson DM (2003) Economic incentives for restoring natural capital in southern African rangelands. Front Ecol Environ 1:247–254. https://doi.org/10.1890/1540-9295(2003)001[0247:EIFRNC]2.0.CO;2

Modiba RV, Joseph GS, Seymour CL et al (2017) Restoration of riparian systems through clearing of invasive plant species improves functional diversity of Odonate assemblages. Biol Conserv 214:46–54. https://doi.org/10.1016/j.biocon.2017.07.031

Mohieldin M, Caballero P (2015) Protect, restore and promote sustainable use of terrestrial ecosystems, sustainably manage forests, combat desertification, and halt and reverse land degradation and halt biodiversity loss. UN Chron 51(4):34–35. https://doi.org/10.18356/f405cab8-en

Moran VC, Hoffmann JH (2012) Conservation of the fynbos biome in the Cape Floral Region: the role of biological control in the management of invasive alien trees. BioControl 57:139–149. https://doi.org/10.1007/s10526-011-9403-5

Morgan JW (1999) Have tubestock plantings successfully established populations of rare grassland species into reintroduction sites in western Victoria? Biol Conserv 89:235–243. https://doi.org/10.1016/S0006-3207(99)00014-2

Mori AS, Furukawa T, Sasaki T (2013) Response diversity determines the resilience of ecosystems to environmental change. Biol Rev 88:349–364. https://doi.org/10.1111/brv.12004

Morris TL, Witkowski ETF, Coetzee JA (2008) Initial response of riparian plant community structure to clearing of invasive alien plants in Kruger National Park, South Africa. S Afr J Bot 74:485–494. https://doi.org/10.1016/j.sajb.2008.01.177

Mostert E, Gaertner M, Holmes PM et al (2017) Impacts of invasive alien trees on threatened lowland vegetation types in the Cape Floristic Region, South Africa. S Afr J Bot 108:209–222. https://doi.org/10.1016/j.sajb.2016.10.014

Mostert E, Gaertner M, Holmes PM et al (2018) A multi-criterion approach for prioritizing areas in urban ecosystems for active restoration following invasive plant control. Environ Manag 62:1150–1167. https://doi.org/10.1007/s00267-018-1103-9

Mugwedi LF, Rouget M, Egoh B et al (2017) An assessment of a community-based, forest restoration programme in Durban (eThekwini), South Africa. Forests 8:255. https://doi.org/10.3390/f8080255

Mugwedi LF, Ray-Mukherjee J, Roy KE et al (2018) Restoration planning for climate change mitigation and adaptation in the city of Durban, South Africa. Int J Biodivers Sci Ecosyst Serv Manage 14:132–144. https://doi.org/10.1080/21513732.2018.1483967

Ndou E, Ruwanza S (2016) Soil and vegetation recovery following alien tree clearing in the Eastern Cape Province of South Africa. Afr J Ecol 54:460–470. https://doi.org/10.1111/aje.12305

Norton DA (2009) Species invasions and the limits to restoration: learning from the New Zealand experience. Science 325:569–571. https://doi.org/10.1126/science.1172978

Nsikani MM, Novoa A, van Wilgen BW et al (2017) *Acacia saligna* 's soil legacy effects persist up to 10 years after clearing: implications for ecological restoration. Austral Ecol 42:880–889. https://doi.org/10.1111/aec.12515

Nsikani MM, van Wilgen BW, Gaertner M (2018) Barriers to ecosystem restoration presented by soil legacy effects of invasive alien N2-fixing woody species: implications for ecological restoration. Restor Ecol 26:235–244. https://doi.org/10.1111/rec.12669

Ntshotsho P, Reyers B, Esler KJ (2011) Assessing the evidence base for restoration in South Africa. Restor Ecol 19:578–586. https://doi.org/10.1111/j.1526-100X.2010.00753.x

Pearson DE, Ortega YK, Runyon JB et al (2016) Secondary invasion: the bane of weed management. Biol Conserv 197:8–17. https://doi.org/10.1016/j.biocon.2016.02.029

Pretorius MR, Esler KJ, Holmes PM et al (2008) The effectiveness of active restoration following alien clearance in fynbos riparian zones and resilience of treatments to fire. S Afr J Bot 74:517–525. https://doi.org/10.1016/j.sajb.2008.01.180

Prins N, Holmes PM, Richardson DM (2004) A reference framework for the restoration of riparian vegetation in the Western Cape, South Africa, degraded by invasive Australian Acacias. S Afr J Bot 70:767–776. https://doi.org/10.1016/S0254-6299(15)30178-2

Raimondo D, Staden VL, Foden W (2009) Red list of South African Plants. South African National Biodiversity Institute, Pretoria

Rebelo AJ, Scheunders P, Esler KJ et al (2017) Detecting, mapping and classifying wetland fragments at a landscape scale. Remote Sensing Appl Soc Environ 8:212–223. https://doi.org/10.1016/j.rsase.2017.09.005

Reyers B, O'Farrell PJ, Cowling RM et al (2009) Ecosystem services, land-cover change, and stakeholders: finding a sustainable foothold for a semiarid biodiversity hotspot. Ecol Soc 14:38. https://doi.org/10.5751/ES-02867-140138

Richardson DM, van Wilgen BW (2004) Invasive alien plants in South Africa: how well do we understand the ecological impacts. S Afr J Sci 100:45–52

Richardson DM, Macdonald IAW, Hoffmann JH et al (1997) Alien plant invasions. In: Cowling RM, Richardson DM, Pierce SM (eds) Vegetation of Southern Africa. Cambridge University Press, Cambridge, pp 535–570

Richardson DM, Bond WJ, Dean WRJ et al (2000) Invasive alien organisms and global change: a South African perspective. In: Mooney HA, Hobbs RJ (eds) Invasive species in a changing world. Island Press, Washington DC, pp 303–349

Richardson DM, Holmes PM, Esler KJ et al (2007) Riparian vegetation: degradation, alien plant invasions, and restoration prospects. Divers Distrib 13:126–139. https://doi.org/10.1111/j.1366-9516.2006.00314.x

Rohr JR, Johnson P, Hickey CW et al (2013) Implications of global climate change for natural resource damage assessment, restoration, and rehabilitation. Environ Toxicol Chem 32:93–101. https://doi.org/10.1002/etc.2036

Rohr JR, Bernhardt ES, Cadotte MW et al (2018) The ecology and economics of restoration : when, what, where, and how to restore ecosystems. Ecol Soc 23:15. https://doi.org/10.5751/ES-09876-230215

Rouwntree K (1991) An assessment of the potential impact of alien invasive vegetation on the geomorphology of river channels in South Africa. South Afr J Aquat Sci 17:28–43. https://doi.org/10.1080/10183469.1991.9631311

Ruwanza S, Gaertner M, Esler KJ et al (2013a) Both complete clearing and thinning of invasive trees lead to short-term recovery of native riparian vegetation in the Western Cape, South Africa. Appl Veg Sci 16:193–204. https://doi.org/10.1111/j.1654-109X.2012.01222.x

Ruwanza S, Gaertner M, Richardson DM et al (2013b) Soil water repellency in riparian systems invaded by *Eucalyptus camaldulensis*: a restoration perspective from the Western Cape Province, South Africa. Geoderma 200–201:9–17. https://doi.org/10.1016/j.geoderma.2013.01.017

Ruwanza S, Gaertner M, Esler KJ et al (2015) Allelopathic effects of invasive Eucalyptus camaldulensis on germination and early growth of four native species in the Western Cape, South Africa. South For 77:91–105. https://doi.org/10.2989/20702620.2014.965985

Ruwanza S, Gaertner M, Esler KJ et al (2018) Medium-term vegetation recovery after removal of invasive *Eucalyptus camaldulensis* stands along a South African river. S Afr J Bot 119:63–68. https://doi.org/10.1016/j.sajb.2018.08.002

Samways MJ, Sharratt NJ, Simaika JP (2011) Effect of alien riparian vegetation and its removal on a highly endemic river macroinvertebrate community. Biol Invasions 13:1305–1324. https://doi.org/10.1007/s10530-010-9891-8

Sher AA, Hyatt LA (1999) The disturbed resource-Flux Invasion Matrix: a new framework for patterns of plant invasion. Biol Invasions 1:107–114. https://doi.org/10.1023/A:1010050420466

Stafford W, Blignaut J (2017) Reducing landscape restoration costs: feasibility of generating electricity from invasive alien plant biomass on the Agulhas Plain, South Africa. Ecosys Serv 27:224–231. https://doi.org/10.1016/j.ecoser.2017.04.008

Stafford W, Birch C, Etter H et al (2017) The economics of landscape restoration: benefits of controlling bush encroachment and invasive plant species in South Africa and Namibia. Ecosyst Serv 27:193–202. https://doi.org/10.1016/j.ecoser.2016.11.021

Standish RJ, Cramer VA, Wild SL et al (2007) Seed dispersal and recruitment limitation are barriers to native recolonization of old-fields in western Australia. J Appl Ecol 44:435–445. https://doi.org/10.1111/j.1365-2664.2006.01262.x

Stanturf JA, Palik BJ, Williams MI et al (2014) Forest restoration paradigms. J Sustain For 33 (Sup1):S161–S194. https://doi.org/10.1080/10549811.2014.884004

Strydom M, Esler KJ, Wood AR (2012) *Acacia saligna* seed banks: sampling methods and dynamics, Western Cape, South Africa. S Afr J Bot 79:140–147. https://doi.org/10.1016/j.sajb.2011.10.007

Strydom M, Veldtman R, Ngwenya MZ et al (2017) Invasive Australian *Acacia* seed banks: size and relationship with stem diameter in the presence of gall-forming biological control agents. PLoS One 12:e0181763. https://doi.org/10.1371/journal.pone.0181763

Suding KN, Hobbs RJ (2009) Threshold models in restoration and conservation: a developing framework. Trends Ecol Evol 24:271–279. https://doi.org/10.1016/j.tree.2008.11.012

Suding K, Leger E (2012) Shifting baselines: dynamics of evolution and community change in a changing world. In: van Andel J, Aronson J (eds) Restoration ecology: the new frontier, 2nd edn. Wiley Blackwell, Oxford, pp 282–292. https://doi.org/10.1002/9781118223130.ch21

Suding KN, Gross KL, Houseman GR (2004) Alternative states and positive feedbacks in restoration ecology. Trends Ecol Evol 19:46–53. https://doi.org/10.1016/j.tree.2003.10.005

Suding K, Higgs E, Palmer M et al (2015) Committing to ecological restoration. Science 348:638–640. https://doi.org/10.1126/science.aaa4216

Temperton VM, Higgs E, Choi YD et al (2014) Flexible and adaptable restoration: an example from South Korea. Restor Ecol 22:271–278. https://doi.org/10.1111/rec.12095

Tererai F, Gaertner M, Jacobs SM et al (2013) Eucalyptus invasions in riparian forests: effects on native vegetation community diversity, stand structure and composition. For Ecol Manag 297:84–93. https://doi.org/10.1016/j.foreco.2013.02.016

Tererai F, Gaertner M, Jacobs SM et al (2015a) Eucalyptus camaldulensis invasion in riparian zones reveals few significant effects on soil physico-chemical properties. River Res Appl 31:590–601. https://doi.org/10.1002/rra.2762

Tererai F, Gaertner M, Jacobs SM et al (2015b) Resilience of invaded riparian landscapes: the potential role of soil-stored seed banks. Environ Manage 55:86–99. https://doi.org/10.1007/s00267-014-0374-z

Török K, Szili-Kovács T, Halassy M et al (2000) Immobilization of soil nitrogen as a possible method for the restoration of sandy grassland. Appl Veg Sci 3:7–14. https://doi.org/10.2307/1478913

Traveset A, Richardson DM (2006) Biological invasions as disruptors of plant reproductive mutualisms. Trends Ecol Evol 21:208–216. https://doi.org/10.1016/j.tree.2006.01.006

Turner SR, Pearce B, Rokich DP et al (2006) Influence of polymer seed coatings, soil raking, and time of sowing on seedling performance in post-mining restoration. Restor Ecol 14:267–277. https://doi.org/10.1111/j.1526-100X.2006.00129.x

Turner KG, Grace K, Anderson S et al (2016) A review of methods, data, and models to assess changes in the value of ecosystem services from land degradation and restoration a review of methods, data, and models to assess changes in the value of ecosystem services from land degradation and resto. Ecol Model 319:190–207. https://doi.org/10.1016/j.ecolmodel.2015.07.017

Turpie JK, Marais C, Blignaut JN (2008) The working for water programme: evolution of a payments for ecosystem services mechanism that addresses both poverty and ecosystem service delivery in South Africa. Ecol Econ 65:788–798. https://doi.org/10.1016/j.ecolecon.2007.12.024

Urgenson LS, Prozesky HE, Esler KJ (2013) Stakeholder perceptions of an ecosystem services approach to clearing invasive alien plants on private land. Ecol Soc 18:26. https://doi.org/10.5751/ES-05259-180126

van Wilgen BW (2018) The management of invasive alien plants in South Africa: strategy, progress and challenges. Outl Pest Manage 29(1):13–17. https://doi.org/10.1564/v29_feb_04

van Wilgen BW, Richardson DM, Seydack AHW (1994) Managing fynbos for biodiversity: constraints and options in a fire-prone environment. S Afr J Sci 90:322–329

van Wilgen BW, Carruthers J, Cowling RM et al (2016) Ecological research and conservation management in the Cape Floristic Region between 1945 and 2015: history, current understanding and future challenges ecological research and conservation management in the Cape Floristic Region between 1945 and 2015. Trans R Soc S Afr 71:207–303. https://doi.org/10.1080/0035919X.2016.1225607

Vilà M, Espinar JL, Hejda M, Hulme PE, Jarošik V, Maron JL, Pergl J, Schaffner U, Yan Sun Y, Pyšek P (2011) Ecological impacts of invasive alien plants: a meta-analysis of their effects on species, communities and ecosystems. Ecol Lett 14:702–208. https://doi.org/10.1111/j.1461-0248.2011.01628.x

Vosse S, Esler KJ, Richardson DM et al (2008) Can riparian seed banks initiate restoration after alien plant invasion? Evidence from the Western Cape, South Africa. S Afr J Bot 74:432–444. https://doi.org/10.1016/j.sajb.2008.01.170

Waller PA, Anderson PM, Holmes PM et al (2016) Seedling recruitment responses to interventions in seed-based ecological restoration of Peninsula Shale Renosterveld, Cape Town. S Afr J Bot 103:193–209. https://doi.org/10.1016/j.sajb.2015.09.009

Walters C, Berjak P (2013) Preservation of recalcitrant seeds. Science 339:915–916. https://doi.org/10.1126/science.1230935

Whisenant SG (1999) Repairing damaged wildlands. Cambridge University Press, Cambridge. https://doi.org/10.1017/CBO9780511612565

Whisenant SG (2002) Terrestrial systems. In: Perrow MR, Davy AJ (eds) Handbook of ecological restoration, Principles of restoration, vol 1. Cambridge University Press, Cambridge, pp 83–105. https://doi.org/10.1017/CBO9780511549984.008

Wilson JR, Gaertner M, Griffiths CL et al (2014) Biological invasions in the Cape Floristic Region : history, current patterns, impacts, and management challenges. In: Allsopp N, Colville JF, Verboom GA (eds) Fynbos: ecology, evolution, and conservation of a megadiverse region. Oxford University Press, Oxford, pp 273–298. https://doi.org/10.1093/acprof:oso/9780199679584.003.0012

Wood AR (2017) Fungi and invasions in South Africa. Bothalia 47:a2124. https://doi.org/10.4102/abc.v47i2.2124

Yelenik SG, Stock WD, Richardson DM (2004) Ecosystem level impacts of invasive *Acacia saligna* in the South African fynbos. Restor Ecol 12:44–51. https://doi.org/10.1111/j.1061-2971.2004.00289.x

Zahawi RA, Dandois JP, Holl KD et al (2015) Using lightweight unmanned aerial vehicles to monitor tropical forest recovery. Biol Conserv 186:287–295. https://doi.org/10.1016/j.biocon.2015.03.031

Zaloumis NP, Bond WJ (2011) Grassland restoration after afforestation: no direction home? Austral Ecol 36:357–366. https://doi.org/10.1111/j.1442-9993.2010.02158.x

Zink TA, Allen MF (1998) The effects of organic amendments on the restoration of a disturbed coastal sage scrub habitat. Restor Ecol 6:52–58. https://doi.org/10.1046/j.1526-100x.1998.00617.x

24

Biological Invasions: Social and Human Dimensions

Ross T. Shackleton, Ana Novoa, Charlie M. Shackleton and Christian A. Kull

Abstract This chapter examines current knowledge relating to the human and social dimensions of biological invasions in South Africa. We do so by advancing 12 propositions and examining the evidence for or against each using South African literature. The propositions cover four broad issues: how people cause invasions; how they conceptualise them; effects of invasive species on people; and peoples' responses to them. The propositions we assess include: (1) intentional introductions were and continue to reflect the social ethos of the time; (2) people go to great lengths to ensure that newly introduced species establish themselves; (3) human-mediated modifications help invasive species to establish; (4) how people think about and study invasive species is strongly shaped by social-ecological contexts; (5) knowledge and awareness of invasive species is low amongst the general public; (6) personal values are the primary factor affecting perceptions of invasive alien species and their control; (7) specific social-ecological contexts mediate how invasive species affect people; (8) research on social effects of invasive species primarily focuses on negative impacts; (9) the negative social impacts of invasive species on local livelihoods are of more concern to people than impacts on biodiversity; (10) people are less willing to manage species regarded as 'charismatic'; (11) social heterogeneity increases conflicts around the management of biological invasions; and (12) engagement with society is key to successful management. By advancing and questioning propositions, we were able to determine what is known, provide evidence for where gaps lie, and thus identify areas for future research.

R. T. Shackleton (✉)
Institute of Geography and Sustainability, University of Lausanne, Lausanne, Switzerland

Centre for Invasion Biology, Stellenbosch University, Stellenbosch, South Africa
e-mail: rtshackleton@gmail.com

A. Novoa
Department of Invasion Ecology, Institute of Botany, The Czech Academy of Sciences, Průhonice, Czech Republic

C. M. Shackleton
Department of Environmental Science, Rhodes University, Makhanda, South Africa

C. A. Kull
Institute of Geography and Sustainability, University of Lausanne, Lausanne, Switzerland

24.1 Introduction

Research that addresses the human and social dimensions of invasion science is crucial for understanding and responding to biological invasions as people are involved in all parts of the introduction-naturalisation-invasion-response continuum (Head 2017; Shackleton et al. 2019a). Despite the need for humanities and social science perspectives in invasion science, to date there has been relatively little work in this area (Le Maitre et al. 2004; Vaz et al. 2017a; Abrahams et al. 2019). Consequently, there are many research gaps and missed opportunities for interdisciplinary collaboration which is necessary to truly advance and address pressing challenges in dynamic and varied contexts (Vaz et al. 2017a; Abrahams et al. 2019). To promote uptake, Shackleton et al. (2019a) recently highlighted four broad issues in which research on the human and social dimensions of invasion science can help to improve understanding and guide management responses. These four areas are: (1) how people cause invasions, (2) how people conceptualise and perceive invasions, (3) the effects of invasions on people, and (4) how people respond to invasions.

In this chapter, for each thematic area, we advance three propositions (statements or proposals for consideration and which can help in asserting a generalisable trend or process—similar to a hypothesis) and examine the evidence in support of each of them in the South African context (Table 24.1). We tried to ensure that the propositions were pertinent to current research topics and trends, and to ensure that they would be useful and relevant to guide future work. We used propositions as a means to move beyond just a summary of current knowledge of the social contributions to invasion science towards a more focussed, analytic and critical stance as the necessary foundation for development of knowledge and theory and future research. The process of examining the available evidence for each proposition fosters in-depth thinking of what evidence is available, where different points conflict and where gaps in research persist. The propositions are not specific to South Africa, but the relative richness of research on biological invasions in the country offers a reasonable first opportunity to investigate them.

24.2 Humans as Causes of Alien Species Invasions

Humans are the primary agents for the deliberate or inadvertent introduction of alien species outside their native ranges, some of which become invasive. In South Africa, work by historians and, to a lesser degree, by researchers in other social science and humanities disciplines, has detailed the role that people play in facilitating biological invasions. In particular, many purposeful introductions have been driven by specific societal mind-sets or ethos, operating in different eras (Carruthers et al. 2011; Kull et al. 2011; Udo et al. 2019), and humans facilitated the establishment of invasions by modifying species and landscapes. In South Africa and globally, understanding the social drivers and processes of species introductions is probably the best researched of the four thematic areas.

Table 24.1 A summary of key findings from the assessment of a set of 12 propositions that examine the social dimensions of invasion science, with notes on the degree to which the proposition can be supported by South African evidence

Proposition	Degree of support	Evidence level	Summary notes
1a. Intentional introductions were and continue to reflect the social ethos of the time	Supported	High	A large body of literature supports this proposition at a broad scale. Further research remains necessary to understand the trends that shape intentional introductions over smaller time scales and the role of different stakeholders
1b. People go to great lengths to ensure that newly introduced species establish themselves	Supported	Moderate	Good evidence in support of this proposition for some invasions. However, some invasions result from accidental introductions that required no or little effort to establish invasive populations
1c. Human-mediated modifications help invasive species establish	Supported	Moderate	Although this proposition is well supported, there is scope for much more work on this topic, especially for better integrating human actions into ecological based models
2a. How people think about and study invasive species is strongly shaped by social-ecological contexts	Supported	Moderate	Although supported for the South African context, it should be tested whether it also holds true for different contexts in other countries
2b. Knowledge and awareness of invasive species is low amongst the general public	Partially supported	Moderate	Generally, public knowledge and awareness of invasive species is low, however, in some regions, it can be high among certain actors or regarding certain species
2c. Personal values are the primary factor affecting perceptions of invasive alien species and their control	Partially supported	Moderate	Personal value systems do affect people's perceptions, but other factors such as utility and economic factors do so as well
3a. Specific social-ecological contexts mediate how invasive species affect people	Supported	Moderate	Although the effects of invasive species are often context specific, the way the effects of a single species can vary when considering different actors still needs to be elucidated
3b. Research on social effects of invasive species primarily focuses on negative impacts	Unsupported	Moderate	There is considerable social research discussing both the benefits and costs or even just the benefits of invasive species. Social research highlights acknowledging trade-offs

(continued)

Table 24.1 (continued)

Proposition	Degree of support	Evidence level	Summary notes
3c. The negative social impacts of invasive species on local livelihoods are of more concern to people than impacts on biodiversity	Unsupported	Moderate	Mixed responses, where in some cases only social effects are mentioned and others where biodiversity effects are ranked higher or as high as many social effects. Context dependence and effects of study approaches and frameworks have great influence on findings
4a. People are less willing to manage species regarded as 'charismatic'	Partially supported	Moderate	We found evidence supporting that charisma can affect people's attitudes towards invasive species management, in South Africa attitudes are also dependent on the economic benefits of invasive alien species and personal value systems
4b. Social heterogeneity increases conflicts around the management of biological invasions	Partially supported	Low	There are too few cases to assess this proposition. A few cases suggest heterogeneity can increase conflict, others showed conflicts in non-diverse areas. More work is needed to evaluate this proposition
4c. Engagement with society is key to successful management	Partially supported	Moderate	Engagement can increase support for management and ensure successful outcomes. However, in some cases it increased public resistance to management and worsened conflict. More understanding on what works and what does not is needed

The 12 propositions correspond to 4 broad issues in which research on the human and social dimensions of invasion science can help to improve understanding and guide management responses: (1a,b,c) how people cause invasions, (2a,b,c) how people conceptualise and perceive invasions, (3a,b,c) the effects of invasions on people, and (4a,b,c) how people respond to invasions

24.2.1 *Proposition 1a: Intentional Introductions Were and Continue to Reflect the Social Ethos of the Time*

A vast number of alien species have been and continue to be purposefully or accidentally introduced into South Africa by people for various reasons (Richardson et al. 2003). Van Sittert (2002) argues that "*biological invasions are thus intrinsically historical processes primarily shaped not by the biology of the invader, but by the shifting cultural values of the invaded society*". We suggest that most motivations for introductions are driven by an ethos that evolves over time and relates closely to social fashions, political-economic circumstances and scientific paradigms (Carruthers et al. 2011). To discuss this, we highlight differences in motivations for introducing alien species during three broad time periods—but we acknowledge there are subtler trends within the broad timelines we outline.

A substantial number of alien species were introduced and established during the colonial period (Bennett and van Sittert 2019)—often with the ethos of making "improvements" to colony landscapes and economies and for botanical interest (Carruthers et al. 2011). Such introductions were strongly influenced by the broad landscape context of South Africa. For example, many tree species were introduced for forestry, linked to economic development, as the country is poorly endowed with natural forests for timber (Bennett and Kruger 2015). Yet other introductions during this period were driven by emotionally related values—in the then Cape Colony *Pinus* species (Pines) were planted by settlers in the 1700s, partly to create a sense of place and familiarity within the treeless landscapes. Similarly, *Oncorhynchus mykiss* (Rainbow Trout) was introduced to improve sense of place and counteract nostalgia for fly fishing by local settler elites from Europe (Alletson 1997; Brown 2013). This phenomenon is emphasised by Thompson (1913) "*The Colonialist, especially of British blood, seemed unable to finally settle down in a new land until many of the animals and plants that minister his pleasure or profit in the homeland had followed him...*". Canavan et al. (2018) showed that even slaves transported to South Africa during the colonisation by the Dutch East India Company (1652–1795) brought useful species, like bamboos, from their native lands with them. During the colonial period, many aesthetically pleasing species were also transferred between colonies, such as *Lantana camara* (Lantana), which was seen as an exotic novelty (Kannan et al. 2013). As a result, many former British colonies share similar issues with invasive ornamental plants brought in by colonial settlers. Collecting exotic plants for newly established public or private botanical gardens was a novelty within the colonies, and was a well-remunerated occupation and promoted by acclimatisation societies (Janick 2007).

By the 1900s appreciation and pride for native flora grew substantially and there was less attachment to species of the homeland by settlers (van Sittert 2003; Bennett 2015). By the mid-1900s, species were often introduced or purposefully dispersed in the context of livelihood development or environmental restoration, and promoted on a mass-scale by the state and non-governmental organisations alike (Carruthers et al. 2011), and less so to fulfil a sense of place for elite settlers or for primary industry than before. This led to the introduction of many so-called "wonder plants" that could yield multiple benefits for people and ecosystems, but that brought many costs once they became invasive (Low 2012; Kull and Tassin 2012). For example, *Prosopis* (Mesquite) was promoted in the apartheid era (mid-1900s) by agricultural departments to solve the effects of drought in the arid Northern Cape (Shackleton et al. 2014). Elderly community members recount stories of how agricultural extension officers distributed *Prosopis* seedlings for planting on private farms and in communal villages (Shackleton and Shackleton 2018). *Leucaena leucocephala* (Leucaena), was promoted by development, agricultural and forestry-focused NGOs during the 1960s to 1980s as a multipurpose tree (Brewbaker 1987). Similarly, a new set of Australian *Acacia* species were introduced for dryland restoration (Carruthers et al. 2011). *Oreochromis niloticus* (Nile Tilapia) was introduced into South Africa in the 1950s for aquaculture, particularly for food security and income generation among poor African communities (Zengeya et al. 2011).

The current ethos (in the post-apartheid democratic area) could facilitate further purposeful introductions. For example, a shift in gardening practices to become less water intensive might lead to the introduction of a new set of non-native species that require little water or care, and yet may become invasive (van Kleunen et al. 2018). Similarly, the rising demand for biofuels and green energy may lead to the promotion of invasive plants such as *Jatropha curcas* (Physic Nut) (Witt 2010; Blanchard et al. 2011). Species of interest to collectors can easily be bought online (e-commerce trade) which is a modern, easy and novel pathway of potential invasive species (Martin and Coetzee 2011; Humair et al. 2015). Simultaneously, the growing ethos of either managing or preventing invasions might lead to fewer purposeful introductions (Carruthers et al. 2011; Udo et al. 2019), although the context of increasing global movement of people and goods could lead to more accidental introductions than in the past (Seebens et al. 2017). For example, biofouling and ballast water has led to the recent introduction of a number of alien marine species along the South African coastline (Faulkner et al. 2017, 2020, Chap. 12).

Overall, evidence from South Africa supports the proposition that there have been clear changes over time in the ethos for introducing and promoting alien species which follow trends in scientific, historical, political and economic contexts (Carruthers et al. 2011). We illustrate this using a very broad set of temporal scales—and there is further need for understanding and analysing changes at finer spatial and temporal scales (e.g. Bennett 2015). Another important aspect would be to analyse the role of different actors—i.e. maybe too much is contextualised under the broad colonial banner and further comparison between British and Dutch settlers might yield useful insights.

24.2.2 Proposition 1b: People Go to Great Lengths to Ensure that Newly Introduced Species Become Well Established

People have put effort into facilitating the establishment and spread of alien species. For example, decades of research went into ensuring that introductions of

Eucalyptus (Gums) for forestry were successful (Bennett 2011). Australian *Acacia* species were planted *en masse* (more than 300 million seeds) to stabilise sand on the Cape flats, and prizes were offered to individuals for successful planting and establishment (van Sittert 2000; Bennett and van Sittert 2019), leading to substantial invasions as a result of this high propagule pressure (Donaldson et al. 2014). Similarly, people made substantial efforts to ensure the establishment of *Prosopis* trees during the mid-twentieth century for silviculture (Shackleton et al. 2015a). Some farmers recount childhood memories of putting *Prosopis* seedlings by the Aga stove in winter and only planting them out in spring to ensure survival. But, in the hot summers, one farmer recounts how during the school holidays he was required to go out and water *Prosopis* seedlings to ensure their survival.

Significant research went into discovering how to introduce fishes for sport and recreation by British colonists, and how to ensure their survival. Tens of thousands of *O. mykiss* and *Salmo trutta* (Brown Trout) eggs were sent from the UK to South Africa in the late 1800s and early 1900s, which failed to establish, but efforts continued until appropriate strategies were implemented to ensure survival during transport (also see Weyl et al. 2020, Chap. 6). In the Cape, special hatcheries were built and a bounty was even established for killing otters that were presumed to pose a threat to the newly introduced fish (Britz 2015). In KwaZulu-Natal, railway sidings were built specifically to ensure that the water did not have time to heat up during transport (Alletson 1997). The time and money that went into ensuring introduction and survival must have helped to promote invasions in the long run.

The available evidence and examples above support the proposition that great perseverance and effort to ensure the survival and establishment of some non-native species was important for facilitating invasions. The degree of human tenacity during the pre-introduction and post-introduction stages is typically poorly accounted for in purely ecological models of invasion dynamics.

24.2.3 Proposition 1c: Human-Mediated Modifications Help Invasive Species to Establish

Humans can modify both species and landscapes, which may facilitate or limit the invasion of some species (Kueffer 2017; Shackleton et al. 2018). Human agency is often not well acknowledged in the biological sciences, often being treated as an "unwelcome extraneous variable", but is actually a key factor in explaining many biological invasion processes (van Sittert 2002).

Le Roux et al. (2013) show how humans have altered the genetic make-up of *Acacia pycnantha* (Golden wattle) through artificial selection both prior to and after introduction, which may facilitate its invasiveness. At a broader scale, human alteration or disturbance of landscapes has facilitated invasions. Initially, many invasive species can only survive and proliferate in human-altered landscapes. For example, the creation of ponds, urban garden microclimates and farm dams has aided the invasion of frogs in South Africa (Davies et al. 2013; Measey et al. 2017). Similarly, the distribution of the invasive bird *Acridotheres tristus* (Common Myna) in South Africa is closely tied to large urban areas (Peacock et al. 2007; Measey et al. 2020, Chap. 5). Environmental degradation and land use change also affects invasions (Kueffer 2017). In South Africa, overstocking of domestic livestock helped *Opuntia ficus-indica* (Mission Prickly Pear) to become a dominant in the Eastern Cape—"*farmers became increasingly aware of their own hand in the gradual transformation of the landscape around them, so opuntia intruded from the margins to the centre*" (van Sittert 2002). In the marine context, *Carcinus maenas* (Green Crab) has established and become invasive only in human-made harbours and small bays nearby because the species cannot maintain a grip on rocks under high wave action (Mabin et al. 2017). In South Africa one of the 26 inter-basin transfer schemes (i.e. channels or tunnels constructed to link different river systems) has aided the spread of at least five non-native species (Ellender et al. 2014).

These few examples show that human modifications of species and landscapes can facilitate invasions, supporting our proposition. This shows some support to the "passenger" part of the driver vs passenger debate—whereby degradation might facilitate invasions rather than invasions causing initial degradation. Despite this, a lot more work can be done on how human modifications facilitate invasions and maybe incorporating this more into models would be useful (e.g. niche modelling, Vimercati et al. 2017; Walker et al. 2017).

24.3 People's Conceptulisation and Perceptions of Invasive Alien Species

A fast-growing body of literature in the field of invasion science considers how people view and conceptualise invasive species, both from a theoretical level to more of a personal and individual level (Kull et al. 2011, 2019; Estévez et al. 2015; Shackleton et al. 2019b). South Africa provides a fertile testing ground for different theories within this topic due to its diverse social and ecological contexts.

24.3.1 Proposition 2a: How People Think About, Value and Study Invasive Species is Strongly Shaped by Social-ecological Contexts

Historical, geographical, ecological, social and institutional contexts help to determine how people think about invasive species. For instance, attention to biological invasions is weak in South America (Speziale et al. 2012), whereas in colonial island landscapes, settings of rapid ecological change caused by the settlers as well as strong scientific interest by the same people, environmental concerns rose to the fore (Grove 1995). Such influence is also apparent in South Africa, where concern over invasions historically arose out of the particular ecological and social context of the Cape Colony (Pooley 2014), an influence that evolved with different periodic ethos (Carruthers et al. 2011), and which continues to this day. As noted by Bennett and van Sittert (2019), "*One fact dominates the history of invasive plants in South Africa. The Cape has consistently led national planning and action on weeds and alien invasive species, especially relating to agricultural weeds and invasive trees.*" We expand on four key aspects relating to this important statement: (1) the focus on Cape biota, (2) the focus on trees, (3) the national spread of the invasion concept, and (4) the institutional structures that embody and permit this dominance.

First, Cape 'exceptionalism', both real and perceived, shapes South African understanding of the phenomena of biological invasions. The distinctive ecology of the Cape, with the Fynbos Biome, combined with its strategic location and early history of European settlement, became the epicentre for species introductions and attracted a lot of scientific attention (Bennett and van Sittert 2019). In this context, scientific concern about the impact of introduced plants on the Fynbos Biome emerged in the late nineteenth century, and this concern continues, fortified with references to the Cape having one of the six global 'Floristic Kingdoms' (Lidström et al. 2016). One consequence is that invasion science research in South Africa has emphasised plants.

Second, a peculiarity of invasion science in South Africa is its focus on trees. The dearth of native forests in the Fynbos Biome, not to mention in the country's vast Grassland and Karoo Biomes, led to particularly strong efforts during the 19th and 20th centuries to introduce and promote trees (Brown 2013; Bennett 2011). These in turn led to highly visible, landscape-transforming invasions and the early catalysts for nascent invasion research and policy in South Africa (Bennett and van Sittert 2019). Possibly as a result, other invasive growth forms, like grasses, have received less attention than they should have (Milton 2004).

Third, the impact of such trees on fynbos landscapes facilitated the early development of interest in biological invasions beyond agricultural weeds in the Cape. This radiated outward to the national level only in the 1980s–1990s, with new policy openings in South Africa after the end of apartheid coming at the same time that an international conception of invasive species was emerging (Lidström et al. 2016; Bennett and van Sittert 2019). The national Working for Water (WfW) programme drew heavily on hydrological and ecological scientific insights from tree invasions in the Cape as motivating evidence for this poverty-relief project (van Wilgen and Wannenburgh 2016).

Fourth, a consequence of the Cape-based origins of thinking, research, and policy-making on invasions is the deep anchoring of this work in Cape-based institutions. The WfW program "*draws its core ideas and leadership from this region*" (Bennett 2014), and urban middle-class whites from the Cape have driven the agenda about invasions (Bennett and van Sittert 2019). The primary institution that has gained agenda-setting authority in the field is the Centre for Invasion Biology (C·I·B), a nationally funded Centre of Excellence established in 2004, based at Stellenbosch University (van Wilgen et al. 2014; Lidström et al. 2016; Abrahams et al. 2019; Richardson et al. 2020, Chap. 30). The C·I·B is not only important at a national level, but also internationally, where it has strongly shaped the development of invasion science (Pyšek et al. 2006; Pouris 2007; Abrahams et al. 2019). It works closely with the South African National Biodiversity Institute, the Council for Scientific and Industrial Research, WfW and many other partners. A recent bibliometric review of research sponsored by WfW shows that people with C·I·B affiliation—closely overlapping with a Stellenbosch affiliation—(co)authored almost half of this work; and that a small number of Stellenbosch-based C·I·B researchers are core authors who play a strong role in maintaining, mediating, and perhaps even controlling relationships and networks in the field (Abrahams et al. 2019). The C·I·B is built upon local expertise in forestry and botany, and previous programs in biological control and plant protection, and as a result has strongly emphasised, until recently, ecological research over other disciplines (Abrahams et al. 2019; Kull 2018)

This illustrates that specific contexts and events can shape the way people think about invasions. South Africa has a unique story where one region, the Cape, has really shaped what is thought and done today. This suggests that maybe some of the knowledge and theory that is accepted as normal might need to be adapted for different contexts to ensure relevance.

24.3.2 Proposition 2b: Knowledge and Awareness of Invasive Alien Species is Low Amongst the General Public

Globally, a growing body of research has focussed on understanding factors that influence people's knowledge and awareness of invasive species. This is an important component in building educational plans and adaptive management strategies (Cole et al. 2019). By knowledge and awareness, we mean what people understand and recognise biological invasions in general but also have knowledge about specific invasive species.

Studies assessing knowledge and perceptions of invasive species in South Africa, display varying results. Despite South Africa being a leading country in terms of policy, outreach and management of biological invasions (Byrne et al. 2020, Chap. 25), 77% of people in a small city (Makhanda (Grahamstown)), did not know that they had one or more listed invasive alien trees in their garden (Shackleton and Shackleton 2016). A similar number could not name a single invasive plant. Of those who did, they mainly knew of *Acacia* species or *Jacaranda mimosifolia* (Jacaranda). This study also suggested that people with higher education and incomes had a broader understanding and knowledge of invasive species. Potgieter et al. (2019), working in Cape Town, highlighted a stark contrast in knowledge levels across different socioeconomic groups, largely as a result of the legacy of apartheid, with more affluent and well-educated citizens having a better knowledge of invasions.

Focusing on *Prosopis,* the second most widespread invasive tree genus in South Africa, Shackleton et al. (2015) highlight greater knowledge and understanding by citizens than the previous two studies. The research did take place in areas of high infestation and where native tree biodiversity is very low, which might explain these results. However, differences in knowledge between social settings and actors was evident. Rural commercial farmers and communal land dwellers who had the greatest exposure to *Prosopis* and whose livelihoods were more closely linked to nature had a much greater knowledge than urban residents. Furthermore, unlike previous studies, Shackleton et al. (2015) revealed greater knowledge of the species amongst poor, urban dwellers (some of whom were reliant on the tree for fuelwood) as compared to affluent urban citizens. Another study in rural villages in the Northern Cape highlighted that knowledge and awareness of invasive species is highly species-dependent (Shackleton and Shackleton 2018). The majority of respondents knew that *Prosopis*, *Eucalyptus* and *J. mimosifolia* were non-native, but very few respondents knew other common invasive alien plants like *Schinus molle* (Pepper Tree), *Tecoma stans* (Yellow Bells), *Melia azedarach* (Seringa), *Morus alba* (White Mulberry) and *O. ficus-indica*, despite them being in the same landscape for similar durations. Differing levels of knowledge across species are likely influenced by factors like species traits, residence time, reasons for introduction, rates of spread and densities of invasions, impacts on people as well as management and outreach efforts.

For this proposition, we suggest knowledge of invasions can be context-dependent, and knowledge is generally low except for a few flagship taxa. Similarly, certain sectors of society are more knowledgeable regarding invasions, such as elites, and those living in rural areas who are likely to be more in contact with invasions and their impacts.

24.3.3 Proposition 2c: Personal Values are the Primary Factor Affecting Perceptions of Invasive Alien Species and Their Control

Various emotionally-related factors can influence people's attitudes and perceptions of invasive species (Urgenson et al. 2013; Shackleton et al. 2019b). For example, the global literature shows that people often "fall in love" with beautiful, cute or charismatic species due to their emotional appeal, while others might be detested based on their ugliness or threat, such as fire ants, wasps or rats (Shackleton et al. 2019b).

Only a few studies have addressed this question in South Africa. Novoa et al. (2017) illustrated that, in South Africa, there was not much difference in the support for management between different taxa, whereas in the UK, the control of a charismatic animal species was less supported than that of an ornamental plant. This might be linked to differences in attitudes by those living in more rural areas (South Africa), but also to the presence and diversity of native flora and fauna.

Linking to the rural-urban debate, people in urban areas were more interested and emotionally attached to charismatic species whereas in natural areas perceptions were more based on species utility. This is mirrored by Dickie et al. (2014) and Gaertner et al. (2016) who highlight that, particularly in cities, people's values relate to aesthetic and recreational benefits, and often take precedent over economic ones. In Cape Town, Potgieter et al. (2019) showed that whilst utilitarian and economic factors play a role in shaping perceptions, cultural values (e.g. aesthetic appearance) also helped to explain people's perceptions—suggesting that multiple interacting factors shape people's attitudes. However, in more rural areas, Shackleton et al. (2015) showed that for *Prosopis* invasions, personal values, whilst important, were not of as much concern as those relating to economic and livelihood benefits and costs. Furthermore, Novoa et al. (2017) indicated that promoting public awareness can assist in changing perceptions and in increasing public support for control, but individuals who are hostile to any invasive species management programs will remain—based on their personal ethical values. Mukwada et al. (2016) discussed how personal or professional values founded on various worldviews and means of living can influence people's perceptions; with park wardens wanting to manage *Acacia* species around Golden Gate National Park and communal villagers having different opinions on the matter as the trees provide utility value for them. Similar socio-political factors relating to biological invasions and different actors' social anxieties are highlighted by Comaroff and Comaroff (2001).

These examples suggest that, overall, multiple factors might influence people's perceptions of biological invasions, including the landscape context, the stakeholder group, their experiences and the species traits. In South Africa, we suspect that people's perceptions are more likely related to the economic and livelihood effects of invasive species, although intrinsic and emotional aspects cannot be discounted, and are likely to be more prominent in urban areas.

24.4 The Effects of Invasive Species on People

Invasive species can affect people positively or negatively in a variety of ways. For instance, invasive species provide both ecosystem services and disservices, which have different implications for livelihoods and human wellbeing (Shackleton et al. 2007; Vaz et al. 2017b; Shackleton et al. 2019c). Seminal works on the role of invasive species on people's livelihoods come from South Africa and have focused on rural settings (de Neergaard et al. 2005; Shackleton et al. 2007).

24.4.1 Proposition 3a: Specific Social-Ecological Contexts Mediate How Invasive Species Affect People

The local socioeconomic and ecological context can greatly influence how biological invasions affect people. For example, invasive *Acacia* species are extremely important for local livelihoods in the high-altitude communal grasslands in South Africa, where other trees are rare. *Acacia* species provide fuelwood, which is a primary source of heating and cooking fuel for almost all rural households (de Neergaard et al. 2005; Shackleton et al. 2007; Aitken et al. 2009). Despite increases in access to electricity in South Africa, *Acacia dealbata* (Silver Wattle) is still highly important for livelihoods today with almost no change in use levels in the past decade (Ngorima and Shackleton 2019). A number of households also earn incomes from selling the wood. However, when they reach high densities, *Acacia* species also have negative implications for water resources (Le Maitre et al. 2020, Chap. 15), grazing (O'Connor and van Wilgen 2020, Chap. 16) and people's health and safety (Shackleton et al. 2018). In the relatively treeless Fynbos Biome, invasive tree species are also an important source of fuelwood and income for rural villagers and foresters (Kull et al. 2011). However, this is a fire-prone biome and *Pinus* species can greatly increase the risk and negative implications of wild fires (Kraaij et al. 2018), highlighting how ecological contexts can result in this unique disservice (Comaroff and Comaroff 2001; Pooley 2014). Shackleton et al. (2015b) show that most local stakeholders from the Karoo and Savanna Biomes prefer to use native trees instead of the invasive *Prosopis* for fuelwood. In these areas, overall use of *Prosopis* for fuelwood is declining with increasing access to and use of electricity, unlike in the high-altitude Grassland Biome.

Research on *Opuntia ficus-indica* invasions in South Africa has provided insights into the complex interaction between invasions and human wellbeing and how benefits and costs are not static in time and space (Novoa et al. 2015a, b; Hill et al. 2020, Sect. 19.3; O'Connor and van Wilgen 2020, Sect. 16.5.5). *Opuntia ficus-indica* was promoted in arid areas to improve agricultural production (Beinart and Wotshela 2012), and at first it greatly benefited commercial farmers and rural villagers (Beinart and Wotshela 2003). Over time, it became invasive and spread over large areas, and its negative impacts started to outweigh its benefits, leading to control measures to reduce its spread, densities and negative impacts. Biological control was highly successful and reduced its spread and population densities, leading to the lowering of costs and the increase of benefits (Zimmermann and Moran 1991). Currently, with lower densities, *O. ficus-indica* has been adopted into society and provides a number of benefits for poor rural people, particularly through the collection and sale of fruits (Shackleton et al. 2007, 2011), showing the important role of spatial and temporal contexts.

Evidence from South Africa therefore supports this proposition and suggests that the effects of biological invasions on people are influenced by specific social and ecological contexts and can be highly dynamic in space and time. It highlights that in some cases some actors benefit more than others or some impacts are more external to people directly in contact with invasions (such as loss of water for cities downstream).

24.4.2 *Proposition 3b: Social Science Research on Invasive Species Primarily Focuses on Negative Effects of Invasive Alien Species for People*

Traditionally, ecological research on biological invasions has focussed on negative effects and ignored benefits of invasive species (Tassin and Kull 2015), and therefore we expect that to be similar in the social sciences and humanities.

Unlike ecological studies, two studies in South Africa discussed only the positive aspects of invasive species for local livelihoods and ignored any negative effects. Shackleton et al. (2011) showed the importance of selling *O. ficus-indica* fruits in rural areas of the Eastern Cape. In the same region the invasive fish *Cyprinus carpio* (Common Carp) benefits livelihoods, as a source of both food and cash income from the sale of fish (Ellender et al. 2010)—as yet no negative livelihood effects have been investigated from this introduction.

Most other studies acknowledge a suite of benefits and costs and often weigh them against each other. For example, the benefits of *O. ficus-indica* and Australian *Acacia* species were viewed as greater than the costs to rural communal land villagers (de Neergaard et al. 2005; Shackleton et al. 2007; Beinart and Wotshela 2012)—although some costs were highlighted as well. For *Prosopis,* both benefits and costs were assessed, and findings indicated that negative impacts outweighed benefits (Wise et al. 2012; Shackleton et al. 2015a). Potgieter et al. (2019) showed that many urban invasives provide both ecosystem services and disservices to people, as is the case for small rural villages in the Kalahari (Shackleton and Shackleton 2018). Harris et al. (2016) argued that work on *Columba* spp. (Pigeons) has traditionally focused on negative issues and control, but their research actually showed the opposite—that people on the University of South Africa's Muckleneuk campus would rather encourage Pigeons.

On a commercial level, the invasive Mediterranean Mussel (*Mytilus galloprovincialis*) is an important aquaculture species in South Africa and benefits are acknowledged (Hecht 1992). Similarly, invasive trees used in forestry provide financial benefits and employment (Tewari 2001; Louw 2004; Bennett and Kruger 2015). At the same time, they have negative implications for other livelihood activities and broader society (Le Maitre et al. 2011). For these kinds of species, economic research commonly applies cost-benefit analyses to estimate both positive and negative impacts (de Wit et al. 2001) and better understand conflicts and trade-offs (van Wilgen and Richardson 2014; Zengeya et al. 2017).

Research that has examined the social effects of invasive species in South Africa is fairly balanced and reports on both positive and negative impacts. This is contrary to our proposition. This also differs substantially from ecological research that tends to focus only on the negative implications (Tassin and Kull 2015). There are many invasive species globally that have mostly negative impacts and a lot of work elsewhere focuses on reporting and quantifying just these negative social effects as opposed to more balanced views (Shackleton et al. 2019c).

This difference in South Africa is likely because of a focus on livelihoods, rather than a species-centred framing. We suggest therefore that more people-centred frameworks could be useful to promote more integrated understanding of invasions and their effects on human wellbeing. For improving further understanding, incorporating larger spatial and temporal frames would be useful. Most studies cover a restricted time period and do not adequately show how longer-term changes in benefits and costs affect people (Shackleton et al. 2007), and what drives the change. Similarly, many studies only focus on one group of actors.

24.4.3 Proposition 3c: The Negative Social Impacts of Invasive Species on Local Livelihoods are of More Concern to People than Impacts on Biodiversity

Most of the work on invasion science in South Africa has focused on biological aspects, with relatively little work on the social aspects (Abrahams et al. 2019), and this bias is common globally (Kull and Tassin 2012; Vaz et al. 2017a). Despite this, some South African research has examined the interaction between invasive species and people's livelihoods—commonly in rural areas (Shackleton et al. 2019c).

Local communities in the Eastern Cape were mainly concerned about the social impacts of *A. dealbata* (Ngorima and Shackleton 2019) and mentioned the impacts of its roots on buildings, effects on cropping activities, issues with crime, and impacts on cultural sites. A handful of villagers mentioned the impacts of *A. dealbata* on water resources but none mentioned its impacts on biodiversity. In the Drakensberg region of the Eastern Cape, over 50% of villagers mentioned the impacts of *Acacia* invasions on crime rates, while 41% mentioned their impacts on water supply and security (although the authors note that "*one may question whether this is a real perception or one borrowed from the WfW programme*" (de Neergaard et al. 2005). Other concerns, voiced by a minority, included their impacts on grazing land and detrimental effects on native species (de Neergaard et al. 2005). In the Kalahari, local communities mentioned eight disservices as a result of *Prosopis* tree invasions (Shackleton and Shackleton 2018). The highest-ranked disservices were social (water resources, human health, infrastructure); the second lowest-ranked (7th) was its effects on species richness. In the same study, the impacts of invasive *O. ficus-indica* on human health were ranked highly (30% of respondents), while only 1% mentioned biodiversity impacts.

Other studies showed that people place importance on the negative impacts of biological invasions on biodiversity, although these impacts are also cited amongst social issues too. For example, communal villages and commercial farmers most commonly mentioned the negative effects of *Prosopis* on water, followed by loss of grazing on commercial farms, and detrimental effects on native plant biodiversity and to a lesser extent a variety of socioeconomic effects, such as reducing property value, revenue loss and effects on human health and infrastructure (Shackleton et al.

2015a). The impacts of tree invasions on fynbos biodiversity were considered by rural communities on the Agulhas Plain as a key threat, because it hampered the collection of wild flowers, with impacts on fire and water viewed as an issue only by a minority of the respondents (de la Fontaine 2013). However, in the same area, farmers considered the impacts of invasive trees on water as a major issue, followed by their impacts on livestock health, and grazing and crop land. In Cape Town, people were mainly concerned about the negative effects of invasive species on native biodiversity, followed by impacts on water supply, fire risk, human health and safety (Potgieter et al. 2019).

In summary, it appears that the literature is quite divergent. In some areas and for some social groups, loss of biodiversity is considered as one of the main impacts of invasive species, while in other areas or for other social groups it is hardly considered at all. Therefore, this proposition is not strongly supported. These differences might also be partly caused by the analytical frameworks used, researcher bias, the methods used and the kinds of questions asked. The findings also suggest that in some cases the concern over biological impacts are a manifestation of social values and therefore in reality biodiversity impacts are a threat to social concepts relating to human values and systems (e.g. preservation, heritage, stewardship, protected areas) (Carruthers 1995). Similarly, biodiversity might underpin many social related services or practices, e.g. flower collection or grazing potential.

24.5 People's Response to Invasive Species

24.5.1 Proposition 4a: People are Less Willing to Manage Species Regarded as 'Charismatic'

Many instances of resistance to invasive species management have arisen within South Africa. Whilst this is undoubtedly a function of the interplay of species traits, stakeholder values and local contexts, resistance is often met due to the charisma of certain species. By "charisma" we mean species that are linked to emotional values such as plants with large, bright flowers or unique growth forms, animals with neoteric features (big eyes and large heads), those that are entertaining and quirky, cute, colourful or those that are majestic, as opposed to less charismatic species which are often thorny, drab, do not have fur, and may sting or bite (Shackleton et al. 2019b). Characteristics that relate to charisma can overwhelm many other considerations, and can hinder or derail proposed management strategies (Dickie et al. 2014; Gaertner et al. 2016).

Control of the charismatic *Anas platyrhynchos* (Mallards) was met with significant resistance from local residents in Cape Town (Gaertner et al. 2016). Residents enjoyed these colourful and friendly ducks that provided them with entertainment and pleasure. In particular, they enjoyed feeding them which is not as easy with native ducks. Also, in Cape Town, there was controversy over the control of *Hemitragus jemlahicus* (Himalayan Tahr), which many residents viewed as majestic (see Davies et al. 2020, Chap. 23, Sect. 23.8.2). However, in this instance control continued because the animals were within a national park and so ecological imperatives overrode public sentiment. In Pretoria, residents successfully opposed the listing of *J. mimosifolia* in the city because it produces an abundance of beautiful purple flowers that residents admire; so much that the city is colloquially known as the Jacaranda City (Dickie et al. 2014). This resonates with the ongoing vocal and active opposition to the control of charismatic *Pinus pinea* (Umbrella Pine) on Table Mountain (Gaertner et al. 2016).

In other situations, charisma is not at the forefront of opposition to the management of particular species and other human values can inspire opposition. For example, a woman in Makhanda said she would refuse the removal of a *M. azedarach* tree from her garden, not because of its beauty or charisma, but because she had fond memories of her children playing in the tree. On the other hand, some see it as ethically wrong to control invasive alien species. The killing of *Columba* species (Pigeons), which some might view as uncharismatic, faced a lot of opposition, as it is an animal and people enjoyed seeing them (Harris et al. 2017). Similarly, citizens in Cape Town have responded "*Everything can be annoying, because its alive... we have to accept it*" and "*You should not kill them, you should move them back to the Eastern Cape*" with regards to the management of *S. gutturalis* (Guttural Toad) that would not traditionally be considered as charismatic (Novoa et al. 2017), and did not come from the Eastern Cape (Telford et al. 2019). Many invasive species provide livelihood benefits, which lead to opposition to their control, and has nothing to do with charisma (Shackleton et al. 2007). The commercial importance of some species, e.g. in forestry or aquaculture, promote resistance to management (van Wilgen and Richardson 2014; Zengeya et al. 2017).

While there was some evidence in support of this proposition, it generally appears that a range of different motivations can catalyse resistance to control efforts, of which species charisma is only one. The relative importance of each motivation can only be tested through carefully chosen experiments with different stakeholders or social groups. Moreover, in socially and culturally diverse countries, like South Africa, the very construct of uniform notions of charisma may be questionable, particularly when more important factors, such as economics, are taken into consideration. Therefore, improved engagement and decision support tools might help to guide and prioritise management in the future (Gaertner et al. 2016; Novoa et al. 2018).

24.5.2 Proposition 4b: Social Heterogeneity Increases Conflicts Around the Management of Biological Invasions

South Africa is colloquially known as the "rainbow nation", reflecting its diversity and mixing of multiple ethnic groups, cultures, races, languages and worldviews. We propose that social heterogeneity increases conflicts around invasive species management. In South Africa, longstanding conflicts include the listing and control of trout as a sport fish (Woodford et al. 2017), managing commercially important species of forestry trees (van Wilgen and Richardson 2014), and managing charismatic species in urban areas (Dickie et al. 2014; Gaertner et al. 2016). From these few, but prominent, cases, it appears that conflicts around management between the state and other actors are driven more often by elite and middle-class individuals and institutions who have resources, power and influence. More often than not, the voices of the poor and marginalised are not heard and therefore are rarely considered.

The literature does however highlight a conflict situation surrounding *Acacia* species in communal lands bordering the Golden Gate Highlands National Park in the Free State (Mukwada et al. 2016), showing the different worldviews among the local villagers and park officials. This has resulted in park-community conflicts relating to perceived threats and management responses, which might also partly be a manifestation of other deeper underlying issues relating to land and resource use, land claims and economic benefit-sharing that result from historic injustices to some marginalised ethic groups. Narratives and experiences by heterogeneous groups are not always accounted for, and can therefore increase complexity and conflict surrounding management (Lidström et al. 2016). These authors describe how environmental engagement does not always transcend, but can sometimes increase, ecological and social inequities in South Africa (Lidström et al. 2016).

Overall, we found this proposition hard to test due to a lack of in-depth information surrounding causes of conflict, although we would suggest it is likely to be true. The results of the few mentioned cases suggest that conflicts are often very context-dependent, relating to different species and different issues (e.g. ethical vs economic debates) and can be a manifestation of other deeper issues often relating to heterogeneity. However, most of the reported conflicts surround elite citizen interests and conservationists rather than between different groups of citizens. However, many studies focus on one specific set of actors or lump all citizens into one group with a lack of comparison between different actors.

24.5.3 Proposition 4c: Engagement with Society is Key to Successful Management

It is increasingly acknowledged that involving people in the development and implementation of management is crucial for the success of natural resource conservation programmes (Ntshotsho et al. 2015; Turner et al. 2016; Novoa et al. 2018). With specific reference to the management of biological invasions, South Africa is unique, because it has the renowned Working for Water programme which has both social and ecological goals. Due to the large variety of social and ecological contexts, as well as the diversity of invasive species in the country, South Africa has been very forward thinking in engaging with different actors and stakeholders to develop policies and management actions aimed at solving contentious issues and ensuring the long-term stability of management through benefiting society (van

Wilgen and Wannenburgh 2016). Research conceptualised in South Africa has led to the development of a stakeholder engagement framework (Novoa et al. 2018).

A number of projects in South Africa have worked with stakeholders to understand how to implement control programs more effectively. Urgenson et al. (2013) highlighted that many private landowners in South Africa understand the need to share management responsibility with the state. However, many of these landowners also emphasised the need for financial and other incentives to improve compliance, but there was a lack of agreement on what these should be. Harris et al. (2017) highlighted that manager's views and methods for control differed very much from the perceptions and desires of staff at the University of South Africa. They suggested that public participation would be needed to develop appropriate management strategies. A number of other projects have engaged with a broad set of actors to help plan and prioritise management strategies. Suggested approaches have included stakeholder workshops, and decision-making tools to develop spatial prioritisation plans (e.g. Roura-Pascual et al. 2009; Forsyth et al. 2012) and adaptation responses (Shackleton et al. 2016). These interventions have led to the development of special prioritisation plans for parts of the Western Cape (e.g. Forsyth et al. 2012), and national taxon-specific plans as in the case of *Prosopis* (Shackleton et al. 2017) and Cactaceae (Kaplan et al. 2017)—although current uptake of these plans is poor (see van Wilgen et al. 2020a, b, Sect. 21.4).

In working through contentious issues, some engagement processes have been highly successful (Novoa et al. 2016), while others have resulted in little change (Woodford et al. 2016). Novoa et al. (2016) showed that, for controlling cacti in South Africa, engagement increased understanding between opposing parties (those promoting different species for ornamental and production purposes, conservationists worried about invasions, and landowners affected by their impacts) and led to better recognition of each other's viewpoints. Furthermore, through engagement, opposing actors came to consensus relatively easily on which taxa in the Cactaceae should be listed as invasive under the National Environmental Management: Biodiversity Act (NEM:BA) regulations (see van Wilgen et al. 2020a, Chap. 1; van Wilgen et al. 2020b, Chap. 21; Box 1.2). Only certain genera and growth forms that are more likely to be invasive were listed, allowing for the use and trade of all others (Novoa et al. 2015a). This allowed for win-win outcomes for both parties. This successful example was due to having a good understanding of Cactaceae ecology, but also having open and bottom-up discussions.

For highly established and profitable industries, such as forestry, engagement has been more difficult (van Wilgen and Richardson 2014; Woodford et al. 2016). These industries have powerful lobby groups and economic interests, making compromise more difficult to achieve. Another big conflict issue has been in listing trout, where engagement has been confrontational and has potentially even worsened the situation. A number of powerful, elite fishers almost derailed the creation of invasive species lists as required under NEM:BA in South Africa. Much of this is to do with elite "recreational selfishness" but there are also economic arguments to be made, with trout fisheries contributing to the economies of a number of small towns across South Africa. The conservation actors were willing to allow mutually beneficial strategies, in which some areas now stocked with trout would be re-prioritised for biodiversity conservation while other localities would allow trout fishing. But proposals for compromise solutions were not accepted by the trout fishing fraternity (Brown 2013, 2016), and because the government is insistent on listing trout, the issue may still end up in court (Ellender et al. 2014). This is driven by a lack of trust in authorities, inadequate communication and understanding (Cox 2013), a similar issue in other regions of the world (Wald et al. 2019). In contrast, Bass (*Micropterus* species) anglers have been more supportive of management suggestions and have engaged with researchers in citizen science projects to control and monitor invasive fish (Weyl et al. 2014). The different outcomes of trout and bass anglers engagement comes down to context. While bass fishing primarily takes place in human-made dams and reservoirs that are already disturbed, allowing for invasives to be present (not needing control), the best trout-fishing takes place in undisturbed mountain rivers with endemic native fishes threatened by trout predation, making regulation desirable (see Davies et al. 2020, Sect. 23.5).

Our proposition finds some support and we have shown instances where engagement has been highly beneficial, but also other cases where it has been difficult. Moreover, most research on management and engagement seeks to improve knowledge as opposed to finding ways to implement it, i.e. there is a research-implementation gap (see Foxcroft et al. 2020, Chap. 27). Involvement of non-research stakeholders can help to swing the balance and thus there is need for better engagement with different actors (Ntshotsho et al. 2015). Often there are social-ecological barriers that can be addressed by meaningful engagement and collaboration (Esler et al. 2010, Shackleton et al. 2016; Angelstam et al. 2017). Having more bottom-up and collaborative management that involves co-design and co-implementation could improve control in the long-term (Reed et al. 2017). Also, better understanding why, where and how engagement works, or doesn't, would benefit planning in the future. South Africa could also learn lessons from elsewhere. For example, from the successful collaboration between parties to better control mink invasions in Scotland (Bryce et al. 2011), and the role of citizen science and volunteering in many parts of Europe (Adriaens et al. 2015; Pagès et al. 2019).

24.6 Conclusions

The chapter has highlighted the important role of human and social dimensions of invasion science in South Africa. Work on this topic in the country has provided better understanding of a number of aspects relating to invasion processes and impacts, and has in some cases improved management responses. South African research has contributed much toward understanding the effects of biological invasions on people's livelihoods, particularly in poor rural areas (e.g. de Neergaard et al. 2005; Shackleton et al. 2007). Historians have theorised the role of past social processes in facilitating invasions and their management (Carruthers et al. 2011; Bennett and van Sittert 2019). Social studies have also helped in developing engagement and management processes to be used in South Africa and elsewhere (Forsyth et al. 2012; Novoa et al. 2015b, 2018). Due to different social-ecological settings within the country, work in South Africa has also shed light on how and why people contextualise biological invasions, and why conflicts arise (Lidström et al. 2016; Shackleton et al. 2015a; Potgieter et al. 2019).

Despite its importance, the volume of work in the social domain still lags behind ecologically-focused research, both in South Africa and elsewhere, and there are many pressing areas and questions where further and deeper contributions are needed. In this chapter we investigated social science and humanities contributions for 12 pertinent propositions, covering four broad topics, using South Africa as a case study region. We found that that substantial evidence was lacking for supporting all but one of these propositions (Table 24.1). Similarly, only five of the 12 propositions could be fully supported (Table 24.1). Those that were unsupported have contradictory information or lack concrete evidence, making further work necessary.

A number of gaps and opportunities were identified relating to the four broad thematic areas outlined in Shackleton et al. (2019a). Regarding people causing invasions, there is need for better examining the role of different sectors of society, and effectively incorporating human actions into models of invasion processes. Concerning people conceptualising invasions, studies have revealed that public knowledge about biological invasions and their effects is low in most, albeit not all, situations, particularly in urban areas. Several techniques from the education and social disciplines could be used to improve public education and awareness, which could lead to more informed and collaborative management efforts in the future (see Byrne et al. 2020, Chap. 25). Better understanding the social-ecological contexts and knowledge systems under which people conceptualise biological invasions would also be beneficial.

Many studies investigating the effects of biological invasion only focus on one group of actors. However, invasions are complex and can have different effects for different groups of people and these can lead to conflicts and trade-offs, and more research is needed to assess this. Most research also lacks deeper considerations of various power dynamics and outcomes relating to the impacts of biological invasions. Such work requires historical nuance, grounding in the social realities of different geographic and institutional contexts, and attention to relationships, practices, perceptions, and discourses. Moreover, a few large-scale cost-benefit analyses have been conducted, but assessing benefits and costs and value chains of invasive species at local levels and between different actors could help to improve understanding and management planning. Linking biological impacts to human wellbeing would also provide further evidence to guide or justify management. Lastly, relating to responding to invasions, increased collaborative research and management planning involving various actors could help to shape better policy, and to effectively control biological invasions in the country (Novoa et al. 2018; Shackleton et al. 2019d). Also, a better understanding of in which forms and the contexts engagement works, and what makes engagement activities fail would help to guide future actions. This might link to improved analysis and understanding of power dynamics and social networks.

In conclusion, structuring this chapter around a series of propositions has highlighted what is known and what is not, whilst simultaneously reviewing meaningful social science theory on biological invasions in South Africa and beyond. Further work in the area should be promoted; South Africa is a superb place to test theories and advance knowledge regarding social dimensions, because of its complex and diverse social-ecological setting and well-developed research capacity (van Wilgen et al. 2020a, b, Chap. 1, Chap. 21).

Acknowledgements RTS acknowledges the Swiss government for funding. CMS's participation was funded by the South African Research Chairs Initiative of the Department of Science and Technology and the National Research Foundation of South Africa (any opinion, finding, conclusion or recommendation expressed is that of the authors and the NRF does not accept any liability). AN acknowledges funding from EXPRO grant no. 19-28807X (Czech Science Foundation) and long-term research development project RVO 67,985,939 (The Czech Academy of Sciences). We thank the reviewers and editors for their valuable inputs which helped to substantially improve the chapter.

References

Abrahams B, Sitas N, Esler KJ (2019) Exploring the dynamics of research collaborations by mapping social networks in invasion science. J Environ Manag 229:27–37. https://doi.org/10.1016/j.jenvman.2018.06.051

Adriaens T, Sutton-Croft M, Owen K et al (2015) Trying to engage the crowd in recording invasive alien species in Europe: experiences from two smartphone applications in northwest Europe. Manage Biol Invasions 6:215–225. https://doi.org/10.3391/mbi.2015.6.2.12

Aitken M, Rangan H, Kull CA (2009) Living with alien invasives. The political ecology of wattle in the eastern highveld Mpumalanga, South Africa. Etudes Océan Indien 42:115–141. https://doi.org/10.4000/oceanindien.769

Alletson DJ (1997) One hundred years of trout in Natal. Piscator 129:80–87

Angelstam P, Barnes G, Elbakidze M et al (2017) Collaborative learning to unlock investments for functional ecological infrastructure: bridging barriers in social-ecological systems in South Africa. Ecosyst Serv 27:291–304. https://doi.org/10.1016/j.ecoser.2017.04.012

Beinart W, Wotshela L (2003) Prickly pear in the Eastern Cape since the 1950s – perspectives from interviews. Environ Hist 29:191–209

Beinart W, Wotshela L (2012) Prickly pear: the social history of a plant in the Eastern Cape. Wits University Press, Johannesburg. https://doi.org/10.18772/22011115300

Bennett BM (2011) Naturalising Australian trees in South Africa: climate, exotics and experimentation. J South Afr Stud 37:265–280. https://doi.org/10.1080/03057070.2011.579434

Bennett BM (2014) Model invasions and the development of national concerns over invasive introduced trees: insights from South African history. Biol Invasions 16:499–512. https://doi.org/10.1007/s10530-013-0601-1

Bennett BM (2015) Margaret Levyns and the decline of ecological liberalism in the Southwestern Cape, 1890-1975. S Afr Hist J 67:64–84. https://doi.org/10.1080/02582473.2015.1019358

Bennett BM, Kruger FJ (2015) Forestry and water conservation in South Africa: history. In: Science and policy. ANU Press, Canberra. https://doi.org/10.22459/FWCSA.11.2015

Bennett BM, van Sittert L (2019) Perceptions of invasive alien plants in South Africa: historicising the national framework. J Environ Manag 229:174–181. https://doi.org/10.1016/j.jenvman.2018.07.029

Blanchard R, Richardson DM, O'Farrell PJ et al (2011) Biofuels and biodiversity in South Africa. S Afr J Sci 107(5/6):186. https://doi.org/10.4102/sajs.v107i5/6.186

Brewbaker JL (1987) *Leucaena*: a multipurpose tree genus for tropical agroforestry. In: Steppler HA, Nair PKR (eds) Agroforestry, a decade of development. International Council for Research in Agroforestry (ICRAF), Nairobi

Britz P (2015) The history of South African inland fisheries policy with governance recommendations for the democratic era. Water SA 41:624–632. https://doi.org/10.4314/wsa.v41i5.5

Brown D (2013) Are trout South African? Stories of fish, people and places. Picador Africa, Johannesburg

Brown D (2016) Native, natural, indigenous, indigenised? Trout in the postcolony. Crit Afr Stud 8:179–195. https://doi.org/10.1080/21681392.2016.1211848

Bryce R, Oliver MK, Davies L et al (2011) Turning back the tide of American mink invasion at an unprecedented scale trough community participation and adaptive management. Biol Conserv 144:575–583. https://doi.org/10.1016/j.biocon.2010.10.013

Byrne MJ, du Plessis D, Ivey PJ et al (2020) Education, training and capacity building in the field of biological invasions in South Africa. In: van Wilgen BW, Measey J, Richardson DM, Wilson JR, Zengeya TA (eds) Biological invasions in South Africa. Springer, Berlin, pp 727–754. https://doi.org/10.1007/978-3-030-32394-3_25

Canavan S, Richardson DM, Le Roux JJ et al (2018) Alien bamboos in South Africa: a socio-historical perspective. Human Ecol 47:121–133. https://doi.org/10.1007/s10745-018-0041-8

Carruthers J (1995) The Kruger National Park: a social and political history. University of Natal Press, Pietermaritzburg

Carruthers J, Robin L, Hattingh JP et al (2011) A native at home and abroad: the history, politics, ethics and aesthetics of Acacia. Divers Distrib 17:810–821. https://doi.org/10.1111/j.1472-4642.2011.00779.x

Cole E, Keller RP, Garbach K (2019) Risk of invasive species spread by recreational boaters remains high despite widespread adoption of conservation behaviours. J Environ Manag 229:112–119. https://doi.org/10.1016/j.jenvman.2018.06.078

Comaroff J, Comaroff L (2001) Naturing the nation: aliens, apocalypse and the postcolonial state. J S Afr Stud 27:627–651. https://doi.org/10.1080/13632430120074626

Cox I (2013) Is this the end of the line for freshwater fishing? Farm Wkly 13036:8–9

Davies SJ, Clusella-Trullas S, Hui C et al (2013) Farm dams facilitate amphibian invasion: extra-limital range expansion of the painted reed frog in South Africa. Austral Ecol 38:851–863. https://doi.org/10.1111/aec.12022

Davies SJ, Jordaan M, Karsten M et al (2020) Experience and lessons from alien and invasive animal control projects in South Africa. In: van Wilgen BW, Measey J, Richardson DM et al (eds) Biological invasions in South Africa. Springer, Berlin, pp 625–660. https://doi.org/10.1007/978-3-030-32394-3_22

de la Fontaine S (2013) Assessing the values and impacts of invasive alien plants on the livelihoods of rural land-users on the Agulhas Plain South Africa. MS thesis, Stellenbosch University

de Neergaard A, Saarnak C, Hill T, Khanyile M, Berzosa AM, Birch-Thomsen T (2005) Australian wattle species in the Drakensberg region of South Africa–an invasive alien or a natural resource? Agric Syst 85(3):216–233. https://doi.org/10.1016/j.agsy.2005.06.009

de Wit MP, Crookes DJ, van Wilgen BW (2001) Conflicts of interest in environmental management: estimating the costs and benefits of a tree invasion. Biol Invasions 3:167–178. https://doi.org/10.1023/A:1014563702261

Dickie IA, Bennett BM, Burrows LE et al (2014) Conflicting values: ecosystem services and invasive tree management. Biol Invasions 16:705–719. https://doi.org/10.1007/s10530-013-0609-6

Donaldson JE, Hui C, Richardson DM et al (2014) Invasion trajectory of alien trees: the role of introduction pathway and planting history. Glob Chang Biol 20:1527–1537. https://doi.org/10.1111/gcb.12486
Ellender BR, Weyl OLF, Winker H et al (2010) Quantifying the annual fish harvest from South Africa's largest freshwater reservoir. Water SA 36:45–52. https://doi.org/10.4314/wsa.v36i1.50906
Ellender BR, Woodford DJ, Weyl OLF et al (2014) Managing conflicts arising from fisheries enhancements based on non-native fishes in southern Africa. J Fish Biol 85:1890–1906. https://doi.org/10.1111/jfb.12512
Esler KJ, Prozesky H, Sharma GP et al (2010) How wide is the "knowing-doing" gap in invasion biology? Biol Invasions 12:4065–4075. https://doi.org/10.1007/s10530-010-9812-x
Estévez RA, Anderson CB, Pizarro JC et al (2015) Clarifying values, risk perceptions and attitudes to resolve or avoid social conflicts in invasive species management. Conserv Biol 29:19–30. https://doi.org/10.1111/cobi.12359
Faulkner KT, Robertson MP, Rouget M et al (2017) Prioritising surveillance for alien organisms transported as stowaways on ships travelling to South Africa. PLoS One 12:0173340. https://doi.org/10.1371/journal.pone.0173340
Faulkner KT, Burnsess A, Byrne M et al (2020) South Africa's pathways of introduction and dispersal and how they have changed over time. In: van Wilgen BW, Measey J, Richardson DM et al (eds) Biological invasions in South Africa. Springer, Berlin, pp 311–352. https://doi.org/10.1007/978-3-030-32394-3_12
Forsyth GG, Le Maitre DC, O'Farrell PJ et al (2012) The prioritisation of invasive alien plant control projects using a multi-criteria decision model informed by stakeholder input and spatial data. J Environ Manag 103:51–57. https://doi.org/10.1016/j.jenvman.2012.01.034
Foxcroft LC, van Wilgen BW, Abrahams B et al (2020) Knowing-doing continuum or knowing-doing gap? Information flow between researchers and managers of biological invasions in South Africa. In: van Wilgen BW, Measey J, Richardson DM et al (eds) Biological invasions in South Africa. Springer, Berlin, pp 827–850. https://doi.org/10.1007/978-3-030-32394-3_28
Gaertner M, Larson BM, Irlich UM et al (2016) Managing invasive species in cities: a framework from Cape Town, South Africa. Land Urban Plan 151:1–9. https://doi.org/10.1016/j.landurbplan.2016.03.010
Grove RH (1995) Green imperialism. Cambridge University Press, Cambridge
Harris E, De Crom EP, Wilson A (2016) Pigeons and people: mortal enemies or lifelong companions? A case study on staff perceptions of the pigeons on the University of South Africa, Muckleneuk campus. J Public Aff 16:331–340. https://doi.org/10.1002/pa.1593
Harris E, De Crom EP, Wilson A (2017) Staff perceptions of pigeons control strategies on the University of South Africa, Muckleneuk campus. J Trans Res S Afr 13:a399. https://doi.org/10.4102/td.v13i1.399
Head L (2017) The social dimensions of invasive plants. Nat Plants 3:17075. https://doi.org/10.1038/nplants.2017.75
Hecht PJ (1992) The current status, future prospects and environmental implications of mariculture in South Africa. S Afr J Sci 88:335–342
Hill MP, Moran VC, Hoffmann JH et al (2020) More than a century of biological control against invasive alien plants in South Africa: a synoptic view of what has been accomplished. In: van Wilgen BW, Measey J, Richardson DM et al (eds) Biological invasions in South Africa. Springer, Berlin, pp 549–568. https://doi.org/10.1007/978-3-030-32394-3_19
Humair F, Humair L, Kuhn F et al (2015) E-commerce trade in invasive plants. Conserv Biol 29:1658–1665. https://doi.org/10.1111/cobi.12579
Janick J (2007) Plant exploration: from Queen Hatshepsut to Sir Joseph Banks. Hort Sci 42:191–196. https://doi.org/10.21273/HORTSCI.42.2.191
Kannan R, Shackleton CM, Shaanker RU (2013) Reconstructing the history of introduction and spread of the invasive species, Lantana, at three spatial scales in India. Biol Invasions 15:1287–1302. https://doi.org/10.1007/s10530-012-0365-z

Kaplan H, Wilson JR, Klein H et al (2017) A proposed national strategic framework for the management of Cactaceae in South Africa. Bothalia 47(2):a2149. https://doi.org/10.4102/abc.v47i2.2149

Kleunen v et al (2018) The changing role of ornamental horticulture in alien plant invasion. Biol Rev 93:1421–1437. https://doi.org/10.1111/brv.12402

Kraaij T, Baard JA, Arndt J et al (2018) An assessment of climate, weather and fuel factors influencing a large, destructive wildfire in the Knysna region, South Africa. Fire Ecol 14:4. https://doi.org/10.1186/s42408-018-0001-0

Kueffer C (2017) Plant invasions in the Anthropocene. Science 358:724–725. https://doi.org/10.1126/science.aao6371

Kull CA (2018) Critical invasion science: weeds, pests, and aliens. In: Lave R, Biermann C, Lane SN (eds) The Palgrave handbook of critical physical geography. Palgrave Macmillan, Cham. https://doi.org/10.1007/978-3-319-71461-5_12

Kull CA, Tassin J (2012) Australian acacias: useful and (sometimes) weedy. Biol Invasions 14:2229–2233. https://doi.org/10.1007/s10530-012-0244-7

Kull CA, Shackleton CM, Cunningham PJ et al (2011) Adoption, use and perception of Australian acacias around the world. Divers Distrib 17:822–836. https://doi.org/10.1111/j.1472-4642.2011.00783.x

Kull CA, Harimanana SL, Radaniela Andrianoro A et al (2019) Perceiving the "neo-Australian" forests of lowland eastern Madagascar: invasions, transitions, and livelihoods. J Environ Manag 229:48–46. https://doi.org/10.1016/j.jenvman.2018.06.004

Le Maitre DC, Richardson DM, Chapman RA (2004) Alien plant invasions in South Africa: driving forces and the human dimension. S Afr J Sci 100:103–112

Le Maitre DC, Gaertner M, Marchante E et al (2011) Impacts of invasive Australian acacias: implications for management and restoration. Divers Distrib 17:1015–1029. https://doi.org/10.1111/j.1472-4642.2011.00816.x

Le Maitre DC, Blignaut JN, Clulow A et al (2020) Impacts of plant invasions on terrestrial water flows in South Africa. In: van Wilgen BW, Measey J, Richardson DM et al (eds) Biological invasions in South Africa. Springer, Berlin, pp 429–456. https://doi.org/10.1007/978-3-030-32394-3_15

Le Roux JJ, Richardson DM, Wilson JR et al (2013) Human usage in the native range may determine future genetic structure of an invasion: insights from *Acacia pycnantha*. BMC Ecol 13:37. https://doi.org/10.1186/1472-6785-13-37

Lidström S, West S, Katzschner T et al (2016) Invasive narratives and the inverse of slow violence: alien species in science and society. Environ Human 7:1–40. https://doi.org/10.1215/22011919-3616317

Louw WJA (2004) General history of the South African forestry industry: 1891 to 2002. S Afr For J 201:65–76. https://doi.org/10.1080/20702620.2004.10431775

Low T (2012) In denial about dangerous aid. Biol Invasions 14:2235–2236. https://doi.org/10.1007/s10530-012-0264-3

Mabin CA, Wilson JRU, Le Roux JJ et al (2017) Reassessing the invasion of South African waters by the European shore-crab *Carcinus maenas*. Afr J Marine Sci 39:259–267. https://doi.org/10.2989/1814232X.2017.1363818

Martin GD, Coetzee JA (2011) Pet stores, aquarists and the internet trade as modes of introduction and spread of invasive macrophytes in South Africa. Water SA 37:371–380. https://doi.org/10.4314/wsa.v37i3.68488

Measey J, Hui C, Somers M (2020) Terrestrial vertebrate invasions in South Africa. In: van Wilgen BW, Measey J, Richardson DM, Wilson JR, van Wilgen BW (eds) Biological invasions in South Africa. Springer, Berlin, pp 113–150. https://doi.org/10.1007/978-3-030-32394-3_5

Measey J, Davies SJ, Vimercati G et al (2017) Invasive amphibians in southern Africa: a review of invasion pathways. Bothalia-African. Biodivers Conserv 47:1–12. https://doi.org/10.4102/abc.v47i2.2117

Milton SJ (2004) Grasses as invasive alien plants in South Africa. S Afr J Sci 100:69–75

Mukwada G, Chingombe W, Taru P (2016) Strifes of the frontier: an assessment of *Acacia mearnsii* related park-community conflicts in the Golden Gate highlands National Park, South Africa. J Integr Environ Sci 13:37–54. https://doi.org/10.1080/1943815X.2015.1130062

Ngorima A, Shackleton CM (2019) Livelihood benefits and costs of an invasive alien tree (*Acacia dealbata*) to rural communities in the Eastern cape, South Africa. J Environ Manag 229:158–165. https://doi.org/10.1016/j.jenvman.2018.05.077

Novoa A, Le Roux JJ, Robertson MP et al (2015a) Introduced and invasive cactus species: a global review. AOB Plants 7:pul078. https://doi.org/10.1093/aobpla/plu078

Novoa A, Kaplan H, Kumschick S et al (2015b) Soft touch or heavy hand? Legislative approaches for preventing invasions: insights from cacti in South Africa. Invasive Plant Sci Manage 8:307–316. https://doi.org/10.1614/IPSM-D-14-00073.1

Novoa A, Kaplan H, Wilson JRU et al (2016) Resolving a prickly situation: involving stakeholders in invasive cactus management in South Africa. Environ Manag 57:998–1008. https://doi.org/10.1007/s00267-015-0645-3

Novoa A, Dehnen-Schmutz K, Fried J, Vimercati G (2017) Does public awareness increase support for invasive species management? Promising evidence across taxa and landscape types. Biol Invasions 19(12):3691–3705. https://doi.org/10.1007/s10530-017-1592-0

Novoa A, Shackleton RT, Canavan S et al (2018) A framework for engaging stakeholders on the management of alien species. J Environ Manag 205:286–297. https://doi.org/10.1016/j.jenvman.2017.09.059

Ntshotsho P, Prozesky HE, Esler KJ et al (2015) What drives the use of scientific evidence in decision making? The case of the South African working for water program. Biol Conserv 184:136–144. https://doi.org/10.1016/j.biocon.2015.01.021

O'Connor T, van Wilgen BW (2020) The impact of alien plants on rangelands in South Africa. In: van Wilgen BW, Measey J, Richardson DM et al (eds) Biological invasions in South Africa. Springer, Berlin, pp 457–486. https://doi.org/10.1007/978-3-030-32394-3_16

Pagès M, van der Wal R, Lambin X et al (2019) Involving volunteers in rationalised nature conservation: challenges and opportunities in the case of non-native species management in Great Britain. J Environ Manag 229:102–111. https://doi.org/10.1016/j.jenvman.2018.06.053

Peacock DS, van Rensburg BJ, Robertson MP (2007) The distribution and spread of the invasive alien common myna, *Acridotheres tristis* L. (Aves: Sturnidae), in southern Africa. S Afr J Sci 103:465–473

Pooley S (2014) Burning Table Mountain: an environmental history of fire on the Cape Peninsula. Palgrave Macmillan, Basingstoke. https://doi.org/10.1057/9781137415448

Potgieter LJ, Gaertner M, O'Farrell PJ et al (2019) Perceptions of impact: invasive alien plants in the urban environment. J Environ Manag 229:76–87. https://doi.org/10.1016/j.jenvman.2018.05.080

Pouris A (2007) The international performance of the South African academic institutions: a citation assessment. Higher Educ 54:501–509. https://doi.org/10.1007/s10734-006-9034-4

Pyšek P, Richardson DM, Jarošík V (2006) Who cites who in the invasion zoo: insights from an analysis of the most highly cited articles in invasion ecology. Preslia 78:437–468

Reed MS, Vella S, Sidoli del Ceno J et al (2017) A theory of participation: what makes stakeholder engagement and public engagement in environmental management work? Restor Ecol 26:S7–S17. https://doi.org/10.1111/rec.12541

Richardson DM, Cambray JA, Chapman RA et al (2003) Vectors and pathways of biological invasions in South Africa - past, present and future. In: Ruiz G, Carlton J (eds) Invasive species: vectors and management strategies. Island Press, Washington DC, pp 292–349

Richardson DM, Abrahams B, Boshoff N et al (2020) South Africa's Centre for Invasion Biology: an experiment in invasion science for society. In: van Wilgen BW, Measey J, Richardson DM, Wilson JR, Zengeya TA (eds) Biological invasions in South Africa. Springer, Berlin, pp 875–912. https://doi.org/10.1007/978-3-030-32394-3_30

Roura-Pascual N, Richardson DM, Krug RM et al (2009) Ecology and management of alien plant invasions in South African fynbos: accommodating key complexities in objective decision making. Biol Conserv 142:1595–1604. https://doi.org/10.1016/j.biocon.2009.02.029

Seebens H, Blackburn TM, Dyer EE et al (2017) No saturation in the accumulation of alien species worldwide. Nat Commun 8:14435. https://doi.org/10.1038/ncomms14435

Shackleton CM, Shackleton RT (2016) Knowledge, perceptions and willingness to control designated invasive tree species in urban household gardens in South Africa. Biol Invasions 18:1599–1609. https://doi.org/10.1007/s10530-016-1104-7

Shackleton SE, Shackleton RT (2018) Local knowledge regarding ecosystem services and disservices from invasive alien plants in the arid Kalahari, South Africa. J Arid Environ 159:22–33. https://doi.org/10.1016/j.jaridenv.2017.07.001

Shackleton CM, McGarry D, Fourie S et al (2007) Assessing the effects of invasive alien species on rural livelihoods: case examples and a framework from South Africa. Hum Ecol 35:113–127. https://doi.org/10.1007/s10745-006-9095-0

Shackleton SE, Kirby D, Gambiza J (2011) Invasive plants – friends or foes? Contribution of prickly pear (*Opuntia ficus-indica*) to livelihoods in Makana Municipality, Eastern Cape, South Africa. Dev South Afr 28:177–193. https://doi.org/10.1080/0376835X.2011.570065

Shackleton RT, Le Maitre DC, Pasiecznik NM, Richardson DM (2014) *Prosopis*: a global assessment of the biogeography, benefits, impacts and management of one of the world's worst woody invasive plant taxa. AoB Plants 6:plu027. https://doi.org/10.1093/aobpla/plu027

Shackleton RT, Le Maitre DC, Richardson DM (2015a) Stakeholder perceptions and practices regarding *Prosopis* (mesquite) invasions and management in South Africa. Ambio 44:569–581. https://doi.org/10.1007/s13280-014-0597-5

Shackleton RT, Le Maitre DC, van Wilgen BW et al (2015b) Use of non-timber forest products from invasive alien *Prosopis* species (mesquite) and native trees in South Africa: implications for management. For Ecosys 2:16. https://doi.org/10.1186/s40663-015-0040-9

Shackleton RT, Le Maitre DC, van Wilgen BW et al (2016) Identifying barriers to effective management of widespread invasive alien trees: *Prosopis* species (mesquite) in South Africa as a case study. Glob Environ Chang 38:183–194. https://doi.org/10.1016/j.gloenvcha.2016.03.012

Shackleton RT, Le Maitre DC, van Wilgen BW et al (2017) Towards a national strategy to optimise the management of a widespread invasive tree (*Prosopis* species; mesquite) in South Africa. Ecosyst Serv 27:242–252. https://doi.org/10.1016/j.ecoser.2016.11.022

Shackleton RT, Biggs R, Richardson DM et al (2018) Social-ecological drivers and impacts of invasion-related regime shifts: consequences for ecosystem services and human wellbeing. Envion Sci Policy 89:300–314. https://doi.org/10.1016/j.envsci.2018.08.005

Shackleton RT, Larson BHL, Novoa A et al (2019a) The human and social dimensions in invasion science. J Environ Manag 229:1–9. https://doi.org/10.1016/j.jenvman.2018.08.041

Shackleton RT, Richardson DM, Shackleton CM et al (2019b) Explaining people's perceptions of invasive alien species: a conceptual framework. J Environ Manag 229:10–26. https://doi.org/10.1016/j.jenvman.2018.04.045

Shackleton RT, Shackleton CM, Kull CA (2019c) The role of invasive alien species in shaping local livelihoods and human well-being. J Environ Manag 229:145–157. https://doi.org/10.1016/j.jenvman.2018.05.007

Shackleton RT, Adriaens T, Brundu G et al (2019d) Stakeholder engagement in the study and management of invasive alien species: a review. J Environ Manag 229:88–101. https://doi.org/10.1016/j.jenvman.2018.04.044

Speziale K, Lambertucci S, Carrete M et al (2012) Dealing with non-native species: what makes the difference in South America? Biol Invasions 14:1609–1621. https://doi.org/10.1007/s10530-011-0162-0

Tassin J, Kull CA (2015) Facing the broader dimensions of biological invasions. Land Use Policy 42:165–169. https://doi.org/10.1016/j.landusepol.2014.07.014

Telford N, Channing A, Measey J (2019) Origin of invasive populations of the Guttural toad *Sclerophrys gutturalis*. Herpetol Conserv Biol 14(2):380–392

Tewari DD (2001) Is commercial forestry sustainable in South Africa? The changing institutional and policy needs. For Policy Econ 2:333–353. https://doi.org/10.1016/S1389-9341(01)00028-4

Thompson WW (1913) The sea fisheries of the Cape Colony. Maskew Miller, Cape Town

Turner BL II, Esler KJ, Bridgewater P et al (2016) Socio-environmental systems (SES) research: what have we learned and how can we use this information in future research programs. Curr Opin Environ Sustain 19:160–168. https://doi.org/10.1016/j.cosust.2016.04.001

Udo N, Darrot C, Atlan A (2019) From useful to invasive, the status of gorse on Reunion Island. J Environ Manag 229:166–173. https://doi.org/10.1016/j.jenvman.2018.06.036

Urgenson LS, Prozesky HE, Esler KJ (2013) Stakeholder perceptions of an ecosystem services approach to clearing invasive alien plants on private land. Ecol Soc 18:26. https://doi.org/10.5751/ES-05259-180126

van Sittert L (2000) The seed blows about in every Breeze': noxious weed eradication in the Cape Colony, 1860-1909. J South Afr Stud 26:655–674. https://doi.org/10.1080/713683604

van Sittert L (2002) 'Our irrepressible fellow-colonist': the biological invasion of prickly pear (Opuntia ficus-indica) in the Eastern Cape c. 1890–c. 1910. J Hist Geogr 28:397–419. https://doi.org/10.1006/jhge.2001.0457

van Sittert L (2003) Making the Cape Floral Kingdom: the discovery and defence of indigenous flora at the Cape ca. 1890-1939. Landsc Res 28:113–129. https://doi.org/10.1080/01426390306529

van Wilgen BW, Richardson DM (2014) Challenges and trade-offs in the management of invasive alien trees. Biol Invasions 16:721–734. https://doi.org/10.1007/s10530-013-0615-8

van Wilgen BW, Wannenburgh A (2016) Co-facilitating invasive species control, water conservation and poverty relief: achievements and challenges in South Africa's working for water programme. Curr Opin Environ Sustain 19:7–17. https://doi.org/10.1016/j.cosust.2015.08.012

van Wilgen BW, Davies SJ, Richardson DM (2014) Invasion science for society: a decade of contributions from the Centre for Invasion Biology. S Afr J Sci 110(7/8):12. https://doi.org/10.1590/sajs.2014/a0074

van Wilgen BW, Measey J, Richardson DM et al (2020a) Biological invasions in South Africa: an overview. In: van Wilgen BW, Measey J, Richardson DM et al (eds) Biological invasions in South Africa. Springer, Berlin, pp 3–30. https://doi.org/10.1007/978-3-030-32394-3_1

van Wilgen BW, Wilson JR, Wannenburgh A et al (2020b) The extent and effectiveness of alien plant control projects in South Africa. In: van Wilgen BW, Measey J, Richardson DM et al (eds) Biological invasions in South Africa. Springer, Berlin, pp 593–624. https://doi.org/10.1007/978-3-030-32394-3_22

Vaz AS, Kueffer C, Kull CA et al (2017a) The progress of interdisciplinarity in invasion science. Ambio 46:428–442. https://doi.org/10.1007/s13280-017-0897-7

Vaz AS, Kueffer C, Kull CA et al (2017b) Integrating ecosystem services and disservices: insights from plant invasions. Ecosyst Serv 23:94–107. https://doi.org/10.1016/j.ecoser.2016.11.017

Vimercati G, Hui C, Davies SJ et al (2017) Integrating age structured and landscape resistance models to disentangle invasion dynamics of a pond-breeding anuran. Ecol Model 356:104–116. https://doi.org/10.1016/j.ecolmodel.2017.03.017

Wald DM, Nelson KA, Gawel AM et al (2019) The role of trust in public attitudes towards invasive species management on Guam: a case study. J Environ Manag 229:133–144

Walker GA, Robertson MP, Gaertner M et al (2017) The potential range of *Ailanthus altissima* (tree of heaven) in South Africa: the roles of climate, land use and disturbance. Biol Invasions 19:3675–3690. https://doi.org/10.1007/s10530-017-1597-8

Weyl OLF, Finlayson B, Impson D et al (2014) Threatened endemic fishes in South Africa's Cape Floristic Region: a new beginning for the Rondegat River. Fisheries 39:270–279. https://doi.org/10.1080/03632415.2014.914924

Weyl OLF, Ellender B, Wasserman RJ et al (2020) Alien freshwater fauna in South Africa. In: van Wilgen BW, Measey J, Richardson DM, Wilson JR, Zengeya TA (eds) Biological invasions in South Africa. Springer, Berlin, pp 151–182. https://doi.org/10.1007/978-3-030-32394-3_6

Wise RM, van Wilgen BW, Le Maitre DC (2012) Costs, benefits and management options for an invasive alien tree species: the case of mesquite in the Northern Cape, South Africa. J Arid Environ 84:80–90. https://doi.org/10.1016/j.jaridenv.2012.03.001

Witt ABR (2010) Biofuels and invasive species for an African perspective – a review. GCB Bioenergy 2:321–329. https://doi.org/10.1111/j.1757-1707.2010.01063.x

Woodford DJ, Richardson DM, MacIsaac HJ et al (2016) Confronting the wicked problem of managing biological invasions. NeoBiota 31:63–86. https://doi.org/10.3897/neobiota.31.10038

Woodford DJ, Ivey P, Jordaan MS et al (2017) Optimising invasive fish management in the context of invasive species legislation in South Africa. Bothalia 47:1–9. https://doi.org/10.4102/abc.v47i2.2138

Zengeya TA, Booth AJ, Bastos AD et al (2011) Trophic interrelationships between the exotic Nile tilapia, *Oreochromis niloticus* and indigenous tilapiine cichlids in a subtropical African river system (Limpopo River, South Africa). Environ Biol Fish 92:479–489. https://doi.org/10.1007/s10641-011-9865-4

Zengeya T, Ivey P, Woodford DJ et al (2017) Managing conflict-generating invasive species in South Africa: challenges and trade-offs. Bothalia 47:1–11. https://doi.org/10.4102/abc.v47i2.2160

Zimmermann HG, Moran VC (1991) Biological control of prickly pear, *Opuntia ficus-indica* (Cactaceae), in South Africa. Agri Ecosyst Environ 37:29–35. https://doi.org/10.1016/0167-8809(91)90137-M

Permissions

All chapters in this book were first published by Springer; hereby published with permission under the Creative Commons Attribution License or equivalent. Every chapter published in this book has been scrutinized by our experts. Their significance has been extensively debated. The topics covered herein carry significant information for a comprehensive understanding. They may even be implemented as practical applications or may be referred to as a beginning point for further studies.

The contributors of this book come from diverse backgrounds, making this book a truly international effort. We would like to thank all the contributing authors for lending their expertise to make the book truly unique. They have played a crucial role in the development of this book. Without their invaluable contributions this book wouldn't have been possible. They have made vital efforts to compile up to date information on the varied aspects of this subject to make this book a valuable addition to the collection of many professionals and students.

This book was conceptualized with the vision of imparting up-to-date and integrated information in this field. To ensure the same, a matchless editorial board was set up. Every individual on the board went through rigorous rounds of assessment to prove their worth. After which they invested a large part of their time researching and compiling the most relevant data for our readers.

The editorial board has been involved in producing this book since its inception. They have spent rigorous hours researching and exploring the diverse topics which have resulted in the successful publishing of this book. They have passed on their knowledge of decades through this book. To expedite this challenging task, the publisher supported the team at every step. A small team of assistant editors was also appointed to further simplify the editing procedure and attain best results for the readers.

Apart from the editorial board, the designing team has also invested a significant amount of their time in understanding the subject and creating the most relevant covers. They scrutinized every image to scout for the most suitable representation of the subject and create an appropriate cover for the book.

The publishing team has been an ardent support to the editorial, designing and production team. Their endless efforts to recruit the best for this project, has resulted in the accomplishment of this book. They are a veteran in the field of academics and their pool of knowledge is as vast as their experience in printing. Their expertise and guidance has proved useful at every step. Their uncompromising quality standards have made this book an exceptional effort. Their encouragement from time to time has been an inspiration for everyone.

The publisher and the editorial board hope that this book will prove to be a valuable piece of knowledge for students, practitioners and scholars across the globe.

Index

A

Agricultural Production, 91, 249

Agricultural Sector, 69, 71

Alien Species, 1-2, 6, 8, 14-18, 22-25, 27-31, 33-34, 37, 141-142, 146, 150, 154-155, 161-163, 205-209, 211-216, 218, 223, 246-247, 250, 253, 258, 261-263

Alien Trees, 1, 17, 28, 41, 50, 67, 73, 108, 142, 158, 163-164, 207, 213, 215, 217-219, 222, 224, 229, 231-232, 246, 260, 263-264

Amphibians, 30-31, 33, 35, 37, 40, 43-44, 49, 114, 133, 167, 182, 198, 261

Apartheid, 56, 67, 72, 242, 245-246

Arundo Donax, 10, 14, 33-34, 48

B

Biodiversity, 2, 18, 21, 25-31, 34, 37, 39, 41, 43-46, 48-51, 181-182, 194, 198-199, 202-204, 206-207, 211-212, 215, 228, 232-233, 251-252, 255-256, 259

Biological Control, 2, 13, 21-23, 25-26, 28, 47, 69, 91-110, 118, 120-121, 126, 129-135, 159, 161-165, 168, 175, 216, 223-224, 249, 260, 265

Biological Invasions, 26-28, 33, 39, 43, 46-47, 49-52, 114, 119-120, 123, 129-133, 145, 161-165, 196-200, 204-207, 244-246, 248-251, 253-254, 256-265

C

Chromolaena Odorata, 9-10, 17, 21, 95, 110, 136, 138, 153, 159, 164, 221, 232

Commercial Farmers, 21, 247, 249, 251

Community Dynamics, 205, 209

Control Agents, 91, 93, 95-98, 101-102, 104, 106-107, 109-110, 121, 129, 131, 135, 138, 154, 216, 224, 234

Control Efforts, 18, 105, 135-136, 139, 144-145, 159, 178, 193, 253

Control Projects, 47, 92-93, 95, 106, 134, 136, 141, 146, 149-150, 153, 157-158, 161, 166-168, 170, 178, 185, 189, 193-194, 259-260, 264

Crop Plants, 60, 62

D

Dispensation, 55-56, 74

Domestic Livestock, 1, 3, 6, 22, 60, 244

Drylands, 22, 99

E

Ecological Restoration, 201-204, 206-213, 220, 226-228, 231-235

Economic Sector, 65, 69

Ecosystem, 2, 18, 24-25, 28, 30-32, 35, 43-48, 51-52, 70, 72, 97, 105-107, 109, 113, 156, 212-216, 218, 220-221, 226-231, 233, 235, 248, 250, 259, 263-264

Extirpations, 34-36, 179

F

Faultless Liability, 55, 61, 65, 68-69, 86, 88

Fencing, 3, 5, 8, 15

Forestry, 2, 25, 45, 55, 66-68, 70, 82, 88-89, 106, 116, 121-122, 124, 130, 134-135, 160, 164, 187, 197, 219, 241-243, 246, 250, 253-255, 258, 261, 263

Freshwater Invertebrates, 40, 67, 114, 127, 132, 169

Freshwater Plants, 39, 67

Fungal Pathogens, 91-92

Fynbos Ecosystems, 162, 213, 215, 225, 230

G

Gastropod, 35, 49

Grazing, 1, 3-8, 15, 17, 19-20, 22-28, 32, 38-39, 49, 64, 156, 202, 249, 251-252

H

Herbaceous Plants, 20, 40, 93

Herbivores, 2, 5-6, 9, 225

I

Invaded Areas, 35, 144, 146, 172

Invasion Management, 55, 70, 83-85

Invasive Alien Plants, 1-2, 6, 8, 16, 18, 23-28, 141, 147, 160-164, 202, 205-206, 208, 211-212, 214, 216, 228, 230-233, 235, 247, 259-264

Invasive Plant Control, 152, 158, 161, 229, 231-232

Invasive Plants, 1, 8, 23, 25, 28, 40, 52, 58, 71-72, 92-93, 95, 97, 104-105, 107, 130, 147-148, 151, 156, 158, 229, 242, 245, 260, 263

J

Jointed Cactus, 68-69, 72, 100, 108, 144, 163

L

Legacy Effects, 201, 205, 208, 214-215, 217, 231, 233

Legislation, 55-56, 58, 62-63, 65-66, 68-69, 73-77, 80, 82, 85-89, 111, 117, 119-123, 125-126, 133, 135, 162, 167, 176, 180, 200, 265

Livestock Production, 1-2, 5, 7, 16-18, 21, 43

Low-impact Species, 40-41

M

Management Plans, 80, 117, 130, 146-147, 162

Marine Invertebrates, 40, 42, 127, 169

Marine Species, 30, 71, 76, 167, 169, 242

Mesquite Trees, 19, 40

N

Native Species, 16, 30-34, 37-38, 41-43, 60, 62, 76, 96, 126, 171, 182, 191, 204-205, 207-210, 215, 217, 219, 223-224, 227, 234, 242-244, 251, 262-263

O

Opuntia Ficus-indica, 11, 17, 21, 23-24, 28, 99-100, 106, 109-110, 244, 249, 263-265

P

Pastures, 22, 60

Plant Species, 1, 6, 8, 10, 14, 23, 25, 27-28, 39, 41-42, 49-50, 91-93, 97-98, 100-102, 108, 144-145, 148, 151, 155, 164-165, 201-202, 214, 228, 232, 234

Plant-feeding Insects, 91-92

Policy Drivers, 56, 58, 70, 73-74, 76, 80

Policy Vacuum, 74, 84, 88

Polistes Dominula, 167, 170, 176-177, 195

Prioritisation, 82-83, 85, 118, 129-130, 132, 134, 159, 161, 168-169, 202, 212, 255, 260

Procambarus Clarkii, 169-170, 198

Prosopis Species, 1, 16-17, 19-20, 27, 50, 140, 152, 164, 263

R

Rangelands, 1-6, 8-9, 14-16, 18, 21-25, 30, 132, 204, 232, 262

Rapid Screening, 114-115

Red-listing Processes, 30-31, 43

Restorative Continuum, 203, 207

Restricted Activities, 78-79, 120, 125-126

Riparian Vegetation, 14, 206, 217, 222, 228-229, 233-234

Risk Analysis, 33, 50, 80, 111-113, 115, 118-126, 129-132

Risk Assessment, 112-116, 118-119, 131-133, 165

Risk Communication, 112-113, 118-119

Riverbanks, 60, 62

S

Sclerophrys Gutturalis, 37, 172, 182-183, 199, 263

Seed Banks, 27, 48, 138, 203, 205-206, 209, 213, 215, 217, 227, 230, 234-235

Shrubland, 1-2, 8, 19, 161, 206, 220, 230

Shrubs, 3, 9, 14-15, 17, 23, 39, 41, 50, 93, 101, 138, 142, 150, 155, 158, 160, 207, 219, 225, 231

Social Ethos, 237, 239, 241

Spontaneous Regeneration, 201, 203, 206-208, 210-211, 214-215, 220-221, 224, 226-227

Stakeholders, 112-113, 116-118, 125-126, 132, 143, 180-181, 189, 191, 193-194, 198, 201-202, 207, 210-213, 227, 233, 239, 249, 253-256, 262

Succulent Shrubs, 9, 14-15

T

Taxonomic Groups, 114, 120, 168

Terrestrial Invertebrates, 35, 40-41, 67, 131, 173, 197

Trogoderma Granarium, 41, 167, 170, 174, 194-195

W

Weedy Species, 209, 215

Wild Fires, 91, 101, 249

Wildlife, 1, 3, 5, 7, 17-18, 25, 46, 49, 73

X

Xanthium Spinosum, 55, 58-65, 68

Z

Zachariades, 13, 28, 91-93, 95-97, 100-101, 104-105, 110, 129, 131, 133